The Hamlyn Guide to the Flora and Fauna of the
Mediterranean Sea

A.C. Campbell

illustrated by
Roger Gorringe
& James Nicholls

HAMLYN
London·New York·Sydney·Toronto

Preface

This book is intended to provide the layman and the student with a simple means of identifying most of the common marine plants and animals in the field. The accurate determination of certain species often depends on the correct use of an identification key in a specialist's monograph. Unfortunately the use of such identification keys frequently presupposes some knowledge of the group of organisms in question and may not therefore be of great help to the beginner. As an alternative, this book provides illustrations and simple accounts of the form, habitats and distribution of a number of species to be found on the shores and in the shallow seas of the Mediterranean. It must be said, however, that a book of this size cannot always provide an exact identification, especially where some of the more obscure groups are concerned. For this reason the reader is directed to other sources at certain points in the text.

It would be impossible to record all the Mediterranean species in a book of this size; those selected for inclusion have been chosen because they are relatively common and because they can generally be identified without the aid of a microscope. In a number of cases the use of a hand lens is called for.

A. C. Campbell

Acknowledgements

The author wishes to acknowledge the great help given by so many friends and colleagues. Special thanks must be given to the following: Mrs L. Campbell, Dr. J. Chatfield, Dr P. Cornelius, Dr A. Demetropoulos, Dr R. Earll, Dr K. Hiscock, Dr S. Hiscock, Mrs L. M. Irvine, Dr R. Ivel, Dr R. Lincoln, Mr R. L. Manuel, Dr T. Norton, Mr A. P. H. Oliver, Mr and Mrs T. Pain, Professor J. D. Pye, Queen Mary College Zoology Department of the University of London, Dr P. S. Rainbow, Royal Scottish Museum, Dr J. Taylor, Dr. T. E. Thompson, Dr P. Tyrer, Mr. A. Wheeler

Illustrations by Roger Gorringe and James Nicholls
Line drawings by James Nicholls, Linda Rogers Associates and Kim Ludlow

Published by The Hamlyn Publishing Group Limited
London · New York · Sydney · Toronto
Astronaut House, Feltham, Middlesex, England

Printed in Spain by Printer industria gráfica sa
Sant Vicenç dels Horts, Barcelona   D.L. B. 29286-1982

Contents

Introduction

The Mediterranean Sea is unique. Not only is it the largest closed sea in the world, but it is the only warm temperate sea. It is not warm enough for coral reefs to grow in, but it is too warm to support many of the familiar marine animals of northern Europe. It has a diverse and fascinating flora and fauna.

For thousands of years it has provided a source of food and a means of trade for the people of southern Europe, Asia Minor and North Africa, and has witnessed many developing civilizations. Today its coastline is shared by twenty or so nations, some of which have played an important role in the development of marine science.

Aristotle was one of the first to describe its marine inhabitants but after his time our knowledge developed slowly. Some indication of the ancient Greeks' knowledge of marine organisms is provided in D'Arcy-Thompson's book *A Glossary of Greek Fishes*. The appearance of Linnaeus's *Systema Naturae* in 1758, in which the scientific means of naming animals and plants was first established, stimulated many naturalists to collect, classify and identify plants and animals. Investigation of the seas followed. In 1841 Edward Forbes published *A Natural History of the European Seas* which included some account of the Mediterranean. In 1859 Darwin's *Origin of Species* was published and this had a profound effect on the future of biological sciences. It may have led Dr Anton Döhrn among others to study the development and growth of animals. Marine animals are ideal for this purpose, and Döhrn required a laboratory where live animals could be kept for study and experiment. He set up such an establishment at Naples in 1874 and the Stazzione Zoologica, as it is known, became the forerunner of marine laboratories throughout the world. Today most countries bordering the Mediterranean have at least one marine laboratory. Many are engaged in academic or applied research and consequently there exists a wealth of literature in the form of books and scientific papers which have been generated by these laboratories. Outstanding in this field are some of the monographs published by the marine laboratories at Naples and in Monaco, which have formed a descriptive series for many groups of Mediterranean animals.

Books written in English specifically on the marine biology of the Mediterranean are scarce, but much general information is contained in *The Open Sea*, Parts 1 and 2, by Sir Alister Hardy (New Naturalist Series, Collins, London); *The Sea Shore* by Sir Maurice Yonge in the same series, and *The Seas* by Sir Frederick Russell and Sir Maurice Yonge (Warne & Co., London). Readers of French and German will find a wider selection mentioned in the bibliography.

The Mediterranean Sea is now visited each year by thousands of tourists; it is an arterial shipping route and many major industrial installations border its coasts. It is very much under threat from man's influence, and being a closed sea it is vulnerable to the accumulation of pollutants. The aim of this book is to provide a ready means of identifying the commoner non-microscopic organisms and to suggest further avenues for study of the inhabitants of this unique sea.

How to use this book

This book describes about 1100 of the most common Mediterranean animals and plants. Because of limitations of space it is impossible to deal with more than a fraction of the total number of species recorded for the area, and emphasis has been placed on the common and more conspicuous organisms. The use of a hand lens will help in many cases.

Many marine plants and animals lack the common names of their terrestrial counterparts, and where such names do exist they may vary from one country to another. In these cases the only name that can be meaningfully used is the scientific one. Each entry in the book begins with the scientific name, written in italics; common names are given only where these have become well established. Sometimes two scientists have independently given an organism different names. Normally one of these names takes precedence by common consent, but where an alternative name persists in use it has been given here in brackets, e.g. *Anseropoda placenta* (=*Palmipes membranaceus*). To help further, the author who first used the name is given after it. If his species has subsequently been transferred to another genus the author's name is placed in brackets, e.g. *Haliclona oculata* (Pallas). In the case of many of the plants the name of the author who made the transfer is also given, hence *Enteromorpha compressa* (Linnaeus) Greville.

If you think you know the name of a plant or animal you have found it can be checked in the index, which contains both scientific and, where they exist, common names. If you wish to identify a species for the first time use the outline key on pages 10–13 to establish where in the book the type of organism you have found is described. It should be emphasized that accurate identification is not necessarily easy since there may be related species which the book does not cover. Where possible, information on the number of similar or related species is provided together with references to more specialist works on the group of organisms in question. It should also be remembered that not all types of organism are equally well documented scientifically; e.g. groups like the sponges and the nemertine worms are not as well known as the fish. In the case of poorly known groups or those with many representatives, a more general description is provided and some typical representatives are shown.

Sometimes habit and habitat will assist with the identification process. The depth distributions in the sea cannot always be regarded as accurate because discoveries are still being made. In general 'shallow' has been used to mean down to about 30m and 'deep' below 30m.

Outline or shape can be particularly important in identification, as can texture and size. Size where given refers to the average recorded; juveniles are often smaller, and then there are always the exceptionally large specimens. Colours in marine organisms are surprisingly variable, and although the artists have taken pains to produce lifelike colour illustrations, all variations and possibilities cannot be included. Mood can drastically affect the colours of many marine animals, especially cephalopods and fishes; they may also change colour after death. Growth form and behavioural characteristics such as locomotion patterns also help with identification in some cases. It should be remembered that the appearance of some animals may be related to the seasons. A number of plants, too, are at their peak in spring and summer, and some even in winter. Certain animals are nocturnal and others, like fish, may migrate. Finally it should be noted that the male and female of some species have a different external appearance. Where a particular sex is illustrated ♂ denotes male and ♀ female.

The Mediterranean Sea

This book covers the Mediterranean area, stretching from the southern part of the Iberian Peninsular and northwest Africa, through the Straits of Gibraltar, the western Mediterranean and the Adriatic to the eastern Mediterranean.

The communications of the Mediterranean Sea to other seas are very restricted. At Gibraltar there is a very narrow strait across which lies a sill, so not only are the straits narrow, but they are also shallow. Because the water of the Mediterranean is slightly more salty than that of the neighbouring Atlantic, having 37–38 parts of salt to 1000 of water rather than 34–35 parts per 1000, it is heavier and there tends to be an outward flow as the deeper heavier water spills over the Gibraltar sill. A very shallow superficial stream flows in from the Atlantic. In the case of the Suez Canal to the East there is a slight northward current taking Red Sea water through the canal and into the Mediterranean. Since the opening of the Suez Canal in 1869 several species of Red Sea fish and invertebrates have been recorded spreading into the Mediterranean.

During the Tertiary period, 15 to 70 million years ago, the geography of Europe was very different from today. Then a huge ocean, known as the Tethys Sea, stretched from the Indo-West Pacific region in the east, passed between Europe and the huge supercontinent of Gondwanaland to the south and passed on westward over the present Atlantic area to connect up with the Eastern Pacific – North and South America not being connected. The Tethys Sea also had connections with what we now know as the Arctic region. In the successive upheavals that have taken place since then, the supercontinent of Gondwanaland split up to form the land masses of Antarctica, South America, Africa, India and Australasia, leaving behind them the Indian Ocean. The eastern basin of the Mediterranean represents the last surviving portion of the Tethys as it was, and the northward movements of the land masses by continental drift have all but closed the Mediterranean off at its western extremity.

Because of its geological history, and the wide connections of the Tethys Sea, the Mediterranean contains species of plants and animals that have descended from ancestors which, in earlier geological periods, inhabited the Arctic, Atlantic, Indian and Pacific Oceans. The unique conditions which prevail in the Mediterranean have enabled many of these organisms to survive there so long as an appropriate habitat was to be found. The western basins of the Mediterranean were a relatively recent development, being formed about 25 million years ago, when small blocks of the European continent moved away from the land mass. These blocks now represent the islands of Corsica, Sardinia and Sicily. The waters below 50 meters depth are cooler than the surface ones and provide a satisfactory habitat for cold water forms, while those above are warmer and allow warm water species to flourish.

As will be discussed later (page 8), there is relatively little tidal movement in the Mediterranean, but at the northern end of the Adriatic in the region of Venice a more marked tidal oscillation is noticeable. The tides in this sea, although slight, are sufficient to produce strong currents, for example that in the Straits of Messina between Italy and Sicily. Here the current may reach 2 meters per second. Despite its relatively small size, the Mediterranean is often rough, particularly when seasonal winds such as the Mistral, Sirocco and Bora are blowing. Waves of 12 meters height have been recorded between Sicily and Tunisia.

Because of its isolation from ocean currents, and the risks of pollution and contamination which are particularly high in an enclosed sea, the surface waters are not as productive in the Mediterranean as they are in some parts of the world. However fish and shellfish are abundant enough to support the fisheries of the various Mediterranean countries provided that excessive quantities are not harvested. Shellfish are an important element in the diet of many Mediterranean people: mussels, lobsters, crabs and shrimps are caught around the rocky coasts, and some deeper water prawns are harvested with special nets. Squid and octopus are great delicacies in many countries. Sardines and herrings are amongst the principal commercial fish taken. These often do not swim very deep and they are effectively captured at night with long nets which are used to surround a shoal. Shoals can be attracted to special lights held aboard the fishing boats. Mackerel, cod, tuna and shark are also taken as well as flatfish. There has been some co-operation between the states bordering the Mediterranean to formulate a fishing policy that will conserve the fish stocks, but there are many problems, not the least of which is that some of the most suitable fish breeding grounds have been lost in the process of land reclamation.

The sea provides other resources. Salt is extracted from it by people of many nations who evaporate the water in shallow pans so that the salt crystals are left behind. In Israel complex plant has been developed so that fresh water can be produced for domestic and irrigation purposes. In a number of countries the commercial sponge is fished; seaweeds are collected in some places for fertilizers or for the industrial extraction of chemicals, and sea-grasses have been used for animal litter in some islands where space for growing arable crops is small.

The following map shows the temperature, and salinity in parts per 1000 of the Mediterranean Sea as well as indicating the extent of coastal waters.

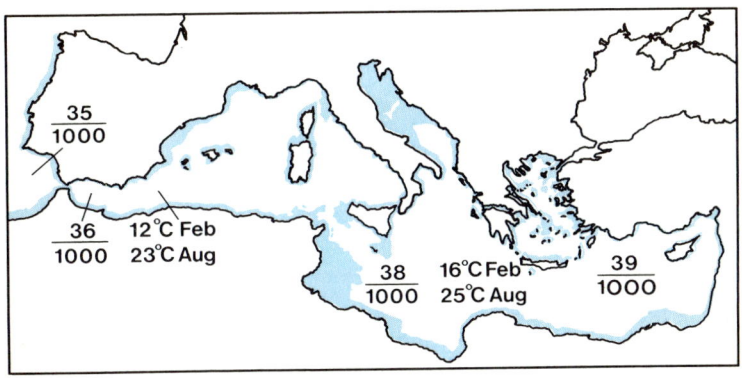

The Mediterranean seashore

The seashore is that region of the coast which lies between the highest point to which the tides flow and the lowest point to which they ebb. All plant and animal life between these two points is subjected to the movements of the tides and their various side effects. Tidal movement is principally brought about by the gravitational pull of the sun and the moon upon the vast water masses of the oceans.

By comparison with the oceans the Mediterranean covers a relatively small area and as such is less influenced by these gravitational forces. Furthermore it is almost cut off from the adjacent Atlantic by the narrow Straits of Gibraltar. Tides in the Mediterranean are therefore restricted, and of much smaller range than those of most of the northwest European coasts. At Gibraltar they rise and fall by a maximum range of a metre, at Naples by a maximum of about 0.5 metres and at Alexandria by a maximum of about 0.6 metres.

For most of the Mediterranean coastline therefore the shore, strictly speaking, is a very narrow strip of rock, sand or pebbles. In stormy weather the influence of the sea can be extended further upwards towards dry land by the effects of salt spray. The spray zone is inhabited by plants and animals of both terrestrial and marine origin which can tolerate salt and spray. The shore proper is alternately covered by seawater and air. In hot weather the rocks or sand may reach a high temperature bringing risks of heat stroke and desiccation. Winds may further enhance the effects of desiccation so that unless the intertidal organisms have a special mechanism to reduce water loss they may die. Because the ranges of the tides in the Mediterranean are less than those of northwest Europe, and because for the most part the temperature ranges are more extreme, shore life as such is considerably less abundant. Most conspicuously absent from many areas of the Mediterranean, (although present in some parts of the Adriatic) are the large brown seaweeds. Animals such as the barnacles, sea-snails and other molluscs with strong shells may be found. Although the large algae on temperate shores provide cover for soft-bodied animals to hide in at low tide, relatively few seashore animals depend upon them as a source of food. The floating planktonic algae are more important to many animals, which are able to strain seawater so as to obtain food at high tide. Animals such as barnacles and bryozoans feed like this. The creeping sea-snails, such as the winkle *Littorina neritoides* may graze on a microscopic film of algae growing on the rocks, or use the seashore lichens as a source of food.

The relative paucity of intertidal life in the Mediterranean is more than compensated for by the abundance of organisms in the shallow water. The use of a snorkel, mask and flippers will enable many people to discover a world of rich plant and animal life just beyond the water's edge.

Conservation and collecting

The Mediterranean Sea is surrounded by about 20 separate states, some of which have a small coastline, e.g. Gibraltar, while others have an extensive coastline, e.g. Italy. Whilst the number of nations surrounding the Mediterranean Sea has advantages in terms of trade and communications, it has some disadvantages as far as establishing a uniform policy on conservation is concerned. It must however be added that in recent years a more unified approach to conservation has emerged. As has already been pointed out, the Mediterranean, being an enclosed sea and somewhat beyond the influence of the Atlantic Ocean, is vulnerable to pollution and over-exploitation. Not surprisingly, different attitudes to exploitation and conservation prevail in different countries, and in some cases there may be legislation affecting what may or may not be permitted in the way of collecting specimens or what type of fishing methods may be used.

It should be remembered that certain species are in peculiar danger from man's activities. Organisms such as semi-precious red coral and large territorial fishes have been greatly over-fished. They are easily caught and exploited. Other species may be far less in danger. It is hoped that this book will be used in the field and not confined to the bookshelf, and that it will not be necessary to remove live specimens for identification at home.

If you are exploring the rocky shore remember that you may have to work hard to discover some animals living in crevices and under boulders. If you have to turn stones over please remember to turn them back afterwards, so that animals which have chosen to live underneath in the dark may continue to do so after your visit.

Snorkel and SCUBA divers are privileged to see animals and plant life 'at home' but should remember that they too can disturb the environment greatly. Some invertebrates like the large and attractive gastropods take years to reach maturity and reproduce. They are vulnerable to curio-hunters. It is also to be hoped that the use of spear guns will be kept to a minimum and certainly that they will not be used against territorial fish which hide in holes and which thus make easy targets.

If you are collecting for a purpose, arrange to look after your specimens carefully. Prevent them from becoming overheated by placing your jars in a rock pool to keep the temperature down. Put the lids on only when actually carrying the jars and add just sufficient water to cover the specimens or let them swim freely. Only take the animals away from the beach if you are certain that you can keep them alive. After you have examined them return them to the spot they came from, or to a similar place. Much of what has been said applies to the diver collecting in the shallow seas. If fishes are being collected from 10m or below, remember that they will have to be decompressed by gradual raising to the surface. Polythene bags are useful for this and they also make good containers for small invertebrates; larger specimens can be carried in string bags.

It is refreshing to see the emergence of groups of individuals dedicated to marine conservation such as the Underwater Conservation Society who produce many pamphlets and guides on particular groups of organisms (see bibliography).

Illustrated key

Plants

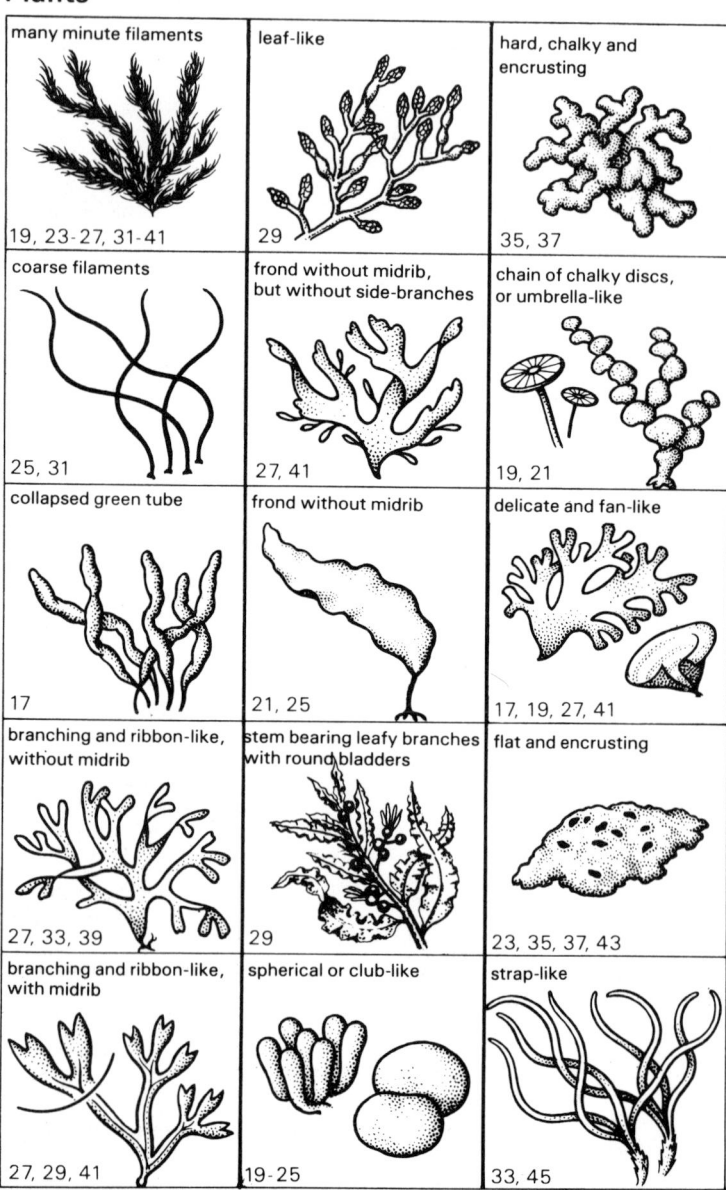

many minute filaments

19, 23-27, 31-41

leaf-like

29

hard, chalky and encrusting

35, 37

coarse filaments

25, 31

frond without midrib, but without side-branches

27, 41

chain of chalky discs, or umbrella-like

19, 21

collapsed green tube

17

frond without midrib

21, 25

delicate and fan-like

17, 19, 27, 41

branching and ribbon-like, without midrib

27, 33, 39

stem bearing leafy branches with round bladders

29

flat and encrusting

23, 35, 37, 43

branching and ribbon-like, with midrib

27, 29, 41

spherical or club-like

19-25

strap-like

33, 45

Plant-like animals

Joint-legged animals

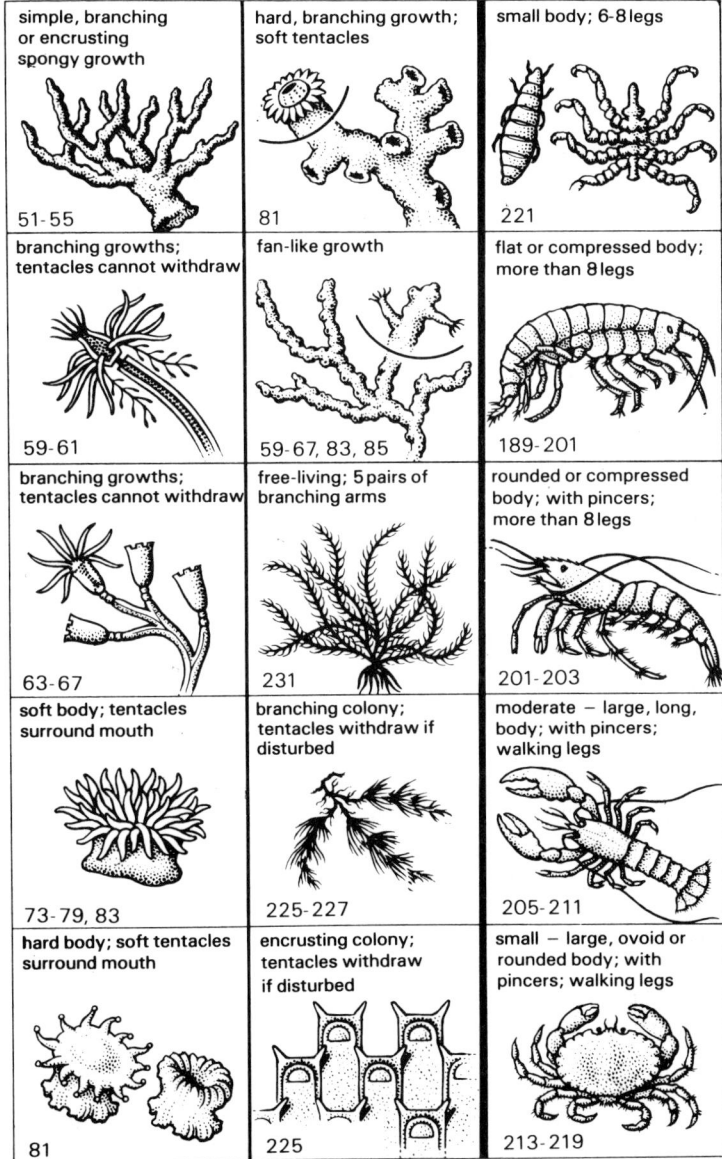

simple, branching or encrusting spongy growth 51-55	hard, branching growth; soft tentacles 81	small body; 6-8 legs 221
branching growths; tentacles cannot withdraw 59-61	fan-like growth 59-67, 83, 85	flat or compressed body; more than 8 legs 189-201
branching growths; tentacles cannot withdraw 63-67	free-living; 5 pairs of branching arms 231	rounded or compressed body; with pincers; more than 8 legs 201-203
soft body; tentacles surround mouth 73-79, 83	branching colony; tentacles withdraw if disturbed 225-227	moderate – large, long, body; with pincers; walking legs 205-211
hard body; soft tentacles surround mouth 81	encrusting colony; tentacles withdraw if disturbed 225	small – large, ovoid or rounded body; with pincers; walking legs 213-219

11

Animals with shells

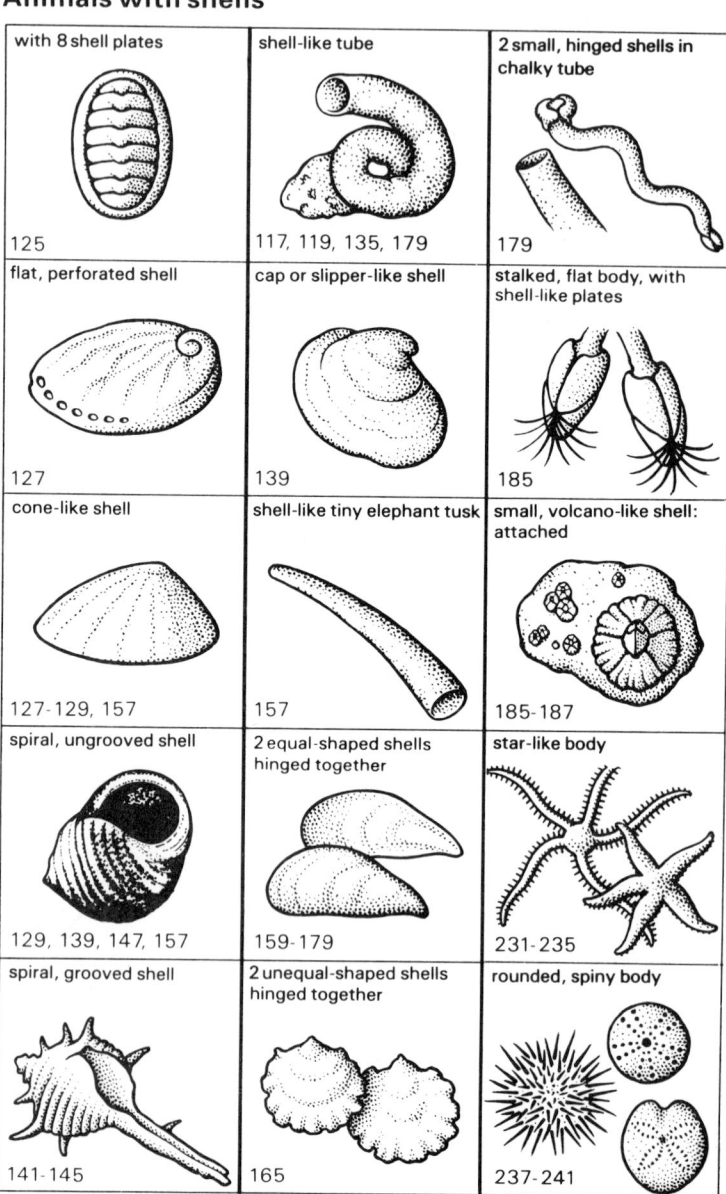

with 8 shell plates	shell-like tube	2 small, hinged shells in chalky tube
125	117, 119, 135, 179	179
flat, perforated shell	cap or slipper-like shell	stalked, flat body, with shell-like plates
127	139	185
cone-like shell	shell-like tiny elephant tusk	small, volcano-like shell: attached
127-129, 157	157	185-187
spiral, ungrooved shell	2 equal-shaped shells hinged together	star-like body
129, 139, 147, 157	159-179	231-235
spiral, grooved shell	2 unequal-shaped shells hinged together	rounded, spiny body
141-145	165	237-241

Animals without shells

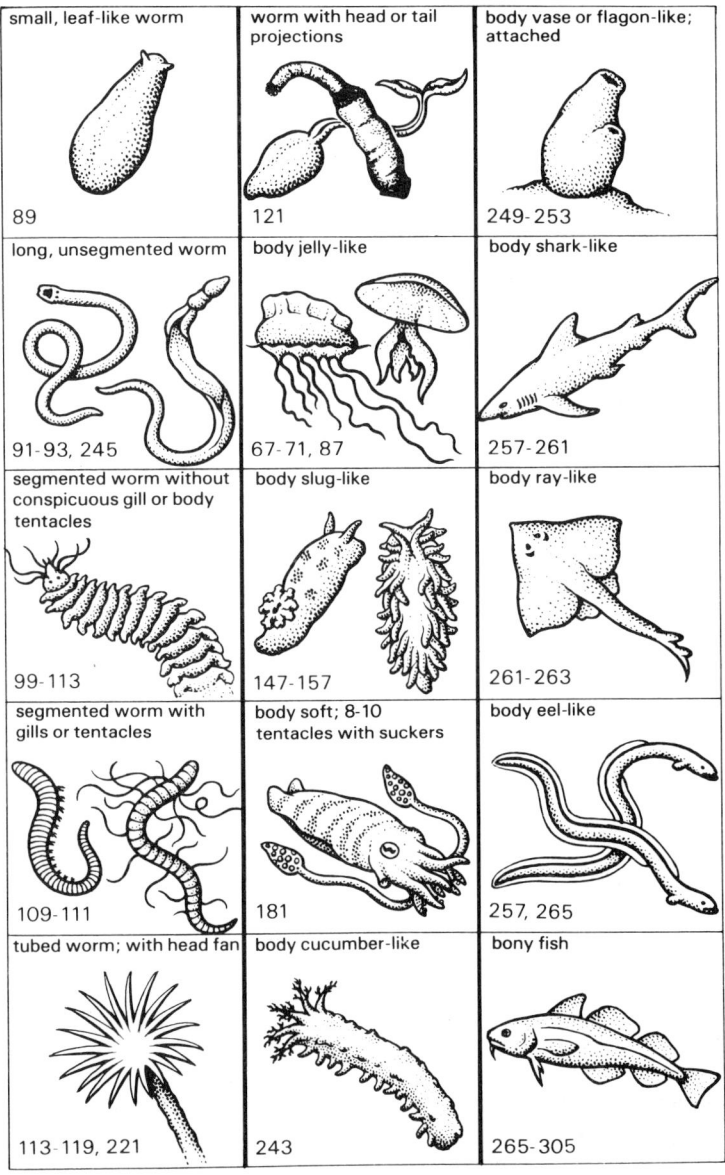

small, leaf-like worm 89	worm with head or tail projections 121	body vase or flagon-like; attached 249-253
long, unsegmented worm 91-93, 245	body jelly-like 67-71, 87	body shark-like 257-261
segmented worm without conspicuous gill or body tentacles 99-113	body slug-like 147-157	body ray-like 261-263
segmented worm with gills or tentacles 109-111	body soft; 8-10 tentacles with suckers 181	body eel-like 257, 265
tubed worm; with head fan 113-119, 221	body cucumber-like 243	bony fish 265-305

Plant Kingdom

The plant kingdom is divided into a number of major groups which range from simple, unicellular organisms that can only be seen with the aid of the microscope, to the great trees which are so familiar on land. The sea provides a satisfactory environment for only a few groups of plants, however, and the majority of these belong to the lowest group of the plant kingdom, the Thallophyta. This group comprises the algae (including the seaweeds), the fungi and the lichens. The only other group that is significantly represented in the sea is the Angiospermae (flowering plants).

Green plants are anabolic organisms, that is, they elaborate organic material from inorganic sources with the aid of photosynthesis. No animal can do this; all animals depend either directly or indirectly on plants for their supply of food. Plants therefore form the first living stage of all *food chains* or *food webs*. (Food chains or food webs are terms used to describe the nutritional relationships existing between organisms and their consumers or predators in a community.)

The marine algae are divided into a number of subgroups of which three will be dealt with here. They are the Chlorophyceae (green algae), Phaeophyceae (brown algae) and Rhodophyceae (red algae). Whilst colour would appear to be their principal distinguishing characteristic, it must be remembered that colour may not only be variable, but it may change considerably if the plants have been removed from the sea for long periods, or if they are dying. Furthermore, many of the red and brown seaweeds are quite similar in colour. These facts make colour alone a rather unreliable criterion for the purposes of identification, and this also holds true for animals. The identification of marine algae is largely based on the form of the plant in question. Precise identification may depend on the accurate recognition of microscopic characters, and to a large extent such aspects of seaweed identification are beyond the scope of this book. Nevertheless with the gross descriptions of form, the colour plates and a little experience, it should be possible for the reader to succeed in what is quite a difficult area of seashore biology.

A typical seaweed is normally composed of a *holdfast* (or attachment organ) and a *frond*. The holdfast and frond together are termed the *thallus*. Many fronds become broken from their holdfasts and washed into positions where they are subsequently found. For this reason the locality in which a plant is found may not be that in which it grew. Not all the illustrations show the holdfast. Unlike the higher plants, many algae show relatively little differentiation within their tissues, and only some, such as the laminarians, have any form of vascular system. Instead, most of the cells are developed in such a way that each can carry out many of the functions of the organism, and any specialized regions that do exist, such as holdfasts and reproductive organs, are reduced to a minimum. The form of the frond and the holdfast are important clues to identification. Holdfasts may be root-like, disc-like or may grow as a shaggy tangle of fine 'rootlets' – these are not roots in the strict sense, however, as they do not provide the seaweed with a special absorptive region. The frond may be rounded or cylindrical, flattened, flattened with a central rib or ribs, or intermediate, i.e. neither round nor flat. The style of branching is also important; it may be dichotomous, alternate, spiral or pinnate. Fig. 1 shows some of these features.

In some species, especially the brown and red algae, attention to the form and disposition of the reproductive organs will assist with identification. In the Phaeophyceae this is true of the order Fucales (e.g. *Fucus*). Here, the reproductive organs generally consist of small pits opening by minute pores. They are usually set in thickened portions of the frond near the tips of the branches. In other orders, such as the Ectocarpales (e.g. *Ectocarpus*), reproductive organs can be identified (with a hand lens) on the sides of the branches or in the angle of a junction between the stem and a side-branch. In some other genera (e.g. *Padina*), they occur on the surface of the frond arranged in groups. Some of the Rhodophyceae also have their reproductive organs set in pits. Other genera, like *Polysiphonia*, carry them on small branches of the frond, but they may be buried in the plant tissue itself. In some red algae the reproductive organs are extremely difficult to discern. The above descriptions do not take into account the various forms or functions of the reproductive organs, only their disposition. Where necessary this will be referred to further in the text. Fig. 1 also illustrates some reproductive organs.

The life histories of marine algae are quite complicated. Essentially two generations alternate with each other. These are known as the *sporophyte* generation, which bears asexually produced spores and which, on the development of these spores, gives rise to the *gametophyte* or sexual generation. The gametophyte generation may consist of plants bearing both male and female organs, or of separate male and female plants. Sexual reproduction results in a new sporophyte generation. In some cases the sporophyte generation closely resembles the gametophyte. If the two generations are similar they are termed *isomorphic*. If they are dissimilar they are known as *heteromorphic*. In some cases the life cycles are less straightforward and a full account of these is beyond the scope of this book.

The occurrence of seaweeds is limited by light, which they require for photosynthesis. Thus they are restricted to depths where sufficient illumination exists. High turbidity in the sea will restrict illumination and thus plant growth, and only certain species will be able to exist in the reduced illumination characteristic of caves, etc. The Mediterranean Sea provides a range of marine habitats in terms of physical conditions and substrate types, but the lack of significant tidal movements limits the distribution of marine plants on the shore. It is in the shallow offshore water that the real variety is to be found, for which reason the plant community is influenced mainly by substrate type. Rocky bottoms are characterized by an abundance of green, brown and red algae, while sandy bottoms may be colonized by extensive communities of the sea-grass *Posidonia*, which may itself form a substrate for algae and sessile invertebrates. Brown algae, although they are plentiful in rocky places, do not exist in the large forms familiar to beachcombers in Northwest Europe. Only in the Adriatic are large browns found.

The literature on the Mediterranean flora is scattered, and because of the number of scientists writing in different languages and using different names it is not easy to give a definite figure for the number of recorded species. An up-to-date list of the names of many British algae is given by Parke, M. and Dixon, P. S. 1976. Many of the green and red algae mentioned in this will be found in the Mediterranean.

This account has paid little attention to the lichens or the marine angiosperms. However, more is said about them on pages 42–45. Here it should be noted that the former group dwell in the splash zone and are more characteristic of the terrestrial habitats. The sea-grasses are highly specialized angiosperms.

Fig. 1 Various patterns of thallus branching and reproductive organs: a = dichotomous; b = opposite; c = opposite with pinnate side-branches; d = alternate with alternate side-branches; e = spiral branching; f = sprig of *Fucus* with reproductive area shaded in black; g = magnified bodies of *Padina*; h = magnified reproductive body of *Polysiphonia*

Subkingdom Thallophyta
Group Algae

These mainly aquatic plants contain the green pigment chlorophyll. Their bodies are not differentiated into true roots, stems or leaves, and they lack a vascular system. An up-to-date check list of names of algae is given by Parke, M. and Dixon, P. S. 1976. Although it is principally concerned with British species, many mentioned in it occur in the Mediterranean.

Class Chlorophyceae Green algae

Algae in which the chlorophyll is not masked by brown or red pigment. Their structure ranges from minute single-celled plants to large thread-like or frond-like plants composed of many cells.

Palmophyllum crassum (Nacc.) Raben Frond rounded, about 300mm long, generally fan-like. Colour olive-green. Habitat between 5 and 60m among stones, coralline algae and *Cystoseira* (see page 28). Similar species few.

Ulva lactuca Linnaeus **Sea lettuce** Frond 150–500mm long; shape variable: lobed, lance-shaped or perforated, normally wider at the top than at the base; stalk if present is solid. Colour translucent green. Habitat on rocks on the shore, in pools and shallow water; may be floating free or get washed up. Similar species few.

Enteromorpha intestinalis (Linnaeus) Link Frond 50mm–1m or more in length; tubular and irregularly inflated and crinkled; generally not branched; may taper continuously along length. Colour pale green. Habitat in rock pools on the shore, often in water with reduced salinity; sometimes washed up. Similar species *E. compressa* and *E. linza* (see below).

Enteromorpha compressa (Linnaeus) Greville Frond up to 300mm long; tubular main frond tapers only at the base and frequently gives off a number of side branches. Colour dark to pale green. Habitat on rocks and pilings on the shore and places with reduced salinity; often washed up, sometimes in estuaries. Similar species *E. intestinalis* and *E. linza* (see below).

Enteromorpha linza (Linnaeus) J. Agardh *(=Ulva linza)* Frond 100–500mm long; superficially flattened, but will be found to be hollow if a transverse section is carefully examined under a microscope, looking especially at the edges; unbranched, tapering slightly from the middle to the apex and base; edges crinkled. Colour bright green. Habitat on rocks on the shore, sometimes washed up. Similar species *E. intestinalis* and *E. compressa*. N.B. precise identification of *Enteromorpha* species is often difficult and may require a microscopic examination.

Cladophora prolifera (Roth) Kützing Frond up to 200mm long with a short stalk, cartilaginous filaments and long rhizoids developing from the first junction of the filaments; branching often into three filaments at the junctions; microscopic examination shows elongated cells becoming shorter further away from the base. Colour brownish-green. Habitat amongst stones and *Cystoseira* (see page 28) down to 20m. Similar species *C. pellucida* and *C. mediterranea*.

Cladophora pellucida (Hudson) Kützing (Not illustrated) Similar to *C. prolifera* above. Frond up to 140mm long, rather rigid, composed of a stalk anchored to substrate by rhizoids; distal region is irregularly branched. Microscopic examination shows frond to be composed of elongated cells which become shorter further away from the base. Colour bright green tinged with red at the base. Habitat in pools on the lower shore.

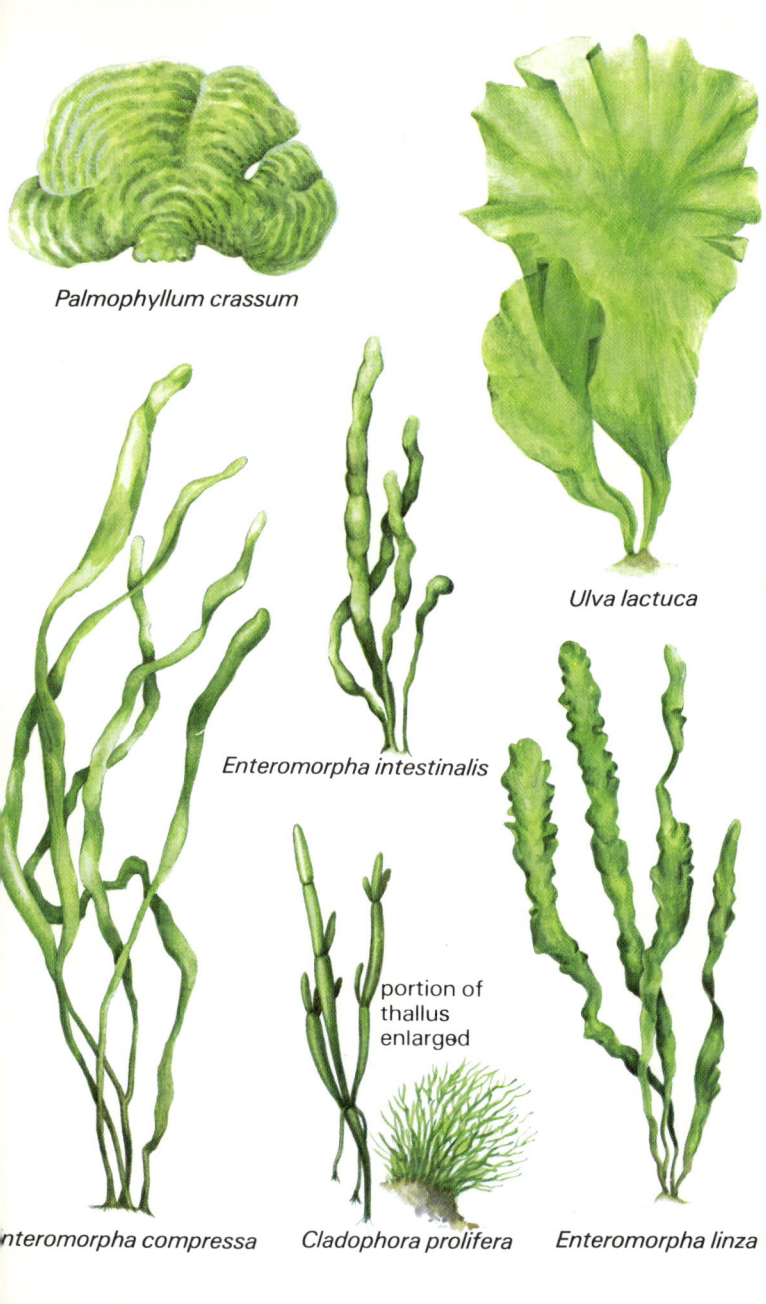

Palmophyllum crassum

Ulva lactuca

Enteromorpha intestinalis

nteromorpha compressa

Cladophora prolifera

portion of
thallus
enlarged

Enteromorpha linza

Dasycladus clavaeformis (Roth) C. Agardh Frond about 30mm long and tapering towards base; several fronds may grow from one holdfast. **Colour** green, with a felty texture. **Habitat** on rocks among sand and gravel in shallow and deeper water. **Similar species** few.

Acetabularia mediterranea Lamouroux **Mermaid's cup** Frond up to 80mm long; stalk is stiffened by calcareous secretion and bears up to 100 leaf-like structures arranged segmentally so that collectively they resemble a disc. **Colour** greenish-white. **Habitat** on rocks and stones on the shore and in shallow water. **Similar species** none.

Derbesia lamourouxi (J.G. Agardh) Solier Frond about 50mm long, growing as a cluster of fine filaments rising from a basal portion with occasional lateral branches. **Colour** bright green. **Habitat** growing on other weeds and mud on the shore and in shallow water. **Similar species** few.

Bryopsis plumosa (Hudson) C.A. Agardh Frond up to 100mm long although often less; feather-like pinnate branches arranged more or less opposite on the main stem, smallest ones towards the top. **Colour** yellow-green in the male and dark green in the female; glossy. **Habitat** on rocks and stones and the sides of pools. **Similar species** *B. balbisana* below.

Bryopsis balbisana (Lamouroux) Frond up to 70mm long with pinnate branches arranged terminally so that long pieces of stalk show free. **Colour** dark green. **Habitat** in shallow water on stones and on other algae, e.g. *Corallina elongata* (see page 34).

Udotea petiolata (Turra) Børgesen Frond about 65mm long, a fan-like growth carried on a stalk; multiple growths of several fans may develop. **Habitat** on sand, stones and rocks down to 60m. **Similar species** few.

Dasycladus clavaeformis

Acetabularia mediterranea

Derbesia lamourouxi

Bryopsis plumosa

Bryopsis balbisana

Udotea petiolata

Valonia utricularis (Roth) **C. Agardh Frond** about 30mm long, often quite bulbous or club-shaped; a number of fronds may arise from a communal holdfast. **Colour** iridescent green. **Habitat** on rocks down to 10m. **Similar species** few.

Halimeda tuna (Ellis & Solander) **Lamouroux Frond** about 90mm long, consisting of a number of calcified discs strung together in chains; there is some irregular branching. **Habitat** on hard substrates down to 20m. **Similar species** none.

Codium tomentosum **Stackhouse Frond** 250–350mm long, tubular and branching dichotomously; texture is felt-like. The holdfast is disc-like, consisting of many entwined threads encrusting the substrate. **Colour** dark green. **Habitat** on mud, sand and rocks down to 20m. **Similar species** none.

Codium bursa (Linnaeus) **C.A. Agardh Frond** 30–200mm across; slender filaments entwine to form a soft sponge-like spherical growth; a mat of interwoven filaments attach this to the substrate. **Colour** dark green. **Habitat** attached to rocks in shallow water, sometimes washed up. **Similar species** *C. difforme* below.

Codium difforme **Kützing** (Not illustrated) **Frond** rounded, encrusting the substrate and up to 60mm long. **Colour** dark green. **Habitat** encrusting hard substrates down to 40m. **Similar species** *C. bursa* above.

Caulerpa prolifera **Lamouroux Frond** broad, up to 150mm long, borne on creeping stems which have root-like attachment organs. **Habitat** on sandy bottoms, often in habours and sheltered places where it may form extensive 'meadows'. **Similar species** none.

Valonia utricularis

Halimeda tuna

Codium tomentosum

Codium bursa

Caulerpa prolifera

Class Phaeophyceae Brown algae

Algae in which the chlorophyll is often masked by the brown pigment fucoxanthin. These are multicellular plants, often large, and they are normally attached to the substrate. Typically they prefer cooler waters, and so although they are conspicuous on the rocky shores of northern Europe, few large brown algae occur in the Mediterranean. Hiscock, S. 1979 gives details of brown algae, some of which occur in the Mediterranean.

Ectocarpus siliculosus/fasciculatus (Dillwyn) Lyngbye Frond 120–300mm long, a tangled growth of fine branching filaments which becomes free towards the tip; branches are variably arranged; holdfast creeping and filamentous. A hand lens may reveal both club-shaped and pointed reproductive bodies carried towards the tips of the branches, usually on short stems. **Colour** yellow-green brown. **Habitat** attached to rocks and stones from the shore down. These species and *E. confervoides* below represent an aggregation of formerly separate species.

Ectocarpus confervoides (Roth) Le Jolis (Not illustrated) Frond up to 50mm long, a tangled growth of fine branching filaments; holdfast is creeping and filamentous. A lens may reveal club-shaped reproductive bodies. **Colour** brownish with a banded appearance due to concentrations of pigment along the fronds. **Habitat** on stones and larger algae. **Similar species** *E. siliculosus* and several others.

Ralfsia species Groups of individuals forming irregular incrustations 20–100mm across; individuals may be rounded and are about 2.5mm thick. In the winter small club-like reproductive bodies grow upwards and may be seen with the aid of a hand lens. Isomorphic generations (see page 15) follow each other in the life-cycle. **Colour** dark brown-black. Habitat attached to rocks and shells, often in exposed places. *R. verrucosa* (see illustration) is typical of the genus.

Stilophora rhizodes (Turner) J.G. Agardh Frond 150–600mm and branching dichotomously; branches taper near tips and are covered with small spots; frond is solid when young, becoming tubular with age. **Habitat** on rocks and on other seaweeds on the shore and in shallow water, often where the salinity is reduced by freshwater streams.

Spermatochnus paradoxus (Roth) Kützing Frond up to 200mm long, alternately branching and rounded; examination under a lens reveals many small darker wart-like growths arranged in twos or threes along its length. **Colour** brownish. **Habitat** from 10–40m on stones and algae.

Stictyosiphon adriaticus Kützing Frond up to 200mm long, much branched, alternate or opposite. **Colour** yellow-brown. **Habitat** attached to stones, shells and to other algae on the shore and in shallow water.

Asperococcus turneri (Smith) Hooker *(=**A. bullosus**)* Frond 150–300mm long, tubular structure carried on a short narrow stalk; often growing in groups; holdfast is small and discoidal. Texture is soft, slightly transparent and membranous, thickening with age. **Colour** olive-green. **Habitat** attached to rocks and large algae on the shore and in shallow water.

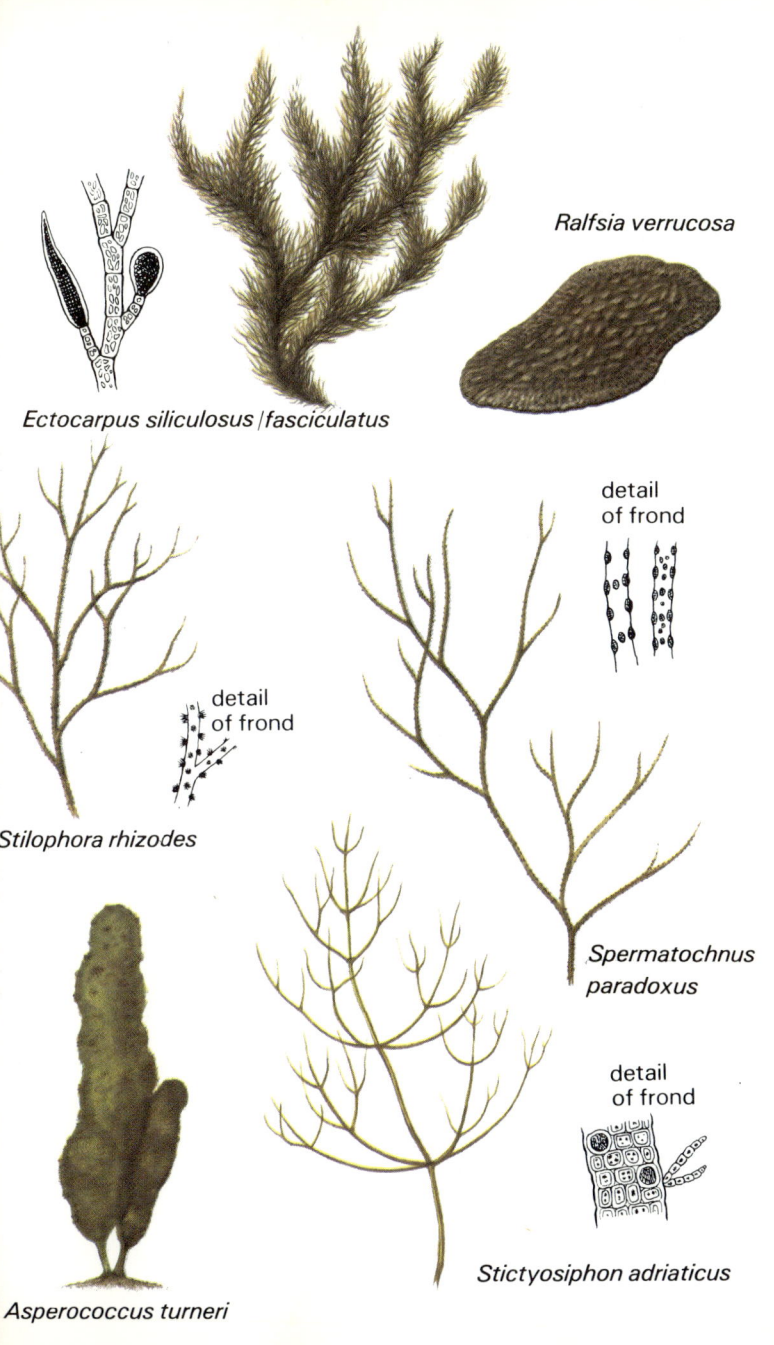

Ralfsia verrucosa

Ectocarpus siliculosus/fasciculatus

detail
of frond

detail
of frond

Stilophora rhizodes

Spermatochnus paradoxus

detail
of frond

Asperococcus turneri

Stictyosiphon adriaticus

Punctaria species Frond 200–400mm long and 75mm across, unbranched an leaf-like, borne on short stalks and dotted with hairs or small spots; may terminat in a wide-angled or narrow-angled point. Habitat on rocks, stones and shells o the shore and in shallow water. *P. latifolia* (see illustration) is typical of the genus

Colpomenia sinuosa (Roth) Derbes & Solier Frond is a thin-walled globula hollow growth up to 200mm in diameter although often smaller, covered all ove with fine brown dots. Habitat in pools and attached to various seaweeds and shell on the shore and in shallow water.

Petalonia species Frond may reach 300mm in length and 60mm across; edge may be frilled; the short stalk widens quickly to form the frond and is much shorte than that of *Laminaria* species. Colour glossy green-brown. Habitat on rocks, ofte those covered with sand, and in pools on the shore. Similar species several. *F fascia* (see illustration) is typical of the genus.

Scytosiphon lomentaria (Lyngbye) Link Frond 150–300mm long, tubula and resembling a miniature chain of sausages; not branched, tapering at the tip Colour green-yellow. Habitat attached to rocks, stones and other algae on the shore and in shallow water, often in exposed places.

Cutleria multifida (Smith) Greville Frond 100–400mm long, forming fla fan-like growths; branched dichotomously, tips of branches divided; texture wher fresh rather springy; usually spotted. Holdfast is disc-like. Colour yellow-green Habitat commonly attached to rocks and shells in shallow water; sometime washed up.

Sporochnus pedunculatus (Hudson) C.A. Agardh Frond 150–450mm long, filamentous growth of a single thread, bearing branches; each branch bears alternately arranged side branches. Colour olive-green. Habitat on rocks in shallow water.

Arthrocladia villosa (Hudson) Duby Frond 150–900mm long, consisting o fine filaments with oppositely arranged branches; these in turn bear branchlets se in whorls; branchlets may carry very fine unbranching filaments which form part o the reproductive bodies and impart a shaggy green appearance. Habitat growing on rocks, stones and *Zostera* (see page 44).

Laminaria rodriguezi Bornet Frond up to 400mm long with a relatively shor stalk giving rise to a wide ribbon-like blade with crinkled edges; several stalks may arise from a single branching holdfast. Habitat on gravelly and harder substrates.

Colpomenia sinuosa

unctaria latifolia

Petalonia fascia

Scytosiphon lomentaria

ria multifida

Sporochnus
pedunculatus

detail of
reproductive
body

aminaria rodriguezi

Arthrocladia villosa

Sphacelaria cirrhosa (Roth) C.A. Agardh Frond up to 30mm long, tufted and branched; branches bear branchlets which are seen under a hand lens to be one cell wide; terminal reproductive bodies. Colour brownish. Habitat living on stones and other algae on the shore and down to 10m. Similar species several.

Halopteris scoparia (Linnaeus) Sauvageau Frond up to 60mm long; main stem arises from holdfast and then divides into several major branches which themselves subdivide into bushy, further subdivided branchlets like small shaving brushes: under a lens these will be seen to be segmented and each several cells wide. Colour green-brown. Habitat on stones and other algae on the shore and down to 5m.

Halopteris filicina (Grattan) Kützing Frond 50–100mm long; main stem bears alternately arranged side branches which themselves carry pinnate branchlets: under a lens these will be seen to be several cells wide. Usually the upper part of the stem bears more branches than the lower. The holdfast is root-like. Colour green-brown. Habitat on rocks, larger seaweeds and shells on the shore and in shallow water.

Cladostephus verticillatus (Lightfoot) C.A. Agardh Frond 100–250mm long; main stem more or less dichotomously branching, with branches bearing whorled spiny branchlets; these may be lacking in the lower regions of the main stem and principle branches. The holdfast is disc-like. Colour dull brown. Habitat on rocks, stones and coralline algae on the shore and in shallow water.

Dictyopteris membranacea (Stackhouse) Batters Frond 100–300mm long, dichotomously branching, flattened, thinner than that of *Fucus virsoides* (page 28); a conspicuous midrib may be the only part of the stem left in the lower regions of older specimens. The membranous edges of the fronds are dotted with groups of minute hairs; tips of branches are rounded and slightly split or notched; the holdfast is disc-like and fibrous. Colour yellowish when juvenile, growing darker brown. Habitat on rocks on the shore and down to 80m. N.B. when freshly collected it has an unpleasant odour.

Dictyota dichotoma (Hudson) Lamouroux Frond about 130mm long, regularly dichotomous, non-transparent, delicate and flattened with rounded notched tips; no midrib; fronds may be covered with groups of minute hair-like reproductive bodies. Colour non-iridescent yellow-olive-brown. Habitat on rocks and on other algae on the shore and in shallow water.

Padina pavonia (Linnaeus) Lamouroux Frond about 100mm long, fan shaped; narrow rounded stalk gives rise to a rounded lamina: when young and smaller this is often quite thin and flat, but as it matures it develops into a characteristic concave fan. Colour outer surface has brown-green stripes, inner surface is lime-green. Habitat on rocks and stones in shallow water.

Taonia atomaria (Woodward) J.G. Agardh Frond 70–300mm long, membranous, transparent and non-shiny, broadening out sharply from the base and branching into wedge-shaped growths; the presence of reproductive bodies and hairs imparts a striated appearance. Colour pale olive-green-brown above, darker below. Habitat usually on stones and rocks down to 20m.

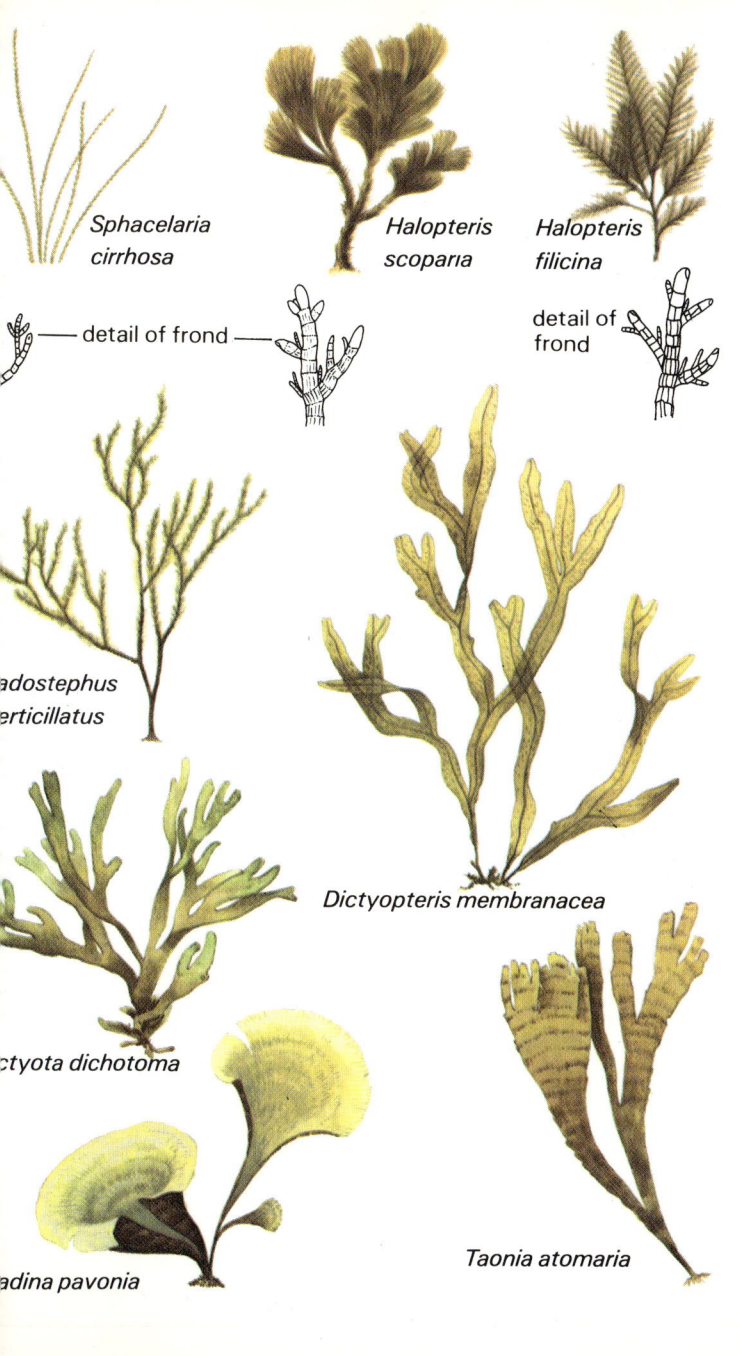

*Sphacelaria
cirrhosa*

*Halopteris
scoparıa*

*Halopteris
filicina*

—detail of frond—

detail of
frond

adostephus
erticillatus

Dictyopteris membranacea

ctyota dichotoma

adina pavonia

Taonia atomaria

Fucus virsoides J.G. Agardh Frond up to 200mm long, tough, strap-like dichotomously branching, without air bladders and lacking marginal serrations conspicuous midrib may have terminal swollen reproductive bodies. Colour olive brown. Habitat on rocks on the shore.

Cystoseira barbata (Goodenough & Woodward) C.A. Agardh Fronc up to 1m long; stout main stem arising from holdfast, sending off side branche which themselves carry 'segmented' branchlets; these side branchlets are dicho tomously arranged and may carry swollen terminal reproductive bodies. Colou brownish. Habitat on hard bottoms down to 50m, often supporting other epiphyti algae. Similar species several.

Cystoseira spicata Ercegovic Frond about 300mm long; main stem divide near the holdfast into a few long branches which may carry alternately arrange pinnate branchlets; reproductive bodies are carried on these. Habitat on har substrates. Similar species several.

Cystoseira abrotanifolia J.G. Agardh Frond about 250mm long; main ster more or less straight, giving rise to alternately arranged side branches; these ar shorter towards the apex of the plant and bear alternately arranged branchlets reproductive bodies are carried subterminally on the branchlets. Habitat on ston bottoms down to about 30m. Similar species several.

Cystoseira tamariscifolia (Hudson) Papenfuss Frond 300–450mm long cylindrical main stem may be branched a few times and bears many alternatel arranged branchlets; branchlets themselves carry many short spines along the length as well as tufted reproductive bodies near their tips; air bladders may occu singly or in groups; overall appearance bushy. Colour olive-brown; iridescer green-blue under water. Habitat attached to rocks on the shore, in pools and i shallow water. Similar species several.

Sargassum vulgare J.G. Agardh Frond 150–300mm long; irregularly branch ing stem bears lance-like 'leaves' in addition to rounded bladders and branchin reproductive bodies which are carried in clusters. Colour brown. Habitat on har substrates down to 30m. Similar species several.

Sargassum linifolium (Turner) C.A. Agardh (Not illustrated) Frond up t 100mm long; main stem bears fewer 'leaves' and bladders than *S. vulgare* or *S hornschuchi*; 'leaves' are narrow and lance-like. Colour brown. Habitat on variou bottoms down to 20m. Similar species several.

Sargassum hornschuchi C.A. Agardh Frond 300–400mm long; main ster bears irregular or alternate crinkled side branches which are leaf-like; near the ti of the plant the side branches also bear rounded bladders and clusters of taperin reproductive organs. Colour brown. Habitat on rocks from 10m down. Simila species several.

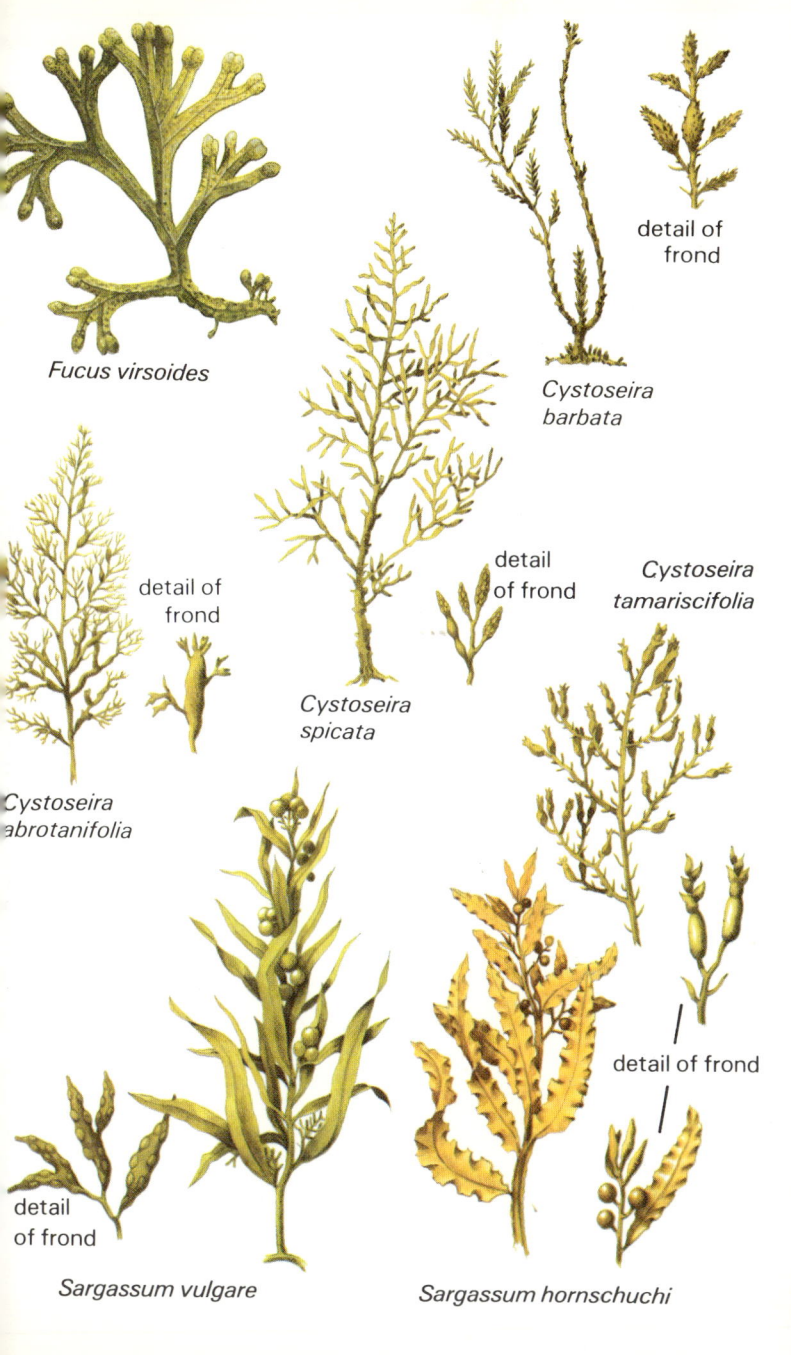

Fucus virsoides

Cystoseira barbata

detail of frond

detail of frond

Cystoseira tamariscifolia

Cystoseira spicata

Cystoseira abrotanifolia

detail of frond

detail of frond

Sargassum vulgare

Sargassum hornschuchi

detail of frond

Class Rhodophyceae Red algae

Algae in which the chlorophyll is masked by the red pigment phycoerythrin. They are always multicellular and usually of small to moderate size. The red algae occur in temperate and warm waters. In the Mediterranean they may be found from the shore down. A number of species occurring in the Mediterranean are described in Dixon, P. S. and Irvine, L. M. 1977.

Gelidium crinale (Turner) Lamouroux Frond about 50mm long; main stem compressed and bearing branches arranged pinnately; branches themselves bear awl-shaped branchlets, especially at the apex; texture cartilaginous. Habitat on sandy and rocky bottoms and in shallow water.

Gelidium sesquipedale (Clemenente) Bornet Thuret Frond up to 200mm or longer; main stem slightly flattened, bearing finer side-branches which taper towards their tips; texture horny and not as pliable as *G. crinale*. Habitat on rocks on the shore and in shallow water. Some authorities regard the various forms of *Gelidium* as a species complex. It may be difficult to distinguish between forms as they can be variable in character.

Pterocladia capillacea (Gmelin) Bornet & Thuret Frond up to 150mm long, generally larger and stronger than *Gelidium* species; somewhat compressed with a tufted appearance; main stem may not bear so many oppositely arranged side branches along its proximal part as distally; branches often taper towards their bases and their free ends. Frond is hollow and has a cartilaginous texture. Habitat on rocks on the shore and in shallow water.

Nemalion helminthoides (Velley) Batters Frond up to 250mm long, worm-like and branching either at the base or dichotomously at various points; although the branches taper along their length they are blunt at the tips. Holdfast is minute and disc-like. Colour brown-red. Habitat on the shore and in shallow water.

Asparagopsis armata Harvey This is the gametophyte generation of **Falkenbergia rufolansa** (Harvey) Smith (see below). When the two plants were named it was not realized that they were heteromorphic generations of the same species (see page 15). Frond up to 200mm long, slender, delicate; main stem bears irregularly placed branches; like the stem these branches are mainly covered with small spirally distributed branchlets giving a tufted appearance. A few side branches lack branchlets and bear alternate barbs or thorns. Plant is attached to substrate by a tangle of 'roots'. Habitat in shady pools on the shore and in shallow water.

Falkenbergia rufolansa (Harvey) Smith The sporophyte generation of **Asparagopsis armata** (above) Frond consists of small tufts of tangled fine filaments attached to other algae, often more abundant than the gametophyte above.

Bonnemaisonia asparagoides (Woodward) C.A. Agardh The gametophyte generation (see page 15) of **Hymenoclonium serpens** (Crouan frat) Batters (not described in this book). Frond up to 230mm long; rounded or compressed stem bears alternate side branches, the longest ones towards the base; the side branches bear alternately arranged sub-branches and branchlets. The branchlets are approximately the same length over the whole plant and are arranged in the same plane. Holdfast is small and disc-like. Habitat in hard substrates in shallow water.

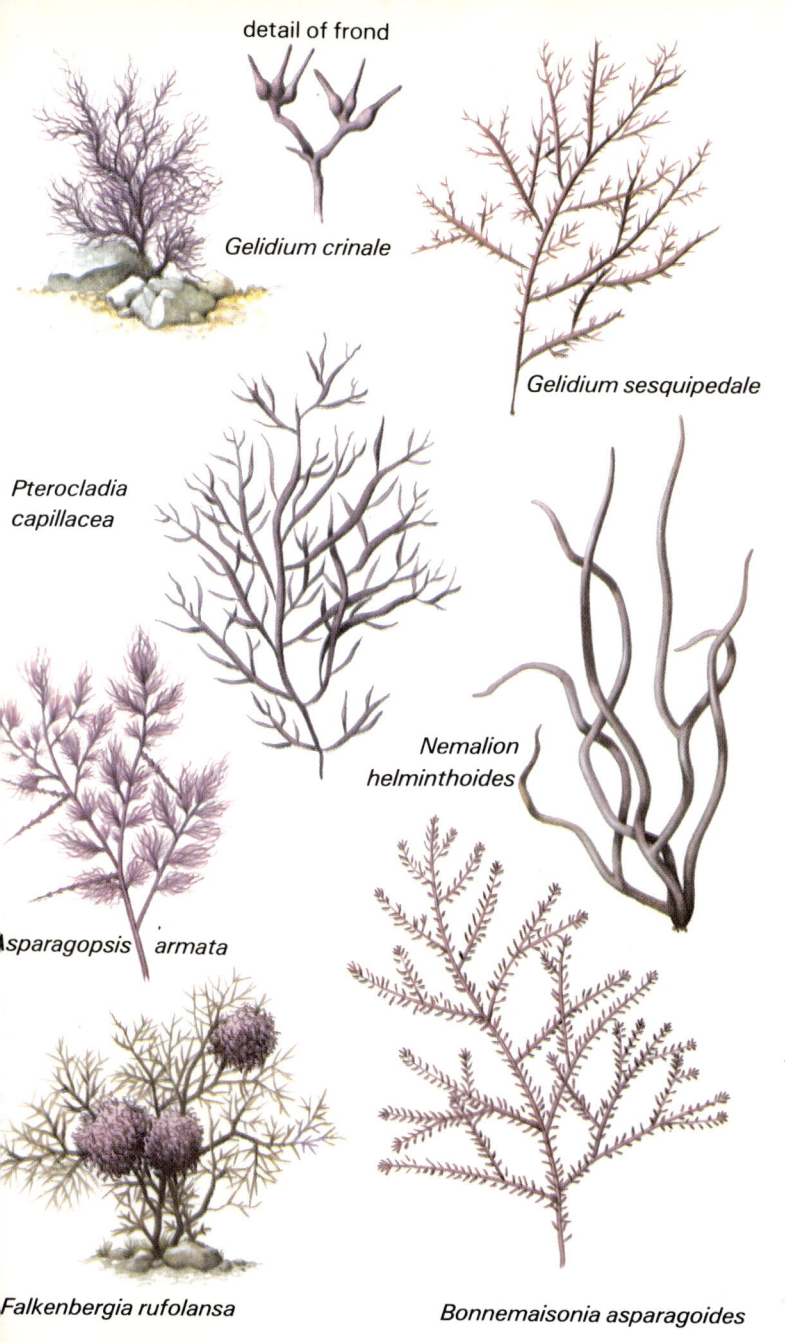

detail of frond

Gelidium crinale

Gelidium sesquipedale

Pterocladia capillacea

Nemalion helminthoides

sparagopsis armata

Falkenbergia rufolansa

Bonnemaisonia asparagoides

Halarachnion ligulatum (Woodward) Kützing Frond 300mm long, strap-like, dichotomously branching; stem bears many side branches which are arranged irregularly or opposite; side branches are much narrower than the stem and often terminate in a notch. Texture is soft and gelatinous. **Colour** pink-red-yellow. **Habitat** on rocks and shells in shallow water.

Catenella caespitosa (Withering) Dixon & Irvine Batters **Frond** up to 30mm long, a moss-like growth of creeping fibres from which tiny irregular branching fronds may arise; may grow in clumps about 50mm across or cover larger areas. **Habitat** on the shore on rocks and in crevices.

Plocamium cartilagineum (Linnaeus) Dixon Frond up to 300mm long, tufted growth with strong main stems which are generally more irregularly branched distally and have fewer or no branches basally; branches carry branchlets the terminal subdivisions of which all occur on the same side; reproductive bodies are almost globular and occur all over the plant. **Habitat** on rocks in shallow water.

Sphaerococcus coronopifolius (Goodenough and Woodward) C. A. Agardh **Frond** up to 200mm high, simple fairly upright growth of fine side branches. **Colour** bright red. **Habitat** on rocks in shallow places. **Similar species** *Plocamium* species.

Gracilaria verrucosa (Hudson) Papenfuss Frond up to 500mm long; stringy stem grows with a number of irregular branches which in turn may carry many slender branchlets; branchlets taper at their bases and at their free ends; small wart-like reproductive bodies may be scattered all over the plant, and several stems may arise from the same fleshy discoidal holdfast. **Habitat** on rocks and gravel on the shore and in shallow water; is one of the few algae which can anchor in sand.

Phyllophora crispa (Hudson) Dixon Frond up to 250mm long; flat ribbon-like stem borne on a very short compressed or cylindrical stipe; branching may be dichotomous; branches generally have blunt tips; texture rigid and crisp. Holdfast very small and disc-like, may coalesce with those of neighbours. **Habitat** usually attached to vertical rocks and pool sides on the shore and in shallow water.

Gigartina acicularis (Wulfen) Lamouroux Frond slender, irregularly branching, up to 80mm long, ends of branches tapering. **Colour** dark green to brown-red or almost black. **Habitat** on stones and on other algae in shallow water.

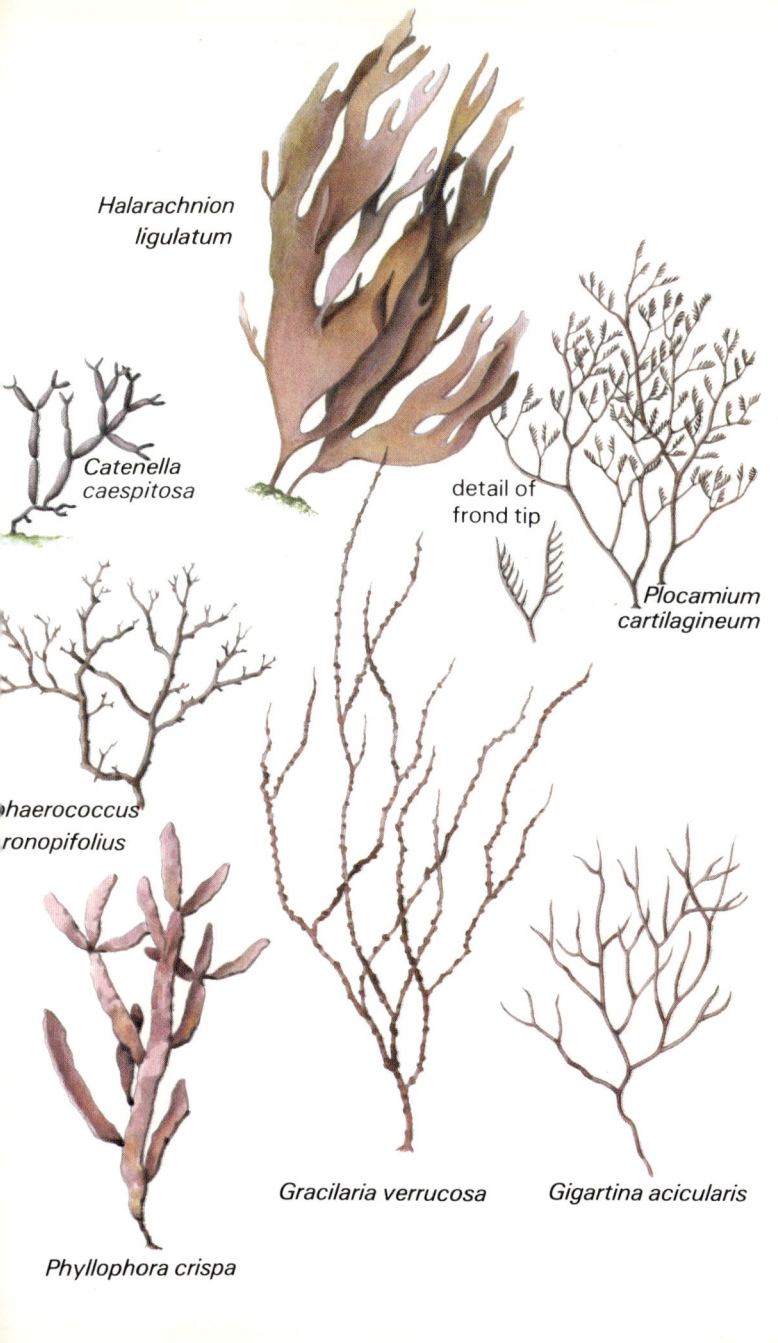

Halarachnion
ligulatum

Catenella
caespitosa

detail of
frond tip

Plocamium
cartilagineum

...haerococcus
...ronopifolius

Gracilaria verrucosa

Gigartina acicularis

Phyllophora crispa

Hypnea musciformis (Wulfen) Lamouroux Frond up to 300mm long; upright stem bears a number of irregularly disposed, almost pinnate branches. Colour black-red-green. Habitat always growing entangled with other weeds in sheltered places.

Peyssonnelia squamaria (Gmelin) (Decaisne) Thallus consisting of leaf-like growths adhering to the substrate by 'roots' developing on their undersides; growths may reach 100mm or more; 'leaves' spread out, upper surfaces may be marked with concentric rings of dark red and red-brown. Habitat on rocks and other weeds on the shore and down to 60m.

Hildenbrandia species Thallus about 30mm across, consisting of a patch of thin tissue closely adhering to a stone or rock; with care it may be peeled off; not chalky; when dry it loses its sheen. Colour pink-brown-red. Habitat on the shore and in shallow water. The status of species hitherto described is questionable. *H. prototypus* (see illustration) is typical of the genus.

Corallina elongata (Ellis & Solander) Frond chalky, up to 80mm long, branching at the base and therefore having a tufted appearance; branches bear pinnate branchlets; under a hand lens the stem appears 'segmented' with segments ovoid or triangular; terminal reproductive bodies bear 'horns'. Colour purple-red-pink-yellow-white. Habitat attached to rocks in pools and shallow water.

Corallina officinalis Linnaeus Frond chalky, up to 120mm long, usually not branching immediately from base; main stem carries branches arranged exactly opposite; branches bear opposite branchlets. The stem appears 'segmented'; a hand lens reveals that segments are longer than they are broad and the joints between them are pliable; terminal reproductive bodies lack 'horns'. Holdfast chalky and encrusting. Habitat on rocks and in pools on the shore and in shallow water.

Jania rubens (Linnaeus) Lamouroux Frond up to 50mm long, chalky and jointed as in *Corallina* but branching dichotomously and not opposite; often grows in dense tufts and in spring bears conspicuous rounded reproductive bodies. Holdfast minute and disc-like. Colour rose-red. Habitat growing attached to other algae.

Lithophyllum incrustans Philippi Thallus comprises a patch of chalky tissue up to 40mm thick; growths irregular in outline, smooth or bumpy in appearance, sometimes overlapping each other or closely encrusting the substrate or shells; the periphery of the thallus is not thicker than the rest of the growth (c.f. *Phymatolithon calcareum*, (see page 36). Colour mauve, purple, red or yellow, darker in shaded places. Habitat in pools on the shore or in shallow water. Similar species several others occur, which are described on page 36. They are often difficult to separate and are best regarded as an aggregation. They are often collected for marl used in building work.

Hypnea musciformis

Peyssonnelia squamaria

Hildenbrandia prototypus

Corallina elongata

Jania rubens

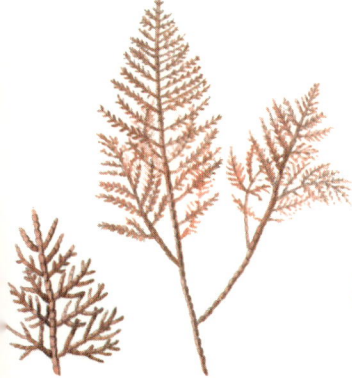

Corrallina officinalis

Lithophyllum incrustans

Phymatolithon calcareum (Pallas) Adey & McKibbon Thallus when young, a patch of chalky tissue up to 40mm thick with an irregular outline and smooth or bumpy appearance, growing over the substrate or other individuals; periphery thicker than rest of growth. When older it becomes erect, nodular and branching, reaching up to 80mm. Colour violet-red. Habitat encrusting rocks and pebbles or free (especially older specimens) on the shore or in shallow water.

Lithothamnion fruticulosum (Kützing) Foslie Similar to *L. calcareum*: thallus with chalky outgrowths up to 30mm thick and up to 100mm across; upper surfaces with short branching processes, either blunt or sharp at the end. Colour violet-red-greyish. Habitat on rocks and softer bottoms down to 80m.

Fosliella farinosa (Lamouroux) Thallus circular or irregular, forming encrusting growths on fronds of other algae, sea-grasses, hydroids and worm tubes. Habitat encrusting other organisms on the shore and in shallow water.

Pseudolithophyllum expansum (Philippi) Lemoine Thallus up to 100mm across, encrusting, chalky, irregularly shaped and adhering to the substrate except at the edges which are free from it, this serving to distinguish this species from *Lithophyllum incrustans*. Colour pink-red. Habitat down to 60m on rocks, calcareous algae and brown algae.

Antithamnion cruciatum (C. A. Agardh) Nägeli Frond up to 50mm long; main stem bears alternate branches and branchlets; tips of branchlets are tufted: under a microscope these will be seen to consist of many single oblong cells joined end to end, each cell bearing two opposite branchlets. Habitat on rocks, often in muddy places. Similar species about four.

Callithamnion corymbosum (Smith) Lyngbye Frond up to 70mm long; main stem bears alternately arranged branches which carry delicate filamentous alternate branchlets. Habitat on rocks and weeds on the shore and down to deep water.

Cryptonemia lomation (Bertol.) J. Agardh Frond up to 80mm long, wide and leaf-like with irregular branching lobes and irregular wavy outline. Colour dark red. Habitat in shallow water, often attached to stones and coralline algae.

Phymatolithon calcareum

Lithothamnion fruticulosum

Fosliella farinosa

Pseudolithophyllum expansum

Antithamnion cruciatum

Cryptonemia lomation

Callithamnion corymbosum

Grateloupia filicina (Lamouroux) C. A. Agardh Frond up to 120mm long; growth tufted and compressed; main stem tapers at free end and at base, leaves alternate, opposite the branches; branches carry alternate or opposite branchlets. Holdfast disc-like. **Habitat** on stones and rocks on the shore or in shallow water, sometimes near freshwater outfalls.

Halymenia dichotoma J. Agardh Frond up to 90mm long, dichotomously branching; branches taper towards their tips. **Colour** blackish-violet to dark green. **Habitat** in shallow water, sometimes in shaded places.

Lomentaria linearis Zanard Frond up to 200mm long, thin, cylindrical, irregularly or dichotomously branching; frond appears as a series of jointed segments. **Colour** pink-red. **Habitat** down to 60m amongst coralline algae and some brown algae.

Rhodymenia corallicola Ardisson Frond up to 190mm long, flattened, branching, twisted over near the junctions of two branches; small rounded holdfast. **Colour** dark red to pink. **Habitat** on hard and soft substrates down to 70m.

Botryocladia botryoides (Wulfen) Feldm. Thallus up to 180mm long, narrow, rounded, alternately branching and bearing spherical growths towards branch tips. **Habitat** from the shore down to about 80m.

Nitophyllum punctatum (Stackhouse) Greville Frond up to 500mm long, broad, wedge-shaped, membranous and growing directly from the small disc-like holdfast; one or more fronds may develop from each attachment point. Branching is fairly regular and dichotomous; the presence of many small terminal branchlets may make the margins look frilly; there are no 'veins'; texture is delicate. **Colour** red-pink, sometimes iridescent near tips. **Habitat** on various seaweeds in pools on the shore and in shallow water.

Dasya hutchinsiae (Harvey) (Not illustrated) Frond up to 100mm long; main stem bears alternately arranged side branches, and these bear fine filamentous branchlets; small flask-shaped reproductive bodies may be carried on stalks by the branchlets. Holdfast consists of 'rootlets'. **Colour** brown-crimson. **Habitat** on rocks on the shore and in shallow water.

Ceramium rubrum (Hudson) C. A. Agardh Frond up to 300mm long; main stem branches dichotomously but unevenly, and arrangement of the side branches is similar. Terminal branchlet tips often point inwards to give forcep-like effect; texture cartilaginous, overall appearance bushy. **Colour** variable, deep red-brown-yellow; under a hand lens stems and branchlets will be seen to be banded with darker pigment. **Habitat** on rocks and weeds on the shore and in shallow water. **Similar species** quite a number. N.B. this species is variable in appearance according to locality.

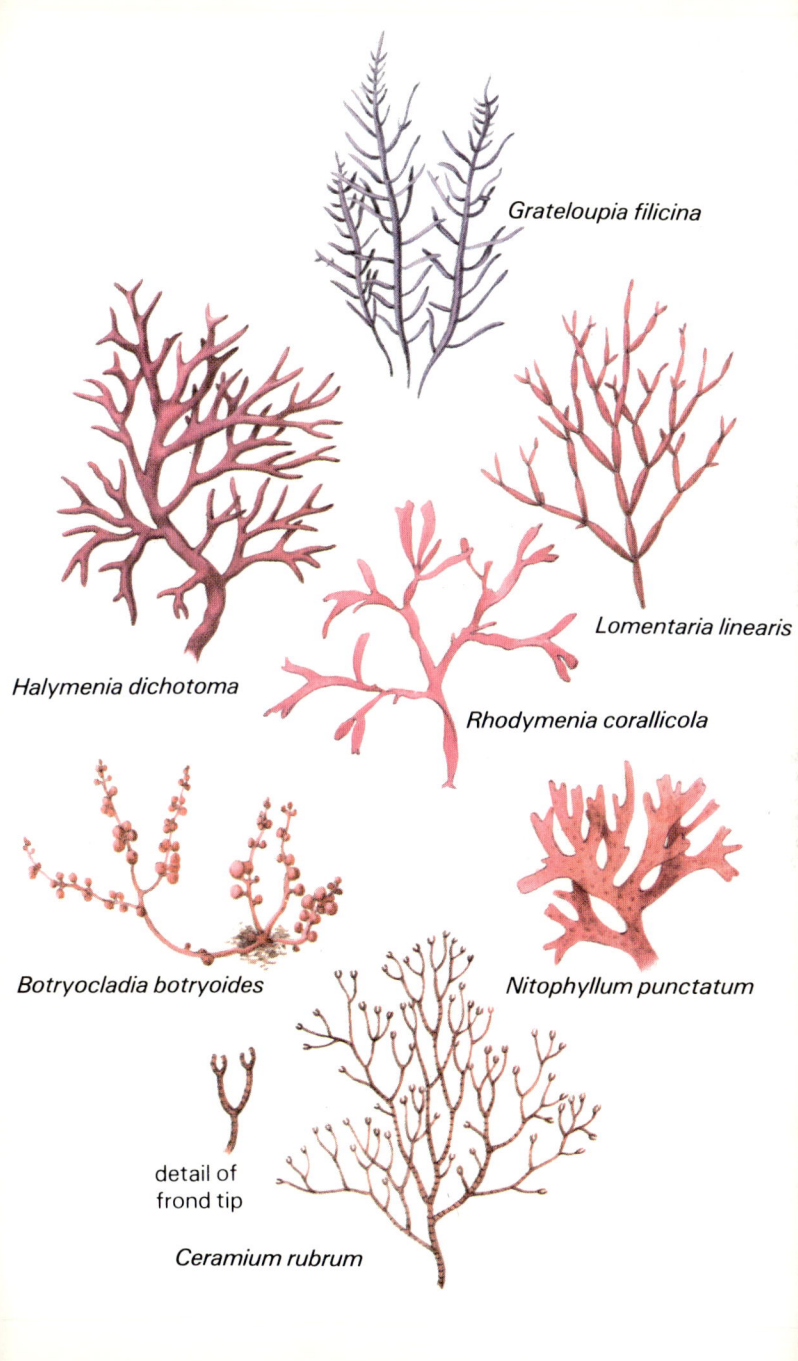

Grateloupia filicina

Lomentaria linearis

Halymenia dichotoma

Rhodymenia corallicola

Botryocladia botryoides

Nitophyllum punctatum

detail of
frond tip

Ceramium rubrum

Laurencia pinnatifida (Hudson) Lamouroux Frond up to 300mm long but often less; main stem flattened, usually well developed and alternately branched; branches subdivided into smaller branchlets; texture cartilaginous. Holdfast disc-like with 'rootlets'. Colour purple-brown to yellow-green. Habitat on the shore.

Laurencia obtusa (Hudson) Lamouroux Frond up to 150mm long, a thick growth often rounded in outline; main stem is rounded in section, bearing alternate or opposite branches which may be arranged on the stem in a spiral fashion, branches and branchlets becoming smaller towards the tips; texture cartilaginous. Holdfast small and disc-like, sometimes with 'rootlets'. Colour variable: purple, pink or yellow. Habitat usually on other seaweeds in pools and in shallow water.

Polysiphonia elongata (Hudson) Sprengel Frond up to 300mm long; main stem clearly defined, overall appearance bushy. Side branches develop about 20mm from the base and are arranged alternately around the stem; sub-branches carry fine, clustered, terminal branchlets, especially in spring; texture is gelatinous. Holdfast consists of fine 'rootlets'. Colour dark red-yellow. Habitat on rocks, stones and shells as well as on some larger algae on the shore and in shallow water.

Polysiphonia sertularioides (Grateloup) J. G. Agardh (Not illustrated) Similar to *P. elongata*. Frond up to 250mm long, main stem less clearly defined; overall appearance bushy; side branches develop at a little distance from the base and may carry small club-shaped reproductive bodies at their tips which end in a tuft of parallel filaments. Colour dark red-brown. Habitat among weeds, stones and sand in shallow water. Similar species several other species of *Polysiphonia* occur in the Mediterranean, and the identification of these is a specialist task.

Polysiphonia urceolata (Dillwyn) Greville Frond up to 250mm long, appears tufted and filamentous; main stem bears alternate branches in all planes around the long axis; several stems may develop from the creeping 'roots' which form the holdfast. Characteristically urn-shaped reproductive bodies may occur on the frond. Colour red-purple-brown. Habitat on stones, weeds and shells down to 20m.

Vidalia volubilis (Linnaeus) J. G. Agardh Frond about 100mm long; stem flat and leaf-like with serrated or slightly branching edges, often turned into a spiral. There may be a few side branches arising from the centre of the frond where there is a midrib; holdfast is disc-like. Habitat on stones and on sandy and muddy substrates down to 80m.

Porphyra umbilicalis (Linnaeus) J. G. Agardh Frond up to 200mm long, irregular, broad, gelatinous, appearing as a membranous growth arranged in 'leaves' all attached at one point. Colour red-purple-green, becoming black when dry. Habitat attached to stones and rocks which themselves may be covered by sand grains.

Porphyra leucostricta Thuret Frond up to 300mm long, leafy, with an elongated oval outline. Colour blue-green, purple-red or sepia-brown. Habitat growing on stones and larger algae.

Laurencia pinnatifida

Laurencia obtusa

Polysiphonia elongata

Polysiphonia urceolata

Vidalia volubilis

Porphyra umbilicalis

Porphyra leucostricta

Group Lichenes Lichens

Lichens have been studied less than algae. They live as dual plants, partly composed of algal tissue and partly of fungal tissue. Formerly they were classified as a discrete group, but recent work has shown that they should be classified with the fungi. Their life processes resemble those of both the algae and the fungi, but they also display particular properties of their own. Many botanical text books provide a good account of their structure and function. Here it may be said that the lichen thallus consists usually of a flat or leafy outgrowth, encrusting or standing up from a substrate of other plants, rocks or shells. The tissue may be brittle or soft, rough or smooth, tufted or flat. Many lichens have fruiting structures which may be important in identification, an operation which is often one for the specialist. Several English language texts are available which include lichens found in the Mediterranean area, but no specific reference work to the Mediterranean lichens exists in English. Reference should be made to Dobson, F. 1979 and Duncan, U.K. 1959 and 1970 and Ferry B.W. and Sheard, J. W., 1969.

Verrucaria adriatica A. Zahlbr. **Thallus** thin and encrusting, covering extensive areas of rock with green fruiting bodies. **Habitat** on rocks around the extreme high water mark.

Lichina confinis (O.F. Müller) C.A. Agardh **Thallus** tufted, branching, erect, about 5mm high, resembling a minute seaweed. **Habitat** on rocks at the extreme high water mark. **Similar species** *L. pygmaea* (below) which is not so richly branched.

Lichina pygmaea (Lightfoot) C.A. Agardh **Thallus** tufted, branching, erect, about 10mm high, resembling a minute seaweed. **Habitat** on the shore where it may be immersed periodically in seawater. Similar to *L. confinis* but is coarser.

Caloplaca marina Weddell **Thallus** flattish, dark orange and encrusting with scattered coarse granules forming patches up to 100mm across; not leafy. **Habitat** on rocks above the shore. **Similar species** *C. aurantia* (Pers.) Helb. which has an extremely flattened thallus and is creamy-orange in colour. In the illustration the centre of the specimen is dying.

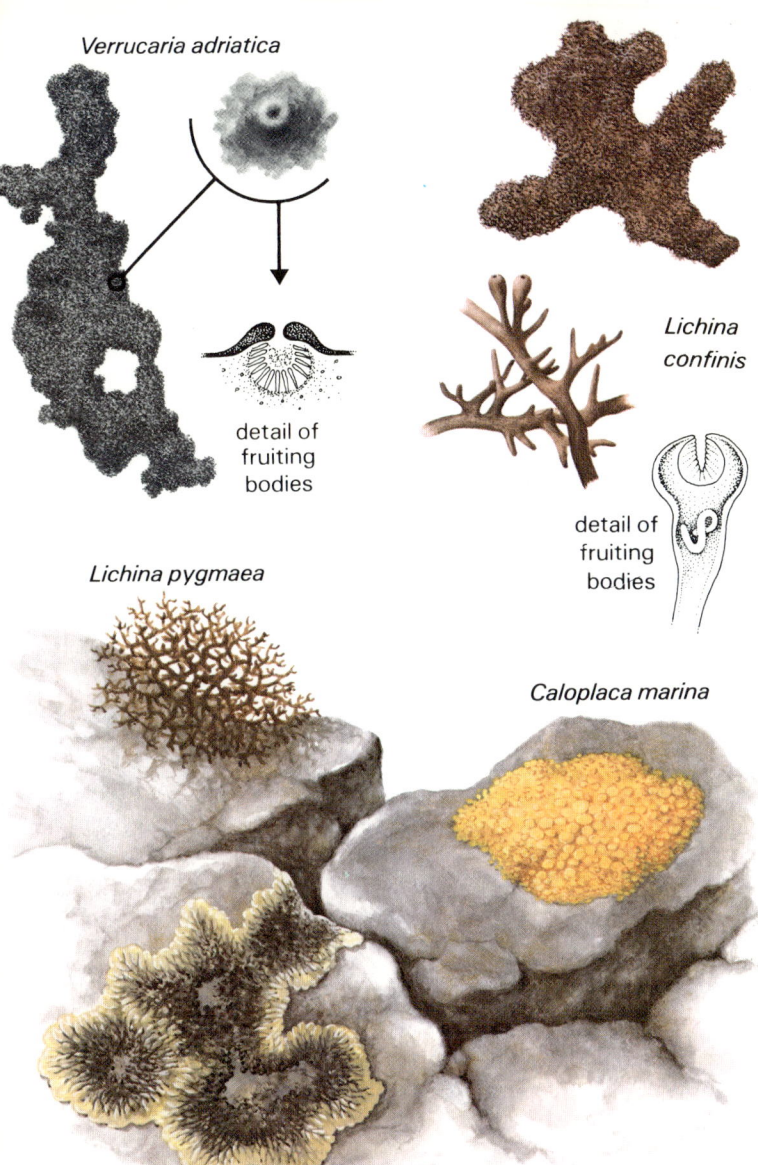

Verrucaria adriatica

detail of
fruiting
bodies

*Lichina
confinis*

detail of
fruiting
bodies

Lichina pygmaea

Caloplaca marina

Caloplaca aurantia

Subkingdom Angiospermae
Marine angiosperms (sea-grasses)

The angiosperms, or flowering plants, are the most familiar representatives of the plant kingdom on land and they include the grasses, herbs, shrubs and trees. Although a number of species are adapted to live partly or wholly in fresh water, very few can tolerate life in the sea. However two groups, the sea-grasses and the mangroves, have evolved that survive and flourish in salt water and may make a conspicuous contribution to marine floras, especially in warmer waters. In the Mediterranean extensive 'meadows' of sea-grasses are not uncommon and provide shelter and the basis of food for major animal communities. Reference should be made to Schrimper. Mediterranean recorded species = 4.

Zostera marina Linnaeus **Eel Grass, Sea-grass** or **Grasswrack** (Only part of leaf illustrated) Flat, long, narrow leaves rise up for 1m, sometimes further; between 5 and 10mm wide; veins of leaves as shown in illustration; inconspicuous flowers, to some extent resembling those of terrestrial grasses, may occur in spring or summer. **Colour** dark green or grass-green. **Habitat** usually on sheltered beaches or estuaries on gravel, sand and mud; **Similar species** *Z.nana* (see below).

Zostera nana Roth **Slender Eel Grass** (Only part of leaf illustrated) Flat, very narrow leaves rise up for 150mm, occasionally further; about 1mm wide; veins of leaves as shown in illustration. **Habitat** on mud banks in estuaries etc., on the shore. **Similar species** *Z. marina* (above).

Posidonia oceanica (Linnaeus) **Neptune Grass** Flat, long, narrow leaves rise up for 300mm, sometimes further; about 10mm wide; base of leaves heavy and shaggy-looking; fairly conspicuous flowers may be found in summer. **Colour** green-yellow. **Habitat** on soft substrates down to 50m; **Note** broken leaves of this plant, when mixed with sand grains and rolled about by the waves, may form brownish soft balls sometimes known as 'sea-balls' (see illustration).

Cymodocea nodosa (Ucaria) Areschoug Narrow, flat leaves rise up to 200mm; thick branching rootstock, leaf bases lacking shaggy appearance of *P. oceanica*. **Habitat** often associated with *Zostera* (above) on mud and sand down to 10m.

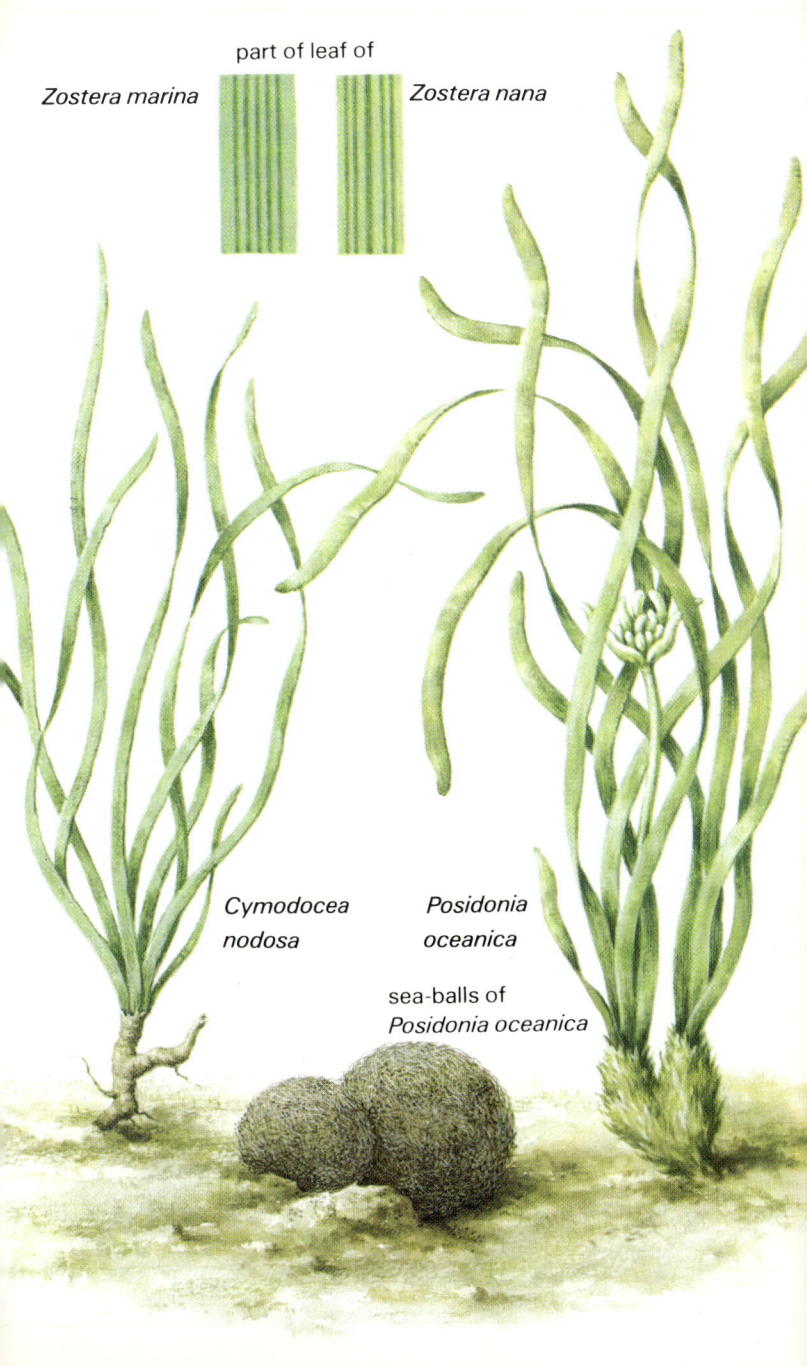

part of leaf of

Zostera marina *Zostera nana*

Cymodocea nodosa

Posidonia oceanica

sea-balls of *Posidonia oceanica*

Animal Kingdom

The animal kingdom comprises an immense variety of organisms ranging from simple, one-celled creatures to complex, multicellular mammals. Between these extremes lies a broad spectrum of forms showing immense diversity.

Animals differ from plants in several fundamental ways. Firstly they are irritable and may respond to stimuli by movements rather than by growth. Secondly they are catabolic rather than anabolic, in other words, for their nutrition they rely on breaking down organic material into its constituent parts rather than elaborating organic compounds from inorganic ones. In this respect animals form themselves into very complex food chains which are founded upon the plants or primary producers. Different animals are often related in the food chain, carnivores preying upon herbivores. In any community there may be a chain or web of energy relationships, and this is as true in the sea as on land. In the marine environment, there is not the diversity of plant life that characterizes terrestrial habitats. The function that is fulfilled by the lowly algae is just as important, however. Whether they are the larger algae of the shallow sea or the unicellular plants of the plankton, they provide food for the herbivorous animals. On the seashore we find weed-eating sea-snails like the periwinkles: in shallow water there are the herbivorous sea-urchins and grazing fishes. Next in the chain are the primary carnivores such as whelks, starfishes, sea-anemones and fishes which in their turn feed on the herbivorous animals. In the ocean, minute crustaceans like copepods form the primary converters of plant protein into animal protein. In addition to finding a food supply, either as large macroscopic seaweeds or minute, suspended food particles, animals need shelter to protect them from their enemies and to provide them with the facilities for reproduction. In this respect other factors are as important as the supply of food in determining whether or not a particular species of animal can survive in a particular location. Different species are able to withstand different circumstances in terms of both the physical and the biological conditions for life. Thus it is that the shore and sea permit such a diversity of life; for the movements of the tides and the consequent alternation of exposure to air and sea provide an almost infinitely graduated range of environments from the terrestrial region immediately above the highest point to which the tides flow, down to the deepest seabed where physical conditions are virtually constant.

The seashore is probably the finest training ground for the zoologist. The principles of ecology, as well as other aspects of interrelationships between and within species, can be appreciated in almost every type of habitat, and also provide examples of the greatest evolutionary interest. All known animal types are represented in the sea. Whilst some groups have conquered fresh water and land with moderate or great success, many have prospered only in the sea. Furthermore some groups, such as the Ctenophora (comb-jellies) and the Echinodermata (starfishes and sea-urchins), are exclusively marine.

The primary divisions or groups in the animal kingdom are known as *phyla* (singular phylum). and all the members of a phylum are considered by zoologists to be related to a common ancestral form by descent. When seeking to classify animals, zoologists must take into account their level of organization and style of organic architecture. In this respect, the following points are important. **1** Is the animal composed of one or many cells? **2** If many cells, are these arranged throughout the body as one type of tissue, or are they differentiated into various tissue types which can then form organs such as glands or muscles? **3** How are the tissues arranged throughout the body? **4** Is there a body cavity inside the animal in addition to the alimentary canal or gut? Another very important feature is the form of symmetry that an animal displays. Some are totally asymmetrical like a few of the sponges, but in fact such examples are rare. More frequent are radially symmetrical animals like the jellyfishes and sea-urchins (as adults, echinoderms display a very special type of radial symmetry based on five radial axes). The great majority of animals display bilateral symmetry (i.e. when divided in two from the head end towards the tail end they produce two halves which are mirror images of each other).

In the descriptions that follow, the text contains, as far as possible, details which complement the illustrations. A quick glance will show that in all cases the type of information that is provided is not exactly equivalent. Where this is the case it is largely a reflection of the state of the information currently available to scientists. In some cases it is known, for example, exactly how deep in the sea a particular species lives. In other cases this information is not known, and it is necessary to rely on a more general statement. It should also be remembered that the collector or beachcomber may find juveniles which are not as large as the adults generally encountered for the species.

Finally, as has already been pointed out, it is not possible in a book of this size to provide details of all the animals which might be encountered on the shore or in the shallow sea. However, as far as possible, references have been provided in the text and in the bibliography which should enable the reader to pursue the line of enquiry for a particular species a good deal further than is possible here.

Phylum Porifera Sponges

These are sessile animals whose bodies consist of a single cavity with a major exhalent opening, and many smaller inhalent openings which are partly lined by special cells called choanocytes. The bodies also contain calcareous or siliceous spicules, or horny fibres, which provide support.

Sponges are the simplest animals treated in this book, yet for all their simplicity they are among the most difficult to describe carefully and scientifically. They are arranged in three classes: the Calcarea, with calcareous spicules; the Hexactinellida or Triaxonida, with siliceous spicules each of which has a six-rayed pattern; and the Demospongiae, with siliceous spicules which are never six-rayed, and/or horny fibres made of material known as spongin. It will be realized from these divisions that the nature and shape of the sponge spicules are of great importance when classifying and identifying sponges. Unfortunately these spicules cannot be seen without the aid of a microscope and the facilities for making the necessary microscopical preparations. However, it is possible to identify sponges to some degree by examining their gross form, texture and coloration, as well as by considering their habitat and distribution. Identifications based on these characters only can be provided here.

Sponges occur on almost all types of seabed from the lower shore down to the ocean depths. The form of their bodies depends to a great degree on the amount of exposure they have to face. Thus, in very exposed conditions sponges will be rounded or flattened against the substrate. In sheltered places they often assume a plant-like growth pattern with erect and delicate branching stems and shoots.

The basic body form of the sponge is illustrated in fig. 2. This diagram also indicates the movement of sea water through the sponge. Water supplies the animal with oxygen and food in the form of minute, suspended particles. This water enters the body by a number of pores and passes along the cavity or *paragaster* to leave by the large exhalent pore or *osculum*. Other sponges are more complex in their organization. Fig. 2 also shows how a compound arrangement may be formed from a number of units which become arranged round a communal paragaster and which share a central osculum. There are many variations on sponge architecture which are beyond the scope of this book, although most general text books on zoology provide good accounts. At first sight the paragaster may be compared with the gut of a higher animal, but it cannot strictly be regarded as such, for it lacks the specializations of tissues associated with a digestive tract. The cells which compose the sponge body are of very few types only, and those which line the paragaster absorb their own food requirements by directly ingesting suitable particles. Because of their level of organization, no organs can be identified within the sponge body which is thus regarded as being developed at the cellular grade. This lack of differentiation of cells into a number of tissue types making up organs means that sponges are capable of regenerating themselves from broken fragments. Despite their simple level of organization, sponges are able to undergo sexual reproduction with the formation of eggs and sperms which, after fertilization, give rise to a free-swimming larva. If the larva settles on a suitable substrate, it will develop into a new sponge.

An interesting feature of sponges is the extent to which they become associated with other organisms. If a large sponge is broken open, a number of other animals such as worms and crabs may be found residing in the cavity. Some sponges are associated with hermit crabs, giving camouflage and protection to the crab which in turn transports the sponge to new feeding areas.

Although a great number of sponge species are known, very few are of commercial importance. One exception is *Spongia officinalis*, the bath sponge, which is extensively collected from some of the Greek island waters.

The sponges are a notoriously difficult group to work on, and there are relatively few books available which help with their identification. Berquist, P. R. 1978 gives a general account of their biology and the classification of calcareous sponges was revised by Burton, M. 1963. Vosmaer, G. C. J. 1933 provides a very technical but well illustrated account of the Naples sponge fauna. Theodor, J. 1964 and Guiterman, D. 1979 (which deals with some sponges occurring in the Mediterranean) should also be referred to.

The layman will find that most of the common Mediterranean sponges are illustrated in this section. There are many less common ones which are not included. Because of the state of research on Mediterranean sponges it is not possible to give an accurate total for the number of species existing in this sea.

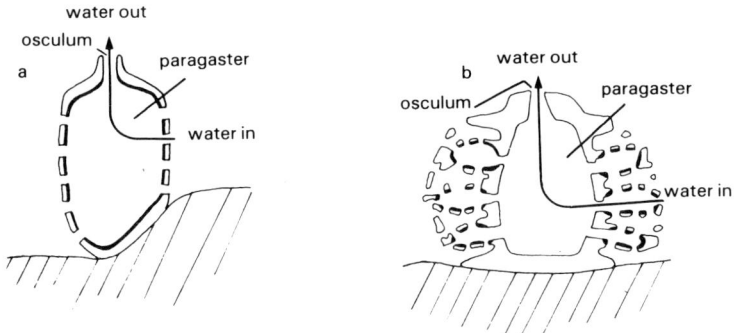

Fig. 2 Diagrammatic sections through (a) simple, and (b) complex sponges to show water currents and position of choanocytes (shown in heavy black)

Class Calcarea

Sponges which contain calcareous spicules only. Their bodies are generally small and cup-shaped or vase-shaped. These sponges are usually pale in colour and prefer to grow in shady places away from the light. They do not compete successfully with algae in illuminated situations but flourish under overhangs, in grottos and dark caves.

Clathrina coriacea (Montagu) Sensu Burton *(=**Leucosolenia coriacea**)* **Form** consists of a branching network of narrow-walled tubes; these are encrusting and do not grow up from the substrate. The network may be several millimetres thick and up to 100mm wide; one osculum is shared between several tubes; texture soft. **Colour** white-grey, occasionally pink or yellow. **Habitat** on clean rocks, often under overhangs in shallow water.

Sycon ciliata (Fabricius) Sensu Burton *(=**S. raphanus**)* **Form** tubular vase-shaped sponges up to 30mm high, outer surface with a shaggy appearance when viewed under a hand lens; a crown of larger stiff spicules may be visible around the osculum at the apex; texture moderately soft to firm. **Colour** cream-grey-brown. **Habitat** on rocks and shells from the lower shore down to about 100m, usually growing in clumps.

Leuconia aspera Schmidt *(=**Leucandra aspera**)* (Not illustrated) **Form** a thick vase-like growth bearing nodules, with a terminal osculum; may occur singly or in numbers, in which case they may be compressed sideways. **Colour** white-brown. **Habitat** in holes, crevices and dark places down to deep water.

Leucosolenia botryoides (Ellis & Solander) Sensu Burton **Form** tubular branching sponge; vertical tubes about 20mm high arise from a mass of creeping root-like canals; oscula large; texture soft. **Colour** whitish. **Habitat** attached to seaweeds on the shore and in shallow water.

Grantia compressa Fabricius **Purse Sponge** **Form** inflated and urn-like when under water, but collapsing into a flat purse-like object when lifted into the air; having a large terminal osculum. **Colour** white-grey-yellow. **Habitat** under overhangs and among seaweeds in shallow water.

Class Demospongiae

Sponges with siliceous spicules (never six-rayed) and/or horny fibres of spongin. Their bodies may be variously shaped and are sometimes quite large. They are often brightly coloured and grow in well illuminated places.

Oscarella lobularis (Schmidt) **Form** a fleshy encrusting growth with rounded nodules reaching up to 100mm in diameter and 3–6mm in thickness; oscula borne on top of nodules; surface texture velvety. **Colour** yellow-brown, occasionally red, green or blue. **Habitat** on stones, rocks and algae in shallow water, usually in well illuminated places.

Chondrosia reniformis Nardo **Form** rounded, oval or lobed, 50–100mm wide and up to 30mm high; skin smooth, texture firm and rubbery; oscula not visible. **Colour** grey-brown or speckled dark and pale with occasional irregular marks; interior white and firm if cut open. **Habitat** on rocks and under stones in shallow water.

Geodia mulleri (Flemming) *(=**G. gigas**)* **Form** rounded and often globular, up to 400mm across; surface firm but often indented; outer skin bears pointed spicules which provide a settlement platform for other organisms e.g. algae, molluscs and worms; inner tissue soft and penetrated by many canals.

Clathrina coriacea

Sycon ciliata

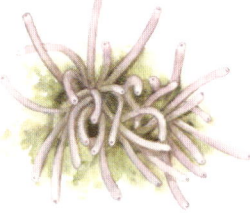

Leucosolenia botryoides

Grantia compressa

Oscarella lobularis

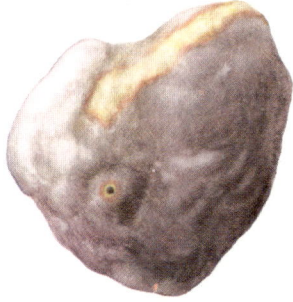

Chondrosia reniformis

Geodia mulleri

Suberites domuncula (Olivi) **Sulphur Sponge** or **Sea-Orange** Form very variable: rounded, globular or elongated; up to 300mm in diameter. Surface is smooth but not velvety; texture fleshy, but will break if unduly stressed; somewhat elastic; when removed from the water will contract to about three-quarters of its original size. **Colour** orange-yellow. **Habitat** from shallow water down to 200m, encrusting or associated with whelk shells occupied by hermit crabs. **Similar species** *S. cavernosus* (not illustrated), which is often stalked and the surface of which is smooth and velvety.

Tethya aurantium (Pallas) Form rounded, globular, up to 100mm in diameter; outer surface covered with conspicuous warts. Examining its internal structure is an aid to reliable identification: fan-like bundles of spicules run at right-angles to the surface and end in a surface wart. Osculum and pores are about the same size. **Colour** pale gold. **Habitat** growing singly or in colonies on stones and rocks as well as in caves from shallow water down to 130m.

Spirastrella cunctatrix Schmidt Form irregular and encrusting, appearing as low masses 50mm across or more; the outer surface may be patterned; texture firm; without papillae; oscula visible. **Colour** variable, orange to blue. **Habitat** on hard substrates.

Cliona celata (Grant) Form massive, rounded or plate-like; spherical growths up to 400mm in diameter and encrusting layers up to 1m long; oscula carried on top of narrow-walled tubes projecting 10mm above the surface; outer texture firm and tough. Young specimens begin life by boring into rock and shells which as a result may have a honeycombed appearance. **Colour** yellowish. **Habitat** encrusting limestone rock, shells and coralline algae, usually in shallow water. N.B. *Cliona* contracts somewhat when removed from the water. The status of the genus is under review and more species may be described.

Axinella polypoides (Schmidt) Form fan-like or erect and pillar-like, with few branches, reaching up to 500mm high. The branches are often oval in cross-section; oscula have shallow surface grooves radiating from them giving a star-like effect; surface velvety and smooth, without ridges or undulations. **Colour** yellow-orange-red. **Habitat** from 30 down to 100m.

Axinella cannabina (Esper) (Not illustrated) Somewhat like *A. polypoides*: oscula are carried on rounded hillocks. **Colour** yellow-orange. **Habitat** often found in muddy places.

Axinella verrucosa (Esper) Form erect and branching, bush-like; up to 250mm; variable in appearance. **Colour** gold-orange-pink. **Habitat** generally on hard substrates between 10 and 100m.

Halichondria panicea (Pallas) **Breadcrumb Sponge** Form very variable, from a low encrusting sponge to a mass or bushy clump of rejoining branches; oscula occur on varied cones in the encrusting forms but are at the tips of the branches in the bushy types; surface texture smooth. Colonies may reach 200mm across and 20mm thick. **Colour** green in the presence of sunlight due to algae living in the tissue; cream-yellow in shade. N.B. there is a red form in the Mediterranean which may be a related species.

Mycale massa (Schmidt) Form massive, round, encrusting; up to 50mm across or larger. **Colour** yellow, grey or pale orange. **Habitat** sometimes attached to shells and coralline algae which may occur on muddy or sandy substrates from 15m down.

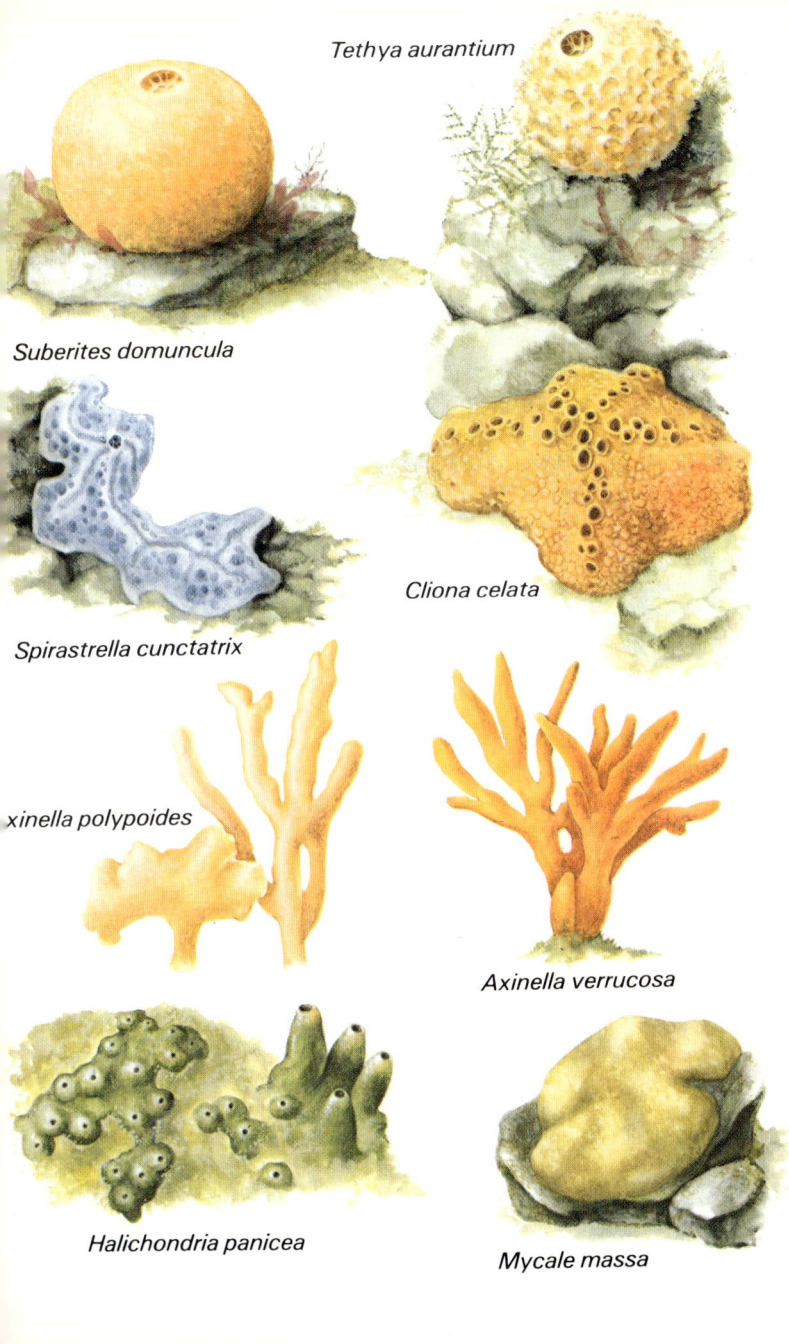

Tethya aurantium

Suberites domuncula

Spirastrella cunctatrix

Cliona celata

xinella polypoides

Axinella verrucosa

Halichondria panicea

Mycale massa

Myxilla incrustans (Johnston) **Form** thick, often massive, encrusting; surface with numerous channels overlain by thin fibres of tissue; up to 150mm across. Texture is soft and elastic; oscula circular, borne at the top of raised hillocks. **Colour** yellowish, sometimes red. **Habitat** on rocks down to 130m, especially in clear water, sometimes encrusting spider-crab shells. N.B. there are several closely related species and precise identification is a specialist task involving use of the microscope.

Hymeniacidon sanguinea (Grant) **Form** encrusting with growths up to 500mm across; there are many small oscula randomly placed; surface is furrowed or smooth. Form may be very variable. **Colour** orange-scarlet to deep red. **Habitat** on rocks on the shore and in shallow water.

Haliclona oculata (Pallas) **Mermaid's Glove** **Form** branching and erect, either with a round axis of constant diameter, or alternatively a mass of rejoining branches (the latter more typical of regions with strong currents). Branch tips are rounded, elastic and soft; branch bases are firmer; there is no strong central core to the branches as in some other sponges; surface texture is soft and slightly velvety (*Axinella polypoides* is much more velvety). Oscula are small, 0.5–2mm in diameter, round and not raised. **Colour** orange-pink. **Habitat** often where there are strong currents; also appears to be able to tolerate sediments.

Petrosia ficiformis Poiret **Form** variable: fig-like with high–standing oscula up to 50mm high, or lattice-like and up to 200mm across with conspicuous oscula, encrusting; hard and paper-like. **Colour** greenish below, brown-violet above. **Habitat** in deep water under rocks and boulders. May be associated with the nudibranch *Peltodoris atromaculata* (see page 152).

Spongia officinalis Linnaeus *(= Euspongia officinalis)* **Bath Sponge** **Form** moderately large, growing in massive irregular or globular shapes, up to 200mm across, occasionally much larger; oscula relatively few, raised from the surface and slightly crater-like. There are no siliceous spicules and the skeleton consists of spongin-like fibres (these remain when the sponge has been cleaned and cured for use in the bathroom). **Colour** red-brown-green. **Habitat** on rocks from shallow water down to 50m or more. **Similar species** few, but see *Hippospongia communis* below.

Hippospongia communis Lamarck *(= H. equinea)* **Horse Sponge** **Form** flat, rounded or lobe-like, reaching up to 600mm across; conspicuous oscula not raised from the surface; there are no siliceous spicules and the skeleton consists of spongin fibres; hard objects often become embedded. **Colour** brown-red-yellow. **Habitat** on rocks from shallow water down.

Dysidia fragilis Montagu **Form** up to 40mm high, rounded and cylindrical with large conspicuous oscula grouped terminally; sometimes encrusting; several may grow side to side. Surface is covered in small pointed outgrowths; texture variable, often soft and elastic; ostia are small and flush with the surface. **Colour** whitish. **Habitat** attached to rocks and often growing in crevices.

Verongia aerophoba (Schmidt) *(= Aplysina aerophoba)* **Form** erect, pillar-like and cylindrical, up to 150mm high; the flattened tips, in which lie the terminal oscula, have a 'sawn off' appearance; individuals join at the base. The skeleton lacks spicules and consists of spongin fibres; hard nodules may be embedded in the tissue. **Colour** yellow-green-black. **Habitat** generally on rocks in shallow water. **Similar species** none.

Ircinia fasciculata (Pallas) *(Hircinia fasciculata)* **Form** irregular, massive, forming growths up to 150mm across; oscula are variable in position and form. **Colour** violet-brown. **Habitat** generally under stone and in crevices in shallow water. **Similar species** several.

Myxilla incrustans

Hymeniacidon sanguinea

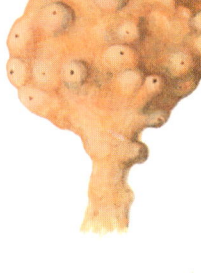

Haliclona oculata

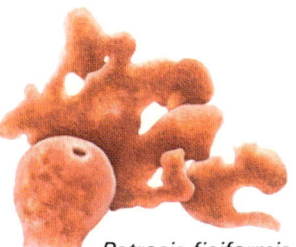

Petrosia ficiformis

Spongia officinalis

Hippospongia communis

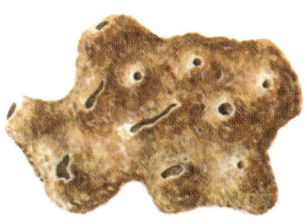

Dysidia fragilis

Verongia aerophoba

Ircinia fasciculata

Phylum Cnidaria

This phylum is composed of generally soft-bodied, flower-like animals, often with a jellyfish stage (*medusa*) in the life cycle.

Cnidarians include some of the most beautiful and abundant marine animals. They are of simple form, having sac-like bodies composed of an inner layer of cells (endoderm) surrounded by an outer layer (ectoderm). These two layers are separated by a jelly-like layer called the mesoglea. The interior of the sac acts as a stomach and opens to the outside via the mouth which also serves for an anus. Around the mouth are rings of tentacles armed with stinging cells. These tentacles grasp the prey and, after it has been immobilized by the stinging cells, force it into the mouth. There are no circulatory or excretory systems, and the nervous system is very simple. Each sessile animal is termed a *polyp* or a *medusa* if it is free swimming like a jellyfish.

Three classes are recognized. Members of the class Hydrozoa include two orders of sea-firs or hydroids which occur in most habitats from the shore to the deep sea, and they are the simplest cnidarians. The polyps often grow close together and are interconnected by tubular extensions of the stomach (see fig. 3). This arrangement frequently leads to the formation of colonies in which some of the polyps are specialized for feeding whilst others are responsible for reproduction or defence. The form and growth pattern of the colony are important identification guides. In some types only the tubular inter-gastric connections are surrounded by a protective skeletal sleeve (the perisarc) – these are the first order or *athecate* hydroids. Elsewhere the perisarc extends up around the gastric region and provides a protective cup (the theca) which houses the whole polyp and into which the tentacles can be withdrawn. Such hydroids form the second order or *thecate* hydroids. Hydroid life cycles are complicated. Adult colonies can grow on stones, shells or seaweeds. Some may develop non-feeding, reproductive polyps from which free-swimming medusae bud off. These are swept away, by tides and currents and can only swim upwards and float downwards. The umbrella-shaped bell (see fig. 4) contracts rhythmically, lifting the medusa in the water. Sex organs are developed by the medusae and when the sperms and eggs are ripe, they are released into the sea where fertilization occurs. A larva is formed which eventually settles on the seabed to form a new colony. Early naturalists did not understand this life cycle and so they gave the hydroids different names from their respective medusae. Both hydroids and medusae generally feed on small organisms which collide with their sting-celled tentacles. A third order in the class Hydrozoa is the Siphonophora. These are large, floating, colonial hydrozoans which differ from other members of the class in that they are free-living for the whole of their life cycle. One specialized polyp forms the *float* and is surrounded by many others developed for feeding, reproduction or defence. Many siphonophores are surface animals relying on wind and water currents for movement. They catch small or larger prey with their long, trailing tentacles, and usually appear in European waters when strong south-westerly winds sweep them in from the Atlantic. Their long tentacles may be broken off when they are cast up on the shore. The small fourth order Chondrophora consists of a few species of free-floating hydroid polyps which produce medusae. There are other hydrozoan orders which are rare or absent from the Mediterranean.

In the class Scyphozoa (jellyfishes) the medusa stage dominates, and individuals spend most of their lives as floating predators catching their prey with their long, trailing tentacles equipped with stinging cells. The ripe gonads of the jellyfish release sperms and eggs and, after fertilization, a small larva forms which settles on the seabed to develop a small, hydroid-like organism (the *scyphistoma*). This stage buds repeatedly, giving rise to miniature jellyfishes which then develop into the characteristic adults. In one group (stalked jellyfishes) the small, trumpet-shaped adults are not free swimming, but live attached to the fronds of seaweeds and stones. Scyphozoans occur at most depths in the seas, and they are less diverse than the hydrozoans.

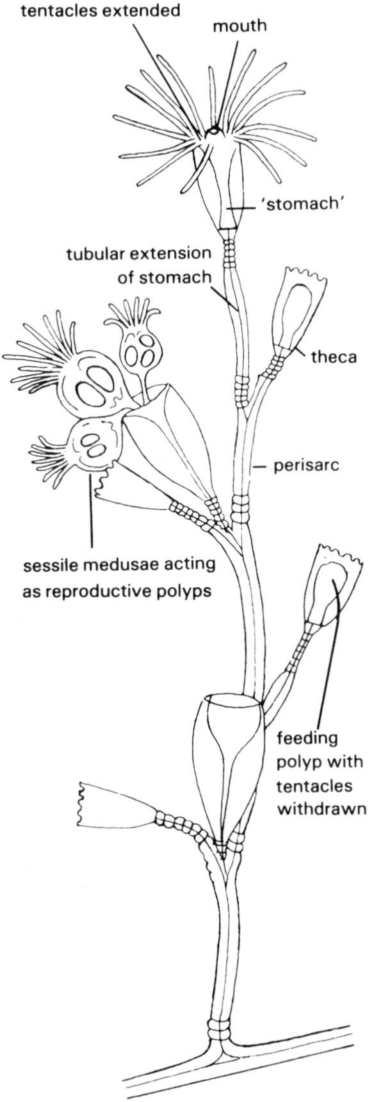

feeding polyp with
tentacles extended

mouth

'stomach'

tubular extension
of stomach

theca

perisarc

sessile medusae acting
as reproductive polyps

feeding
polyp with
tentacles
withdrawn

The class Anthozoa (sea-anemones and their allies) is another diverse group. Its members lack a medusa stage. The polyps are more sophisticated than the hydrozoan type and can either burrow in soft substrates or live attached to rocks and shells. The best-known anthozoans are the sea-anemones which belong to the order Actinaria. Examples are to be found on the shore as well as in deeper water. Like other cnidarians they are carnivorous, catching their food with tentacles. They reproduce asexually by division, or sexually when a fertilized egg develops into another polyp – usually via a larval stage. The order Antipatharia comprises the black corals. These have colonial polyps spread out over a hard, horny, tree-like skeleton. The Ceriantharia and Zoantharia resemble sea-anemones. The former are rather worm-like, and live in slimy tubes buried in mud or sand, where their long tentacles can protrude; the latter are small and, in most cases, colonial, and are usually found encrusting rocks and stones. The true corals (order Madreporaria) are represented by several species in the Mediterranean. They differ from anemones because they have hard, chalky skeletons which support and protect the lower regions of the polyp, and into which the tentacles can usually be drawn. Many are colonial, unlike the European representatives of the closely related Corallimorpharia which, although they resemble coral polyps in many ways, lack the characteristic skeleton of limestone. The orders Alcyonacea, Gorgonacea and Pennatulacea are closely related and usually have polyps with branching tentacles. They are always colonial and have a variety of growth forms.

Fig. 3 Branch of *Gonothyraea loveni* to show the form of feeding and reproductive polyps. In this species medusae are not liberated, but remain attached to the colony.

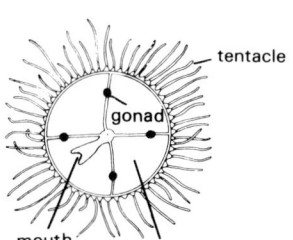

tentacle

gonad

mouth

underside of bell or umbrella

Fig. 4 Free medusa of *Obelia geniculata*

Class Hydrozoa
Order Athecata

Hydroids in which the horny perisarc does not surround the polyps. The polyps can always be seen even when disturbed. Please note that the overall size and appearance of colonies may be quite variable. Hydroids have recently been shown to be good indicators of water quality, see Stebbing, A. R. D. 1979. For a fuller coverage of hydroids together with more details for identification see Hincks, T. 1868, Brinkmann–Voss, A. 1970 and Cornelius, P. F. S. 1975a, b, 1979 and 1982.

Corymorpha nutans (M. Sars) Solitary polyp with thick stem and thin delicate perisarc rising to about two-thirds of the way up the main stem; about 20mm high. Head of polyp bears an outer ring of longer tentacles and an inner ring of shorter tentacles from between which reproductive bodies may arise. **Habitat** anchored in sand or mud by root-like threads. **Similar species** none.

Tubularia indivisa Linnaeus Colonial, with creeping root-like stolons and erect, seldom branching stems, often slightly plaited; up to 180mm high. Terminal flask-shaped polyps bear an outer ring of drooping white tentacles and a shorter stiffer inner ring, occasionally with reproductive bodies resembling miniature bunches of grapes; perisarc yellowish and often striped longitudinally. **Habitat** in pools on the shore, usually well shaded, and down to quite deep water on rocks and wrecks. **Similar species** several.

Coryne pusilla Gaertner Colonial, with creeping root-like stolons and irregularly branching stems; may reach 10mm high or more. Terminal cylinder-shaped pink-coloured polyps bear club-shaped tentacles; occasionally found with spherical reproductive bodies. **Habitat** on rocks in pools and down to quite deep water. **Similar species** few.

Zanklea costata Gegenbaur Colonial, with creeping root-like stolons from which arise unbranching stalks each bearing a long rod-like polyp armed with club-shaped tentacles; reproductive bodies may arise on the sides of the polyp among the tentacles; polyps about 5mm high; perisarc extends part way up the stem only. **Habitat** on stones and rocks from the shore down to deep water. **Similar species** few.

Cladonema radiatum Dujardin Colonial, with creeping root-like stolons from which arise unbranching stalks each bearing a small polyp; the polyp has four short basal tentacles and four longer club-shaped tentacles near the mouth; reproductive bodies may arise on the sides of the polyp near the lower tentacles; polyp about 5mm high; perisarc inconspicuous. **Habitat** on rocks and pebbles on the shore and in shallow water. **Similar species** none. N.B. often appears in marine aquaria where its relatively large medusae (3.5mm) may develop and be found swimming or adhering to the glass.

Eleutheria dichotoma Quatrefages Colonial, with creeping root-like stolons from which arise vertical polyps armed with 6–8 tentacles near the mouth; reproductive bodies arise at the base of the polyp just above the upper limit of the perisarc. **Habitat** growing on rocks in pools and in shallow water. **Similar species** few. N.B. This appears as a line-drawing to emphasise the structure.

Halocordyle disticha Goldf. Colonial, with creeping root-like stolons from which arises a main stem with alternately arranged branchlets, giving an overall feather-like plume effect; there are up to 5 polyps on each side branch; colony is up to 100mm high. Individual polyps have a ring of longer tentacles around the base of the tapered mouth piece and short tentacles up the sides of the mouth piece. **Habitat** on algae and rocks in exposed places. **Similar species** few.

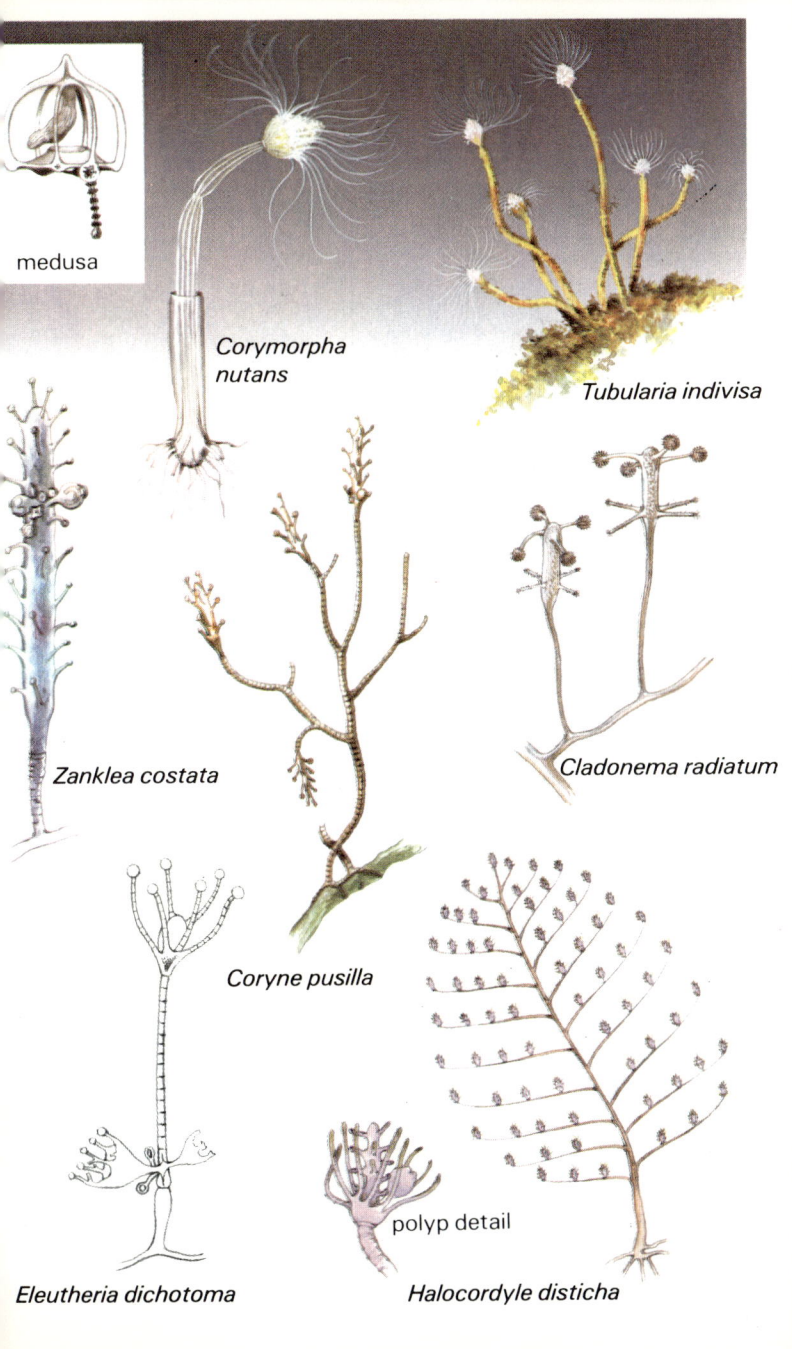

medusa

*Corymorpha
nutans*

Tubularia indivisa

Zanklea costata

Cladonema radiatum

Coryne pusilla

Eleutheria dichotoma

polyp detail

Halocordyle disticha

Tubiclava fruticosa Allman Colonial, with creeping root-like stolons from which branching stems arise bearing the polyps; perisarc extends up to polyp bases; polyps are elongated with about 8 tentacles. Habitat in submarine caves and grottos. Similar species few.

Podocoryne carnea M. Sars Colonial, with encrusting horny basal mass armed in placed with thorn-like tubercles; from this arise unbranched trumpet-shaped polyps and occasional parallel-sided tentacle-like structures. Polyps bear about 10 slender tentacles surrounding the conical mouth piece; reproductive bodies are borne about half way up the polyp stalks. Habitat in shady and pebbly places.

Hydractinia echinata (Flemming) Colonial; densely growing with an encrusting perisarc; spindle-shaped white-brown-red polyps rising on stems up to 15mm high. Habitat disused gastropod shells now inhabited by hermit crabs, occasionally growing on stones in shallow water. Similar species few.

Bougainvillia ramosa (Van Beneden) Colonial; upright alternate or irregularly branching stems arise from an encrusting base; polyps are borne on the tips of the branches and branchlets. Colony is up to 25mm high; polyps themselves bear about 10 tapering tentacles; reproductive bodies may grow subterminally on branchlets. Habitat growing on calcareous algae and on rocks and stones. Similar species few.

Perigonimus repens Wright Colonial; unbranched or branching stems arise from base; tubular perisarc extends up to polyp bases and the small club-shaped polyps can partially withdraw into these tubes. Colony is about 7mm high; reproductive bodies are borne below polyp bases on stems. Habitat growing on other hydroids, on opercula of some gastropods and on other shells down to quite deep water. Similar species several.

Eudendrium rameum (Pallas) Colonial and bush-like; up to 150mm high with creeping roots; strong stem with brown perisarc; flask-shaped pink polyps with one circlet of tentacles just at the base of the mouth piece; reproductive bodies form at the bases of the polyps. Habitat in water usually deeper than 10m, on rocks and in caves. Similar species *E. racemosum* and *E. ramosum*.

Eudendrium racemosum Cavolini Colonial and bush-like; up to 100mm high with creeping roots and a strong stem; flask-shaped polyps have two circlets of tentacles, one at the base of the mouth piece and one towards the polyp base itself. Habitat usually in deep water on rocks and under overhangs. Similar species *E. rameum* and *E. ramosum*.

Eudendrium ramosum (Linnaeus) Colonial and bush-like, but delicate; up to 120mm high; main stem strong; flask-shaped polyps have one circlet of tentacles at the base of the mouth piece. Habitat on stony bottoms and among coralline algae down to 35m. Similar species *E. rameum* and *E. racemosum*.

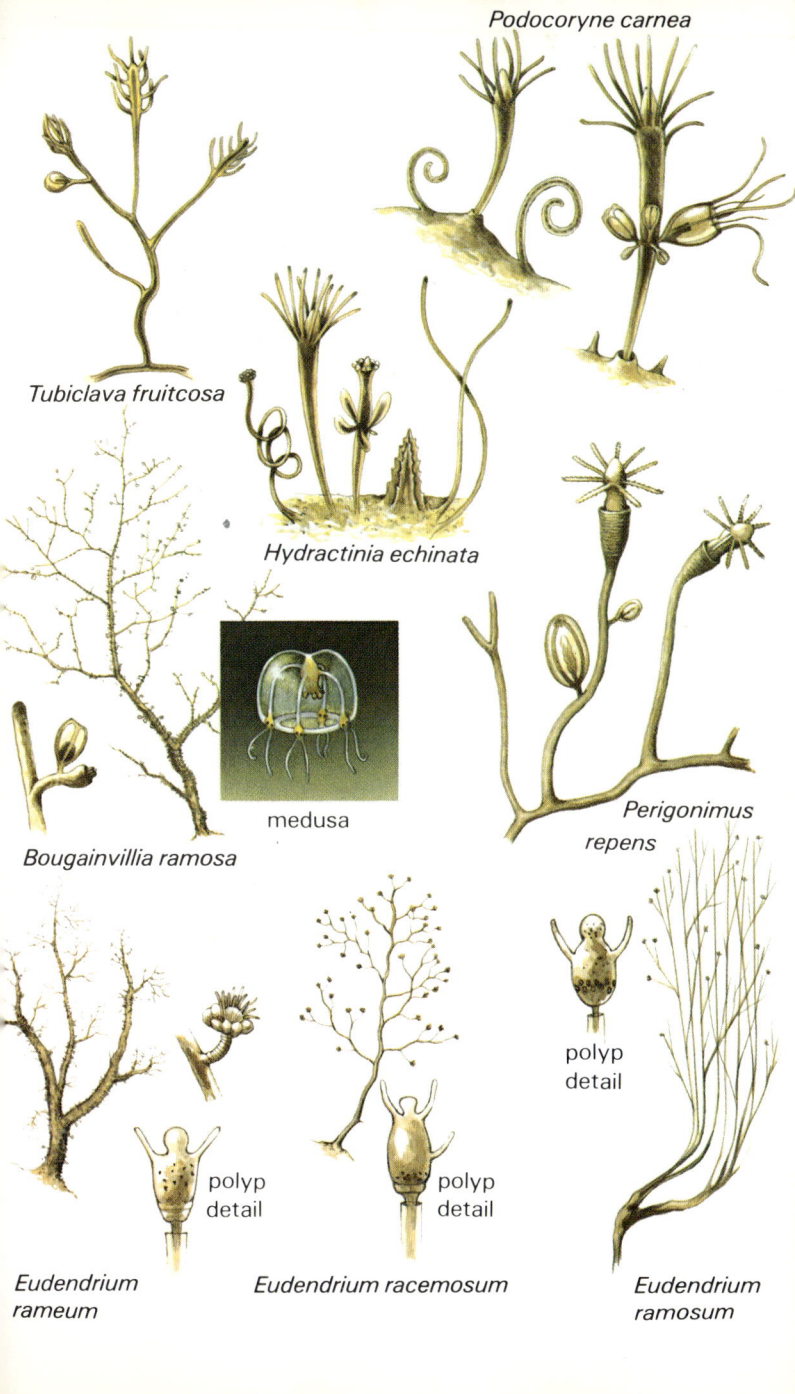

Podocoryne carnea

Tubiclava fruitcosa

Hydractinia echinata

medusa

Bougainvillia ramosa

Perigonimus repens

polyp detail

Eudendrium rameum

polyp detail

Eudendrium racemosum

polyp detail

Eudendrium ramosum

Order Thecata

Hydroids in which the horny perisarc extends round the polyps so providing a cup into which they can withdraw from view when disturbed. Most species have a characteristic growth pattern, but care must be taken in some cases to avoid confusion with bryozoa (see pages 222–227). For further references see under Order Athecata (page 58).

Cuspidella grandis Hincks Colonial; small with a creeping habit and no erect stem; tube-like filaments connect the polyps, which are small, up to 0.5mm long with about eight tentacles. **Habitat** encrusting the stems of other hydroids. **Similar species** several.

Campanulina species Colonial; upright growths arising from creeping root-like stolons; polyp stems themselves dividing; polyps cylindrical with webbed tentacles often held together to give a tapering appearance to the polyp. Large reproductive bodies are produced on side branches well down towards the stem bases; colony is up to 5mm high. Several species are known and because of the difficulty of separating them a general description only is given.

Campanularia hincksi Alder Colonial, with conspicuous elongated polyp cups which are castellated round the margin and supported on a long branch; up to 5mm high from the base of the stem, which creeps over the substrate. Elongated reproductive polyps may arise from the base stem; these are ringed, tapering towards the opening at the top. **Habitat** on shells, other hydroids and bryozoa from 10–60m deep. **Similar species** several.

Orthopyxis caliculata (Hincks) Colonial; simple unbranched stems growing up from a basal mass; a well marked ring occurs at the base of the polyp cup, only faint annulations occurring below this. The polyp cup has a smooth lip; colony is up to 4mm high. **Habitat** on hard substrates and red algae down to moderately deep water. **Similar species** few.

Clytia johnstoni (Alder) Colonial; long unbranched or slightly branched stems arise from a creeping stolon; polyps homed in a polyp cup with a conspicuous ring at the base; the cup lip has a number of pointed teeth but is not castellated. Polyps are up to 5mm high; urn-shaped reproductive bodies are on separate short stalks. **Habitat** often growing on red algae and leaves of *Zostera*. **Similar species** several.

Obelia geniculata (Linnaeus) Colonial, with creeping root-like stolons and erect zig-zag branching stems reaching up to 40mm high. The bell-shaped polyp cups are supported on a red-brown perisarc which is characteristically annulated between polyp cups and main stem, urn-shaped reproductive polyps are borne on alternate sides of the stem. **Habitat** on seaweeds on the shore and in shallow water. **Similar species** *Laomedia flexuosa* (below).

Gonothyraea gracilis (M. Sars) Colonial; similar to *Clytia johnstoni* (above), but with branching erect stems and several annulations at the base of the polyp cup; there are also separate nets of annulations at the stem joints. Colony is up to 15mm tall; reproductive bodies are in flask-shaped cups on separate stalks. **Habitat** in grottos and submarine caves and on hard bottoms in deeper water. **Similar species** few.

Laomedia flexuosa (Hincks) Colonial, like *Obelia geniculata* (above); colony up to 15mm high, not so markedly zig-zag and growing on rocks and stones and pool sides. **Similar species** several.

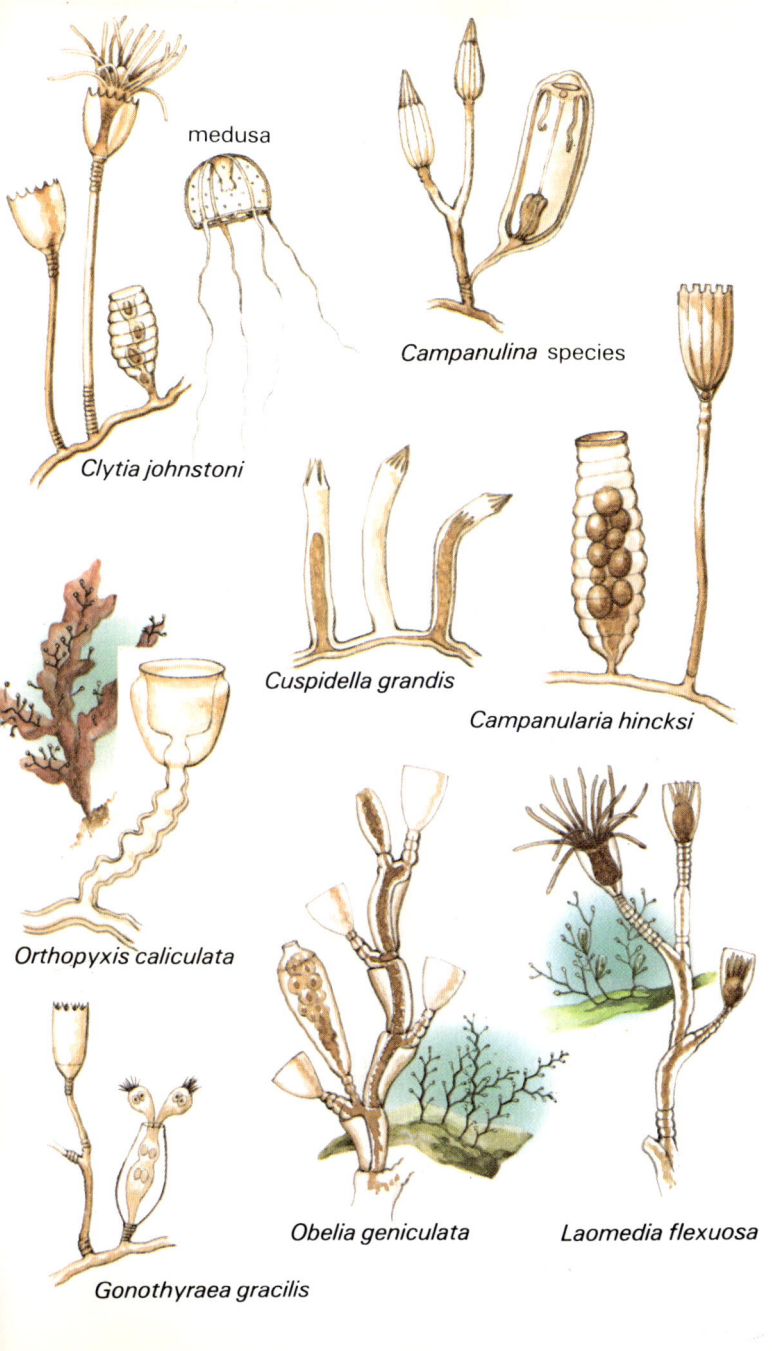

medusa

Campanulina species

Clytia johnstoni

Cuspidella grandis

Campanularia hincksi

Orthopyxis caliculata

Obelia geniculata

Laomedia flexuosa

Gonothyraea gracilis

Lafoea dumosa (Flemming) Colonial and small; creeping stolons forming a delicate network from which arise small tube-like polyp cups reaching about 5mm high. There is a more thickly branched, stronger looking variety. **Habitat** on shells, worm tubes and other hydroids down to 300m; it may occur in great profusion, especially on other hydroids. **Similar species** several.

Halecium halecinum (Linnaeus) Colonial; large feathery growths reaching 120mm tall or more, several arising from a common position. The main stem bears side branches on which the polyps are situated; polyp cups are funnel–like, often several funnels superimposed. **Habitat** growing on stones and shells and on worm tubes such as those of *Chaetopterus* from shallow water down.

Dynamena pumila (Linnaeus) Colonial; stems arising from horizontal creeping root-like stolons; up to 12mm polyps borne on opposite sides on upright stems; polyp cups are cowl-shaped. **Habitat** on rocks and weed on the shore and in shallow water, often very common. **Similar species** several.

Sertularia cupressina (Linnaeus) Colonial, with tall slender stems bearing alternate branches from which arise branchlets; arranged so as to form a long feathery growth; the polyps and the reproductive bodies are borne on the branchlets. The polyp cups occur in 2 rows and are longish and cylindrical; growths may reach 450mm long. **Habitat** on shells and other hydroids down to quite deep water.

Sertularia gayi (Lamouroux) Colonial, with erect branching stems; side branches arranged alternately; polyp cups with lips drawn out into four slight teeth; up to about 220mm high. **Habitat** growing on clean sandy and shell gravel bottoms down to quite deep water. **Similar species** several, including *S. polyzonias* (reproductive polyp only illustrated).

Kirchenpaueria pinnata (Linnaeus) *(=**Plumularia pinnata**)* Colonial; upright branching stems reach to 100mm or more, arising from root-like holdfast; polyps borne on side branches in shallow cups; reproductive bodies borne on main stems. **Habitat** on various substrates, pilings, rocks, other hydroids, worm tubes and shells from shallow water down.

Plumularia halecioides Alder *(=**Ventromma halecioides***) Colonial; fine plume-like growths rising to 20mm from creeping root-like stolons; polyps are borne on side branches in shallow cups. The cups are quite similar to those of *Kirchenpaueria pinnata* above, but they have a distal accessory tube; reproductive bodies are quite large and are borne on the main stems. **Habitat** epizoic on *Plumularia setacea* and on *Nemertesia* spp., and growing on algae.

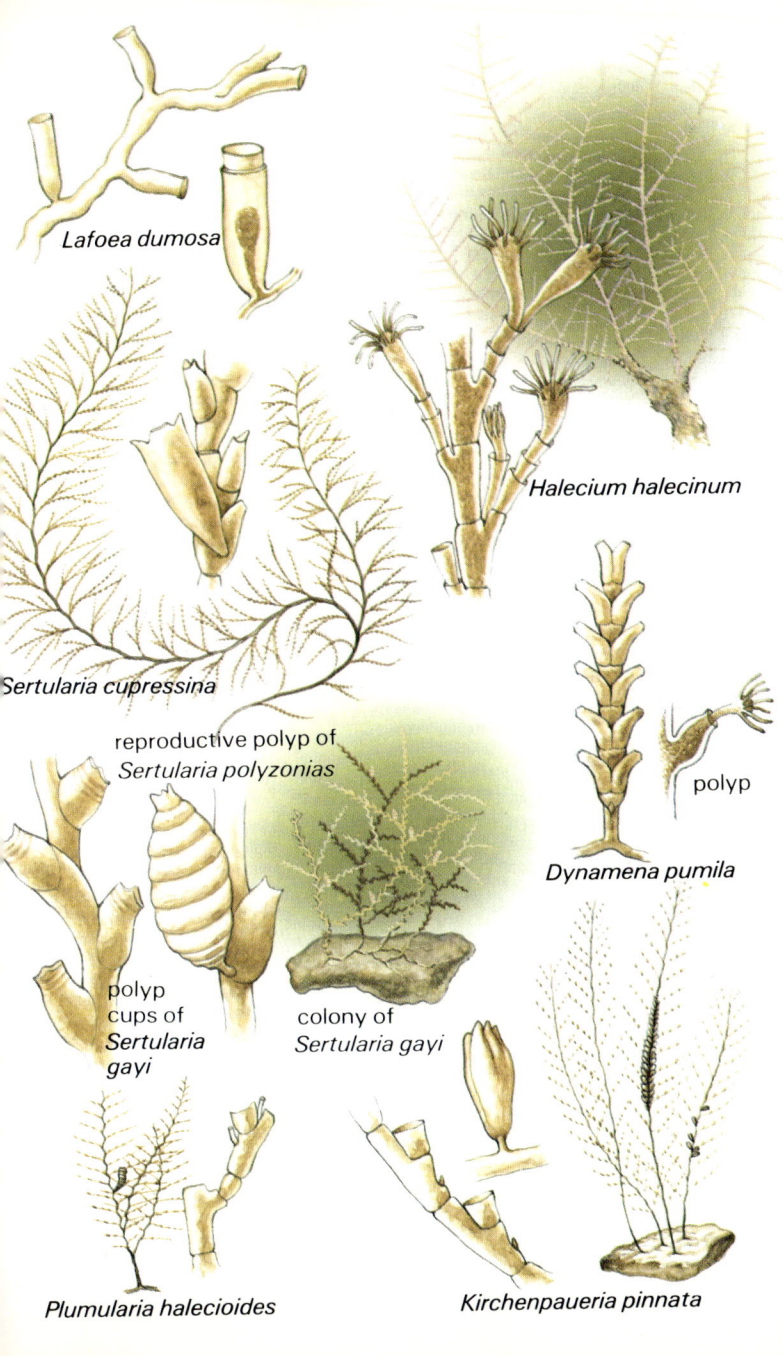

Lafoea dumosa

Halecium halecinum

Sertularia cupressina

reproductive polyp of
Sertularia polyzonias

Dynamena pumila

polyp

polyp
cups of
*Sertularia
gayi*

colony of
Sertularia gayi

Plumularia halecioides

Kirchenpaueria pinnata

Plumularia setacea (Ellis & Solander) Colonial; delicate wavy stems arise from creeping stolons, reaching about 30mm high; polyps are borne on side branches which are alternately arranged; polyp cups quite deep and smooth-lipped. Slender pod-like reproductive bodies may be borne in the angle of the joint of the side branch and main stem. **Habitat** on the shore, on rocks and in pools and in shallow water. **Similar species** *P. catharina*.

Plumularia catharina Johnston Colonial, with creeping root-like stolons and branching stems which reach up to 100mm high; side branches are exactly opposite, polyp cups are quite deep with smooth lips and are borne on the main stems as well as on the side branches. **Habitat** on stones, shells and the tunics of sea-squirts.

Polyplumularia frutescens (Ellis & Solander) Colonial and bush-like; tapering upright stems rise to 140mm; main stem tubular and irregularly branched; branches bear alternately arranged branchlets which carry the polyps. The polyp cups are set hard against the branchlets, deeper than their diameter; reproductive bodies are pear-shaped on short stalks, also carried on the branchlets. **Habitat** on shells and stones in deeper water.

Thecocaulus diaphanus (Heller) Colonial, with erect stems arising from creeping stolons and reaching 150mm high; stems bear alternately arranged branches on which the polyps are carried; there is a basin-like polyp cup at the base of each side branch; reproductive bodies are cowl-shaped. **Habitat** on hard bottoms in deeper water.

Nemertesia antennina (Linnaeus) *(=Antennularia antennina)* Colonial; clusters of upright stems reaching 250mm high; yellowish; horny texture. The stems arise from a matted filamentous mass which serves as a holdfast; the whorled short branchlets are wider at the base, curved in, bearing polyps which are housed in vase-like polyp cups with a smooth rim; reproductive bodies occur in the angle of the main stem and branchlet. **Habitat** usually attached to rigid objects e.g. pebbles and empty shells, although often where sand accumulates.

Nemertesia ramosa (Lamouroux) Colonial; the thick main stem arises from a holdfast of matted fibres and divides and subdivides irregularly; the long tapering outcurved branchlets are hairy and closely set and are arranged in whorls; the polyp cups are small and vase-like; the reproductive bodies are pear-shaped and face in towards the stem. **Habitat** as for *N. antennina* above.

Aglaophenia pluma (Linnaeus) Colonial, with a feathery appearance; main stem rises to 70mm or more from creeping root-like stolons; the side branches are arranged alternately; polyp cups fairly shallow and with a finely toothed lip, turned out at the edges; reproductive bodies pod-like. **Habitat** on algae, shells and rocks on the shore and in shallow water. **Similar species** *A. tubulifera* (below).

Aglaophenia tubulifera (Hincks) Colonial; plume-like growths arising from root-like stolons and reaching 80mm or more high; polyp cups are deep, curving in above at the front, margin of lip very slightly turned out; reproductive bodies are ovoid, carried on short stalks and often arranged in a double row. **Habitat** on weeds, shells and other organisms, usually in deeper water than *A. pluma* above.

Thecocarpus myriophyllum (Linnaeus) *(=Lytocarpa myriophyllum* or **Aglaophenia myriophyllum)** Colonial, very feathery; main stems usually clustered and rising to 300mm or more, thickening at intervals into knobs; side branches may themselves be slightly branched into branchlets; the polyp cups are quite deep, cylindrical, with the lip slightly crenated and bearing a single large tooth in front; reproductive bodies resembling miniature mussel shells are borne in pairs near the bases of the polyp cups, and are protected by long-toothed curved processes on the outer edge. **Habitat** in deeper water.

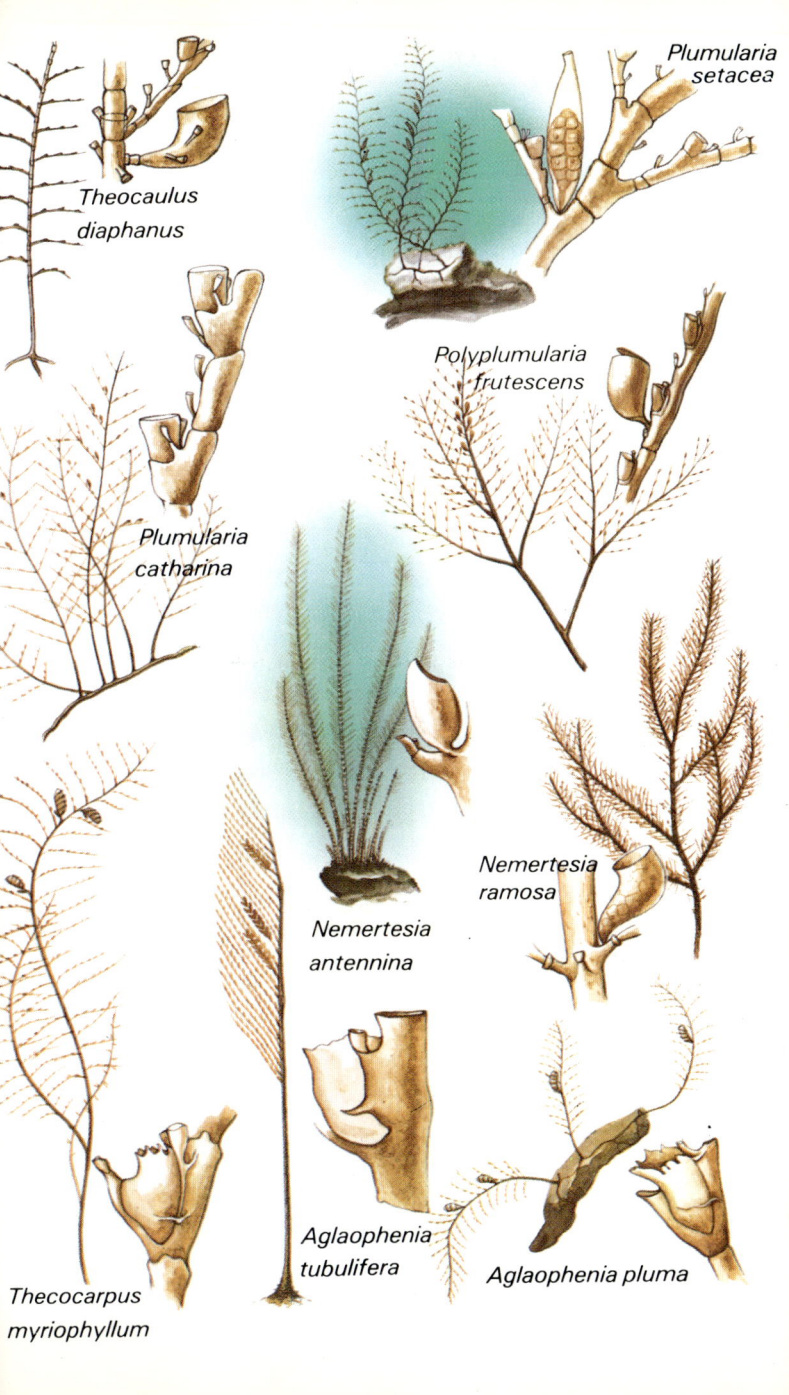

*Theocaulus
diaphanus*

*Plumularia
setacea*

*Polyplumularia
frutescens*

*Plumularia
catharina*

*Nemertesia
ramosa*

*Nemertesia
antennina*

*Thecocarpus
myriophyllum*

*Aglaophenia
tubulifera*

Aglaophenia pluma

Order Siphonophora

Colonial, free floating or swimming hydrozoans with many individuals variously modified to provide floats, stems and feeding, defensive and reproductive polyps. Some species float on the surface; others swim in deeper water. They are generally oceanic but are occasionally washed into coastal areas by rough weather. A benthic polyp stage has been abandoned and its role has been taken over by part of the floating colony. Mediterranean recorded sp = about 20. See Hardy, A.C. 1970.

Physalia physalis (Linnaeus) **Portuguese Man-o-war** Large bag-like float or pneumatophore reaching up to 300mm long by 100mm wide with conspicuous sail attached; many other individuals suspended below the float in a complex association of polyps; a short stem attached to the base of the float supports several large fishing and defensive tentacle-like polyps known as dactylozooids, smaller dactylozooids, bunches of feeding individuals or 'gastrozooids' without tentacles but with mouths, and many branched reproductive individuals or gonodendra; the gonodendra may release medusae which form part of the life-cycle like the medusae of other hydroids. **Colour** pneumatophore silver-blue tinged with red; rest of colony blue-purple. **Habitat** pelagic; surface dwelling. N.B. when rarely cast ashore the pneumatophore may be broken from the remaining part of the colony which thus appears to be missing. **Beware** dangerously powerful stinging cells. **Similar species** none precisely the same but a number may appear superficially similar, especially after being cast ashore.

Muggiaea atlantica Cunningham Helmet-shaped swimming bell up to 20mm long; transparent retractile stem hangs from the bell and supports a great number of combinations of individuals; each group is known as a cormidium and comprises 3 members: a feeding zooid equipped with 1 tentacle, a reproductive zooid and a small bract (like a medusa) which enables the reproductive zooid to lead a free existence. **Colour** transparent. **Habitat** found swimming at various depths. **Similar species** several, which may only be distinguished by a careful examination of bell shape and stem morphology.

Lensia conoidea (Keferstein & Ehlers) (Not illustrated) Similar to *Muggiaea atlantica*; 2 swimming bells each about 10mm long; front bell slightly larger than the rear; retractile stem bears numerous cormidia. **Colour** transparent. **Habitat** found swimming at various depths, often in shoals. **Similar species** several may only be distinguished by examination of bell shape and stem morphology.

Physophora hydrostatica Forskål Small apical float or pneumatophore below which hangs a stem polyp about 60mm long supporting 5 pairs of swimming bells arranged in 2 rows; below these trail feeding, defensive, fishing and reproductive polyps. **Colour** predominantly yellow-pink-red. **Habitat** swimming at various depths. **Similar species** *Halistemma rubrum* is larger, with 9 pairs of swimming bells.

Halistemma rubrum Vogt (Not illustrated) Similar to *Physophora hydrostatica* but overall length 200mm or more; 9 pairs of swimming bells on the stem polyp. **Colour** red. **Habitat** found swimming at various depths. **Similar species** see above.

Order Chondrophora
Free-floating colonial hydrozoans

Velella velella (Linnaeus) **By-the-wind-sailor** Bluish horny oval or disc up to 80mm in diameter enclosing a float system and equipped with a crescent-like sail; when alive the sail and the disc are covered with soft tissue and the sail projects above the surface of the water to catch the wind and aid dispersal. A large feeding zooid under the disc is encircled by a ring of reproductive zooids; at the periphery is a ring of tentacle-like fishing zooids. **Habitat** pelagic surface-dweller; sometimes in shoals. **Similar species** *Porpita umbella* (see below) lacks the sail.

Porpita umbella Otto (Not illustrated) Similar to *Velella velella* above but without the characteristic sail; blue-green disc reaching up to 80mm in diameter. **Habitat** pelagic surface-dweller; often in shoals. **Similar species** see above.

Physalia physalis

Velella velella

Physophora hydrostatica

Muggiaea atlantica

Class Scyphozoa Jellyfishes

Cnidaria in which the medusa stage is dominant, but which normally pass through a small polyp stage during the life-cycle; this polyp is termed the scyphistoma. The medusae are mostly pelagic but a few are sessile. Russell, F.S. 1970 gives a detailed account of the scyphozoa. Mediterranean recorded species = about 6.

Aurelia aurita (Linnaeus) **Common Jellyfish** Saucer-shaped 'umbrella' reaching up to 250mm in diameter with 4 frilly mouth arms longer than the numerous peripheral short tentacles; 8 sense organs arranged regularly around the periphery of the 'umbrella' and 4 conspicuous purple-violet reproductive organs visible through the body wall; these are horseshoe-shaped as seen from above. **Colour** transparent, tinged blue-white. **Habitat** pelagic. **Similar species** when stranded several may be confused, but when swimming *A. aurita* does not closely resemble other species. N.B. inset illustration shows scyphistoma (see above) which is found attached to rocks and seaweeds in pools and shallow water; from this small larvae with bilobed arms bud off; they are known as ephyrae.

Chrysaora hysoscella (Linnaeus) **Compass Jellyfish** (Not illustrated) Saucer-shaped 'umbrella' up to 300mm in diameter, and bearing 24 tentacles alternating with 8 sense organs at the periphery; 4 mouth arms longer than the tentacles. **Colour** yellow-white with 16 characteristic brownish radiating bands on top of the 'umbrella'. **Habitat** pelagic. **Similar species** see under *Aurelia aurita*.

Pelagia noctiluca (Forskål) Mushroom-shaped 'umbrella' reaching up to 100mm in diameter with 4 arms around mouth and 8 slender trailing tentacles around periphery; tentacles longer than mouth arms when fully extended; 8 small wart-like sense organs alternate with tentacles; **Colour** transparent but tinted yellow-red. **Similar species** see under *Aurelia aurita*.

Charybdea marsupialis (Linnaeus) **Mediterranean Sea-wasp** Box-shaped 'umbrella' up to 60mm high trailing 4 long tentacles reaching 300mm or more. **Colour** transparent with yellow-red tints. **Habitat** pelagic. **Similar species** see under *Aurelia aurita*. N.B. can inflict painful stings.

Rhizostoma pulmo (Macri) *(=R. octopus)* Dome-shaped 'umbrella' up to 900mm in diameter; no peripheral tentacles but 96 edge lobes; 16 sense organs and 8 fused mouth arms. **Colour** blue-white-yellow with yellow or blue-red mouth arms. **Habitat** pelagic, sometimes associated with young fish, e.g. *Boops, Seriola* and *Trachurus*. **Similar species** see under *Aurelia aurita*.

Cotylorhiza tuberculata Agassiz Saucer-shaped 'umbrella' with pronounced central dome above and tapering below; about 200mm in diameter; 16 peripheral lobes and many tentacles of various lengths, some terminating in frilly tips; 8 sense organs on periphery of 'umbrella'; 8 mouth arms. **Colour** green-brown, the green colour being caused by commensal algae. **Habitat** pelagic, sometimes in shallow water associating with young fishes, e.g. *Boops, Seriola* and *Trachurus*. **Similar species** see under *Aurelia aurita*.

Nausithoë punctata Köll Relatively flat 'umbrella' from which arise 16 conspicuous peripheral lobes, 8 tentacles and 8 sense organs; mouth cross-shaped surrounded by 4 small flaps about 12mm in diameter. **Habitat** pelagic. **Similar species** see under *Aurelia aurita*.

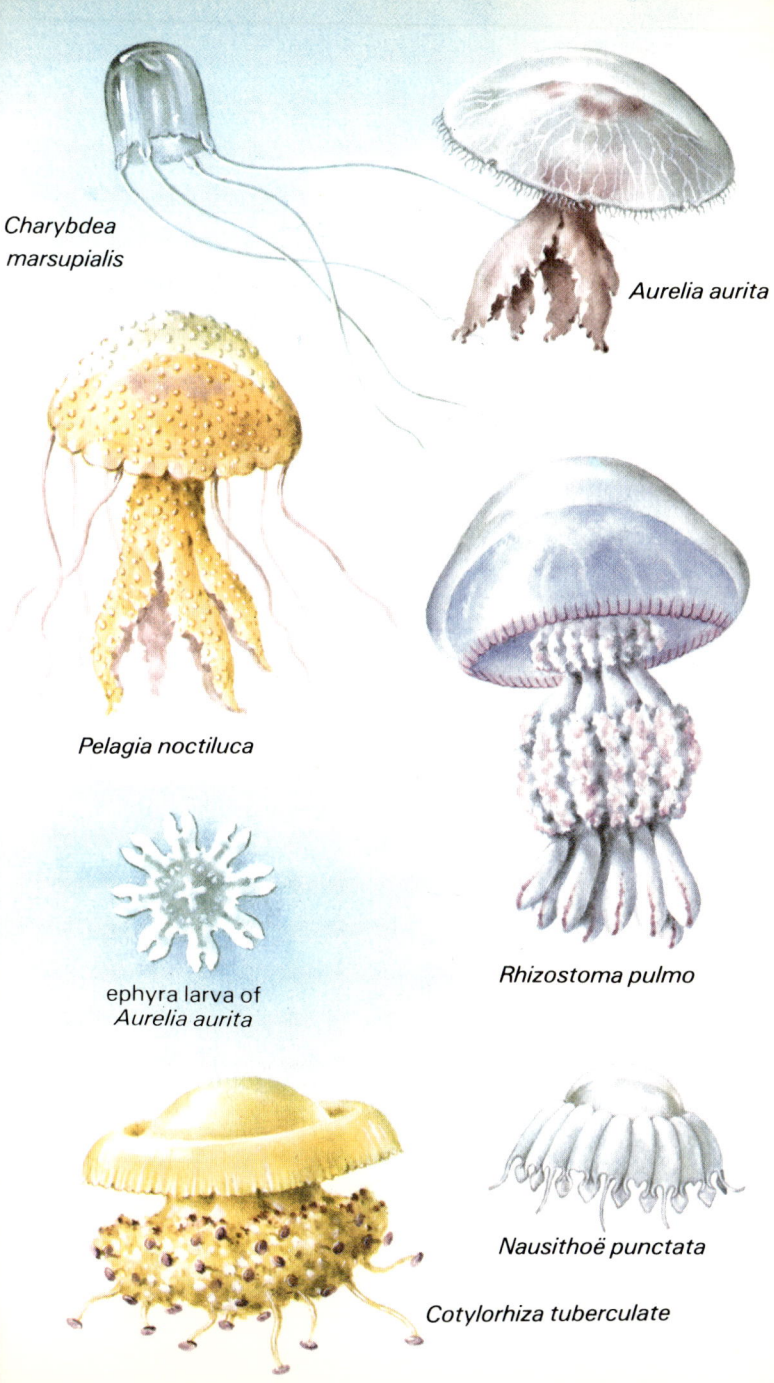

Charybdea marsupialis

Aurelia aurita

Pelagia noctiluca

Rhizostoma pulmo

ephyra larva of *Aurelia aurita*

Nausithoë punctata

Cotylorhiza tuberculate

Class Anthozoa Sea-anemones, corals and their allies

Coelenterates which lack a medusa stage in their life cycle. The polyps may be solitary or colonial and are often large and conspicuous; a chalky skeleton may or may not be present.

Order Antipatharia Black corals

Thorny black branching skeleton surrounded by softer tissue bearing polyps which cannot retract their tentacles; tentacles unbranched. Mediterranean recorded species = 2.

Antipathes subpinnata (Ellis & Solander) **Black Coral** Colonial; thorny black-brown skeleton reaching up to 1m high; white-grey outer tissue with small bilaterally symmetrical polyps up to 1mm high, each bearing 6 tentacles. **Habitat** on muddy substrates with stones between 10 and 250m. **Similar species** none.

Order Ceriantharia

Solitary polyps living in thick, felt-like mucous tubes buried in sand or mud so that the crown of unbranched tentacles, which are arranged in two whorls, or cycles, can protrude. There is no adhesive basal disc as there is in the true anemones. Mediterranean recorded species = 2.

Cerianthus membranaceus (Spallanzini) Large solitary polyp up to 350mm high with about 100 tentacles arranged in 4 rings around the mouth; the tentacles are up to 200mm in length and may be erect, giving the animal a fountain-like appearance; polyp may withdraw rapidly into mucous tube which usually has grains of mud incorporated into it. **Colour** variable, often brownish, but sometimes violet, white or green and fluorescent. **Habitat** burrowing in muddy sand. **Similar species** one other of doubtful status. N.B. beware of confusion with burrowing anemones (see page 74): these lack the felt-like tube of mucus, and have fewer shorter tentacles. The polychaete *Myxicola* can resemble small specimens.

Order Zoantharia

These polyps are usually colonial, and individuals within a colony are linked by an encrusting stolon which may or may not be seen easily; they encrust surfaces such as rocks or shells, or live organisms like sea-squirts. They have smooth slender tentacles which sometimes appear to have minute terminal expansions or knobs, and which are arranged in 2 distinct whorls or cycles. The zoanthids are rather small and may require a lens for correct identification. Mediterranean recorded species = about 5. For further details see Pax, F. and Müller, I. 1962 and Manuel, R. L. 1980 and 1981.

Epizoanthus arenaceus (Delle Chiaje) (Probably synonymous with *E. couchii*) Colonial; stolons not easily seen, being thin and narrow; grey-brown-pinkish polyps up to 10mm high with 25–35 slender whitish-translucent tentacles reaching 5mm long and arising from serrated parapet at top of column; tentacles have white tip. **Habitat** on rocks, stones, pebbles and shells from low water downward, sometimes encrusting quite extensive areas. **Similar species** *E. paxi*.

Epizoanthus paxi Abel Colonial, with rather obscure stolons; dark ginger-brown polyps have a narrow dark violet band encircling just below the origin of the tentacles at the parapet of the column; up to 35 tentacles. **Habitat** often in quite shallow water. **Similar species** *E. arenaceus*.

Parazoanthus axinellae (Schmidt) Colonial, with thin plate-like stolons connecting the yellowish polyps which reach 20mm high; sides and bases of the polyps may be encrusted with sand; about 34 golden-yellow tentacles arise from the serrated parapet; mouth may be orange. **Habitat** on other animals such as sea-squirts, corals and worm tubes as well as on cave walls, rocks and shells from 6 to 100m. **Similar species** pale specimens may resemble *Epizoanthus* spp.

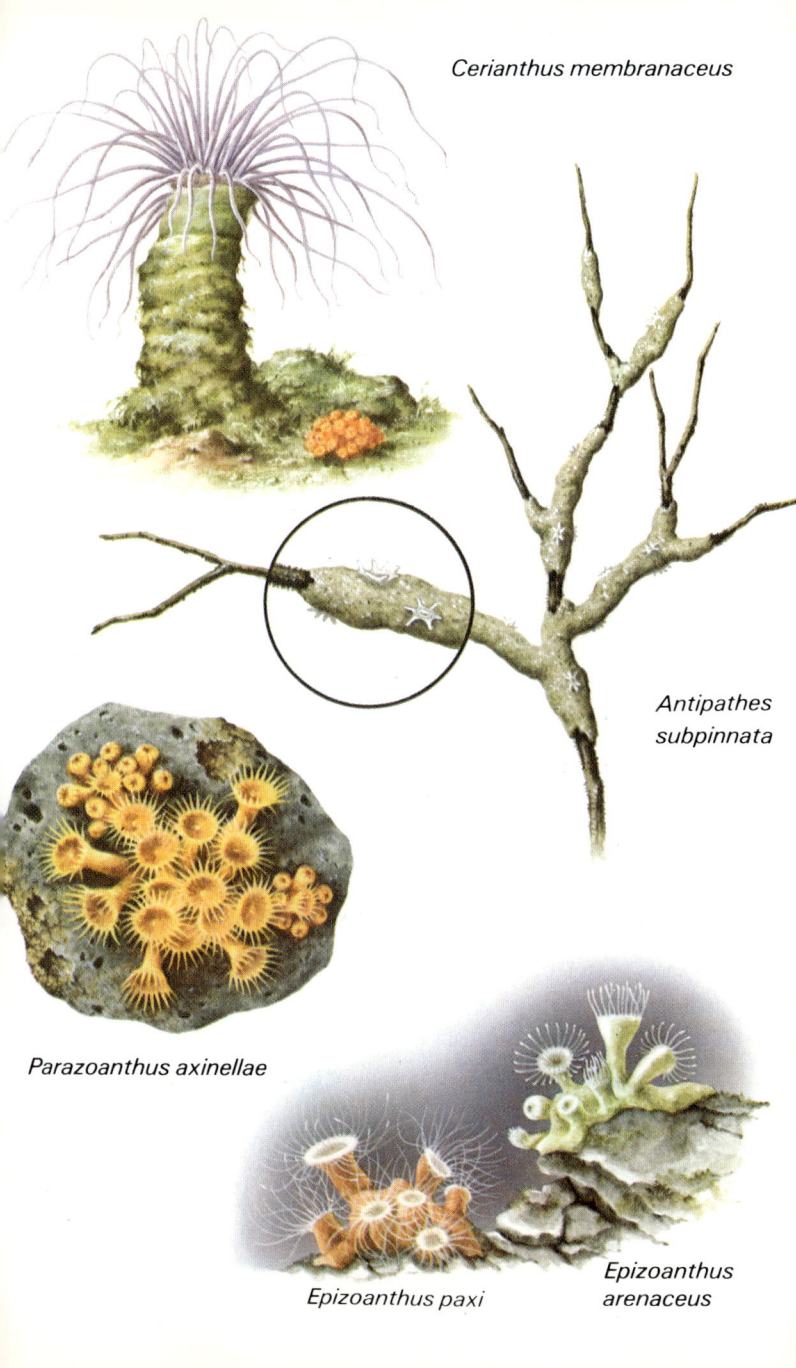

Cerianthus membranaceus

Antipathes subpinnata

Parazoanthus axinellae

Epizoanthus paxi

Epizoanthus arenaceus

Order Actinaria Sea-anemones

Solitary, often conspicuous anthozoans with no calcareous hard skeleton. They are sedentary, with the polyp base modified either for burrowing or for adhering to shells and rocks by means of a sucker-like disc. The tentacles are simple and not branched. For more details see Manuel, R.L. 1981 or Schmidt, H. 1972.

Actinia equina Linnaeus **Beadlet Anemone** Base adhesive and sucker-like, up to 50mm in diameter; column smooth and up to 70mm high; about 200 densely packed retractile tentacles up to 20mm long and with a total span when fully extended of up to 70mm; tentacles are arranged in 5–8 circlets which are quickly folded in when the animal is disturbed. There are 24 conspicuous blue spots arranged on the periphery of the oral disc just outside the tentacles. If stranded above the water level this anemone appears as a blob of jelly up to 30mm high. **Colour** variable: brown, red, orange and green. **Habitat** on rocks and in crevices, usually in shallow water down to 8m. **Similar species** *A. cari.*

Actinia cari Delle Chiaje Base adhesive and sucker-like, up to 70mm in diameter; column smooth, up to 50mm high, noticeably cone-shaped when the tentacles are withdrawn; about 190 retractile tentacles reaching up to 20mm long, which fold in rapidly when the animal is disturbed; clear blue spots arranged on the periphery of the oral disc outside the tentacles. **Colour** brown, blue-green, green or yellow with contrasting darker encircling lines. **Habitat** on rocks, stones and piers from 0.5–1.5m. **Similar species** *A. equina.*

Anemonia viridis Forskål **(=*A. sulcata*)** Base lightly adhesive and sucker-like, up to 70mm in diameter; column smooth and variable in height up to about 100mm; up to 170 wavy tentacles reaching to 150mm long are arranged in up to 6 circlets; span of tentacles may reach 180mm; tentacles cannot be fully retracted out of sight when the animal is disturbed. **Colour** column brown, grey or greenish; tentacles similar without purple tips; characteristically 2 white lines run from opposite sides of the disc to the mouth. **Habitat** in shallow water attached to rocks, algae and sea-grasses, often in well illuminated places. **Similar species** *Aiptasia mutabilis* which has many whitish lines running from tentacle bases to mouth.

Bunodactis verrucosa (Pennant)**Gem Anemone** Base adhesive, up to 25mm in diameter; column tall and cylinder-like with small wart-like protuberances arranged in up to 48 vertical rows; column may reach up to 50mm high; up to 48 tentacles present, each up to 15mm long. **Colour** column pinkish; rows of warts mostly grey but interspersed with some white rows; tentacles transparent and mottled green, grey or pink. **Habitat** in rock pools and crevices in shallow water, often well exposed to light. **Similar species** *Anthopleura rubripunctata* which is often larger.

Anthopleura rubripunctata (Grube) Similar to *Bunodactis verrucosa* above. Base adhesive, up to 70mm in diameter; column up to 70mm or more tall with up to 48 vertical rows of wart-like suckers; up to 96 tentacles, not long, round at the ends. **Colour** column yellowish to olive-green; warts may be red; tentacles yellowish. **Habitat** in shallow water.

Anthopleura ballii (Cocks) Base adhesive and sucker-like, up to 50mm in diameter; column tall and widening towards the top when fully extended, may reach to about 60mm; warts arranged in rows, the longest ones at the top; about 48 tentacles up to 15mm long and tapering at the tips. **Colour** column pink-orange or yellow-brown; tentacles translucent, mottled grey, pink or brown and sometimes with a greenish tinge. **Habitat** in crevices and out of strong light, usually in shallow water. N.B. this species is variable in coloration and appearance.

Condylactis aurantiaca (Delle Chiaje) Base adhesive and sucker-like; big, up to 70mm in diameter; column up to 400mm tall; many tentacles. **Colour** column white background with orange-red longitudinal stripes and white wart-like suckers; tentacles green with violet tips. **Habitat** burrowing in sand or gravel and attached to an embedded rock, stone or shell so that the tentacles may be the only conspicuous part. (See also illustration on back cover.)

Anemonia viridis

Bunodactis verrucosa

Actinia equina

Actinia cari

Condylactis aurantiaca

Anthopleura rubripunctata

Anthopleura ballii

Phymanthus pulcher Andres *(=**Ragactis pulchra**)* Base adhesive, up to 30mm in diameter; column up to 65mm high, 70mm when in full expansion, at which time it tapers slightly towards the top; disc wider and saucer-like, fringed with up to 96 small short tentacles in 5 cycles, longer tentacles on the inner ring. **Colour** column dark, striped reddish and white; tentacles dark. **Habitat** to 30m.

Aureliana heterocera (Thompson) Base adhesive and up to 70mm in diameter, often much wider than the column; column tapers slightly and is smooth; disc flat and carrying up to 150 tentacles arranged in 4 equal cycles; tentacles short, club-shaped and knobbed. **Colour** column yellow or red, often with a marbled effect and sometimes with yellow and white marks; disc reddish; tentacles whitish, sometimes dotted white and grey. **Habitat** attached to rocks and shells or buried in mud, sand or gravel; sometimes free; down to very deep water.

Aiptasia mutabilis (Gravenhorst) *(=**A. couchii**)* Base adhesive and sucker-like, narrower than the column; column up to 200mm tall, widening to the disc; about 100 long tapering tentacles. **Colour** pale yellow to brown; disc usually has many whitish lines running from the base of the tentacles to the mouth. **Habitat** on rocks and shells, often in great numbers in shallow water. **Similar species** *Anemonia viridis*.

Aiptasiogeton pellucidus (Hollard) Base adhesive, up to 10mm in diameter; column tapers slightly or noticeably from base towards the top according to the degree of extension; may reach 20mm high; there are about 100 slender long tapering tentacles. **Colour** column translucent white to pale brown with longitudinal opaque lines, sometimes with orange tints; disc translucent and colourless to orange; tentacles translucent pale pink to magenta or orange with magenta tips. **Habitat** attached to rocks and in crevices on the shore and in shallow water. N.B. this species is often confused with small examples of others, e.g. *Sagartia elegans* and *S. troglodytes*.

Haliplanella lineata (Verill) *(=**H. luciae**)* Base adhesive, up to 25mm in diameter; column wrinkled when retracted, smooth and pillar-like when fully extended; may reach up to 40mm tall, often less; tentacles number up to 100 and may reach 100mm in length; they are retractile and delicate. **Colour** dark green-brown with up to 20 vertical orange stripes; tentacles green-grey. **Habitat** on rocks, stones, pebbles and shells, wooden piers etc., in sheltered bays, lagoons etc.

Sagartia elegans (Dalyell) *(=**S. rhododactylos**)* Base strongly adhesive, wider than the column, up to 30mm in diameter. The column has suckers, can be trumpet-shaped in full expansion; disc bears up to 200 tentacles which may span 40mm or more. **Colour** most have a brown-orange column with paler suckers, a variable disc and orange tentacles with paler markings. **Habitat** down to 100m.

Sagartia troglodytes (Price) Base strongly adhesive and up to 50mm in diameter, often wider than the span of the tentacles; column is with suckers, up to 120mm in full extension; disc bears many tentacles with a span of 120mm or more. **Colour** very variable; there are two established varieties: in variety *decorata* the column is yellowish, becoming paler and greyer above with vertical lines; mouth disc and tentacles intricately patterned. In variety *ornata* the column is dark olive-green with slightly paler vertical lines and spots; mouth disc and tentacles intricately patterned; base of the disc is generally about 10mm in diameter. **Habitat** var. *decorata* on the shore and down to 50m in very dirty muddy conditions as well as sand. Var. *ornata* on the shore in cleaner places than *decorata*.

Cereus pedunculatus (Pennant)**Daisy Anemone** Base strongly adhesive and wider than the column; height of column up to 120mm, bearing prominent suckers on the upper region; disc bearing up to 750 tentacles arranged in up to 9 cycles in a large specimen; disc often puckered. **Colour** column variable, paler below, often fleshy-grey-brown; suckers grey-white; tentacles and disc variable, often brownish and dotted with paler colours, sometimes tinted with pink. This species may often be cryptic. **Habitat** on the shore in pools or in shallow water; occurs on rocks and attached to hard objects buried in sand.

Phymanthus pulcher

Aureliana heterocera

Aiptasiogeton pellucidus

Aiptasia mutabilis

Haliplanella lineata

Sagartia elegans

Cereus pedunculatus

Sagartia troglodytes
var. *ornata*

Sagartiogeton undatus (Müller) Base strongly adhesive, usually wider than the disc and reaching up to 60mm in diameter; column smooth, reaching up to 120mm tall, but full extension may occur only in the dark or when the animal is buried; disc bears up to 192 tentacles which in expansion are generally held in a graceful fashion; the tentacles themselves are quite long and gradually tapering. **Colour** column pale brown-yellowish often with brown spots, and bears vertical cream stripes; disc patterned, translucent brown to pale grey; tentacles translucent pale grey. **Habitat** normally attached to stones or shells buried in sand and mud. **Similar species** *Sagartia troglodytes*. See illustration on page 79.

Hormathia coronata (Gosse) Base broad, up to 40mm in diameter and moderately adhesive; column with a thin outer skin, up to 50mm high, tapering slightly and bearing small warts arranged in 12 vertical rows; just before the top of the column is reached these give way to 12 faintly developed ridges; disc small, bearing up to 96 short tentacles arranged in up to 5 cycles. **Colour** the entire column except for the uppermost ridged part is brown-red-orange; uppermost part is dark brown-purple with a terminal white line, the ridges orange-brown; disc and tentacles orange to light brown, sometimes patterned, with a grey region surrounding the mouth. **Habitat** generally growing on worm tubes, shells or rocks, sometimes on rocks and pebbles submerged in mud. **Similar species** none.

Calliactis parasitica (Couch) **'Parasitic Anemone'** Base strongly adhesive, up to 80mm in diameter and a little wider than the column; column stout and pillar-like, reaching up to 100mm high with no warts or suckers but having a granular appearance; disc quite wide, bearing up to 700 moderately long tentacles. **Colour** column pale brown, yellow or light brown with an overall effect of paler vertical stripes; disc and tentacles translucent yellow-cream, sometimes tinted orange. **Habitat** not a parasite but usually found as a commensal on the gastropod shells occupied by *Pagurus bernhardus* and *Dardanus arrosor* (see page 210). This anemone appears to defend the crab against the attacks of *Octopus* and may benefit by obtaining some food from the crabs' feeding activities in return. Rarely found on rocks and stones.

Adamsia carciniopados (Otto) **Cloak Anemone** A commensal anemone almost always associated with *Pagurus prideauxi* or *P. excavatus*. Base and column highly modified to form an adhesive investment around the body of the crab, which itself is usually contained inside the shell of a small gastropod: as the crab grows the base of the anemone secretes a horny substance which effectively extends the shell. The base may reach 100mm across if measured free from the crab and along the longer axis; column squat, giving way almost immediately to a tentacle disc with many small tentacles having a span of up to 50mm, although often less. **Colour** base brown-yellow, usually with red spots or blotches; disc and tentacles white and translucent; when disturbed the anemone may eject fine purple-lilac threads. **Habitat** on sandy and muddy substrates down to 200m, always with a hermit crab.

Amphianthus dohrni (Von Koch) Base elongated, up to 25mm along its long axis, adhering to the substrate which comprises either a gorgonian colony such as *Eunicella* (see page 84) or a hydroid colony such as *Tubularia* (see page 58). Column low, may get wider towards the disc and may be cylindrical; disc flat or saucer-like, up to 10mm across with a rim and bearing up to 80 short irregular tentacles. **Colour** orange-red-pink-buff, sometimes with irregular streaks or patches. **Habitat** always growing on another coelenterate colony, down to 1000m. **Similar species** none. N.B. this species was formerly common in many parts of the Mediterranean and elsewhere but numbers are now diminishing.

Peachia cylindrica (Reid) *(=P. hastata)* No adhesive basal disc; column translucent and worm–like, up to 300mm long and 25mm in diameter but often less; 12 tentacles with a span of 40mm, usually appearing short and stout. **Colour** buff-brown and streaked; tentacles with characteristic arrow-shaped markings. **Habitat** in sandy mud, sand and shell gravel down to about 50m. N.B. this species and the two which follow are true burrowers and have a polyp base modified for burrowing and not for adhesion.

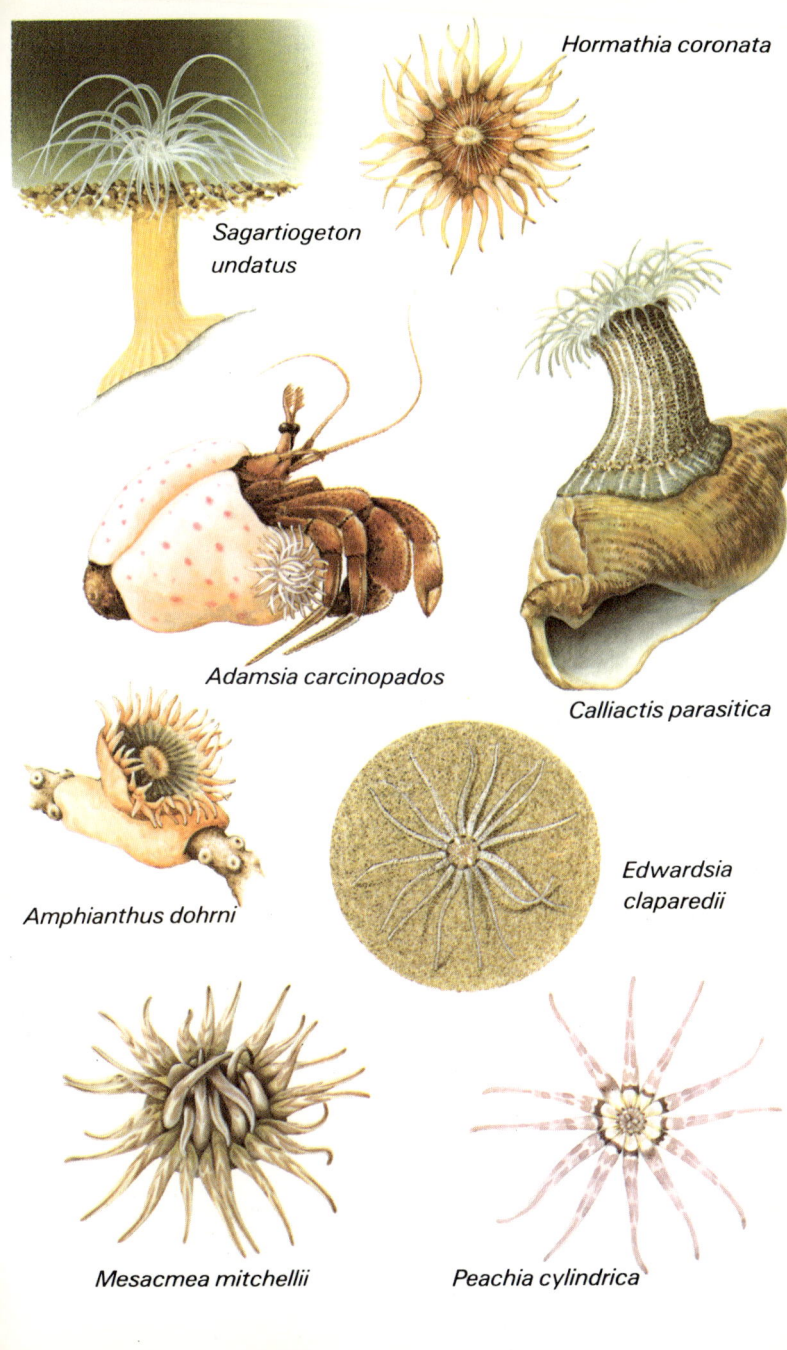

Hormathia coronata

Sagartiogeton undatus

Adamsia carcinopados

Calliactis parasitica

Amphianthus dohrni

Edwardsia claparedii

Mesacmea mitchellii

Peachia cylindrica

Mesacmaea mitchellii (Gosse) Base can be adhesive but is not sucker-like; column pear-shaped when extended and rounded when contracted; may reach 80mm long and 50mm in diameter; up to 36 long tapering tentacles arranged in 3 cycles. **Colour** column brown-orange or reddish below, pale grey-white just below disc which itself is grey-brown-cream with a ring around the mouth; tentacles grey-brown. **Habitat** burrowing in sand or gravel between 15 and 100m. (See illustration on page 79.)

Edwardsia claparedii (Panceri) Base not adhesive; column worm-like and up to 120mm long in full extension; disc small and bearing 16 tentacles arranged in 2 cycles; tentacles long when extended, having a span of about 35mm. **Colour** column translucent pink; disc yellow-buff; tentacles transparent and colourless but finely spotted red-brown and opaque cream. **Habitat** burrowing in muddy sand and gravel, often among sea-grasses, down to about 10m. **Similar species** *Scolanthus callimorphus* (Gosse) not dealt with here. (see Manuel, R. L. 1981 page 204). (See illustration on page 79.)

Order Madreporaria True corals

Anthozoans with hard calcareous skeletons into which the polyps can almost, if not completely, withdraw when disturbed. Often colonial. N.B. avoid confusion with calcified ectoprocts (see pages 222-227). Further information in Best, M. B. 1969 and Manuel, R. L. 1981.

Caryophyllia smithii Stokes & Broderip **Devonshire Cup Coral** Solitary, with stout brown-white skeleton up to 15mm high and with conspicuous ridges (septae); polyps variable in colour: white, pink, brown or green, often with contrasting lips e.g. red or green; tentacles variable in colour, terminating in a small knob. **Habitat** on rocks or stones from the shore down to 100m. **Similar species** two: *Balanophyllia regia* and *Caryophyllia clavus* Sacchi, which was originally thought to be a different species, is now thought to be synonymous with *C. smithi.* This species is sometimes associated with the small barnacle *Boscia anglicum* (see page 186) which may be found growing on the periphery of the coral cup.

Balanophyllia regia Gosse **Scarlet-and-gold Star Coral** Solitary, with cylindrical skeleton up to 10mm high; perforated septae not visible in life (they are in *Caryophyllia smithii*); brilliant scarlet-orange body, tentacles transparent with gold flecks and lacking a knob on the top. **Habitat** from the shore down to 100m on rocks. **Similar species** *Caryophyllia smithii.*

· **Cladocora cespitosa** (Linnaeus) Bushy colonies with branching tubular skeletons reaching up to 100mm high, with stems 20 to 50mm in diameter; polyps brown. **Habitat** on rocks and shells from 1 to 70m.

Dendrophyllia ramea (Linnaeus) Tree-like colonies up to 500mm high with side-branches and polyps arranged in 2 rows; thin-walled skeleton outwardly ribbed; polyps light yellow in colour. **Habitat** on rocks from 30m downward.

Lophelia pertusa (Pallas) Colonial, with irregularly branching, yellow-white skeletons reaching up to 500mm in height; pinkish polyps loosely scattered over the skeleton. **Habitat** on rocky substrates in water from 60–600m.

Leptopsammia pruvoti Lacaze-Duthiers Solitary, coral with stout skeleton reaching up to 60mm high and up to 17mm in diameter; skeleton tapers slightly towards attachment point: skeleton stoutly constructed with thick walls. Polyp with 96 or more tentacles. **Colour** polyps yellow-orange. **Habitat** substrates dominated by coralline algae and under overhanging ledges, in caves and grottos. Similar species *B. regia*, but this has a more delicate skeleton and up to 48 tentacles.

Astroides calycularis (Pallas) Colonial, colonies convex or cushion-like comprising a number of similar individuals: individual skeletons delicate, about 35mm high and 8mm in diameter, densely packed and fused to their neighbours for almost all their height, delicate septae radiate from a central point in each. Polyps have about 30 tentacles. **Colour** polyps orange-red. **Habitat** on rocks and pebbles in shady places down to about 50m.

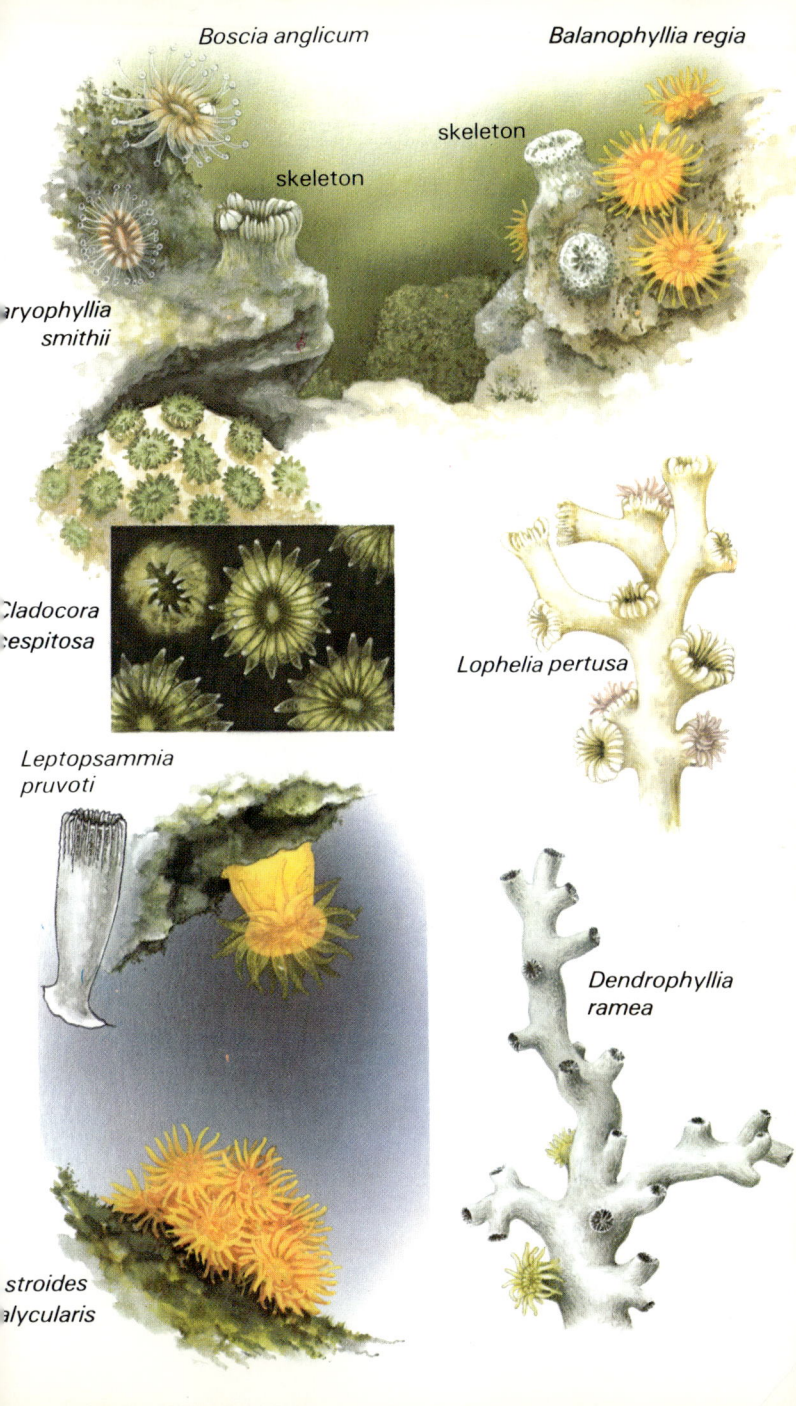

Boscia anglicum

Balanophyllia regia

skeleton

skeleton

Caryophyllia smithii

Cladocora cespitosa

Lophelia pertusa

Leptopsammia pruvoti

Dendrophyllia ramea

Astroides calycularis

Order Corallimorpharia

Anthozoans without a hard skeleton. The polyps have tentacles which terminate in a small knob.

Corynactis viridis Allman **Jewel 'Anemone'** Solitary, small polyp with brilliant colours (which may fade in the aquarium); body 3-5mm in diameter; broad adhesive base; tentacles arranged in 3 circlets; mouth borne on a minute cone. Habitat on rocks on extreme lower shore and down to 100m.

Order Pennatulacea Sea-pens

Feather-shaped anthozoan colonies with a horny or chalky skeleton supporting the central column. Body divided into an upper region bearing polyps on branches and a lower one lacking them which may be buried in the substrate to a variable extent. Mediterranean recorded species = about 5.

Pennatula phosphorea Linnaeus **Phosphorescent Sea-pen** Feather-like colony with relatively slender central column reaching up to 200mm high, occasionally more; region bearing polyps a little longer than the region which does not; polyps borne in rows on relatively narrow side branches; polyps about 1mm tall. Colour red-brown with white polyps. Habitat growing in sand and clay from 20 to 100m. Similar species *Pteroeides spinosum.*

Pteroeides spinosum (Ellis) Feather-like colony with fleshy, relatively stout-looking central column reaching up to 150mm high; region bearing polyps about two-thirds of total length; polyps borne in rows on flap-like spiny side branches. Colour polyps whitish, side branches greyish-yellow to brown, stem brown to orange. Habitat on muddy bottoms from 30 to 250m. Similar species *Pennatula phosphorea* above.

Veretillum cynomorium (Pallas) Finger-like colony with stout central column reaching 150mm; region bearing polyps about four-fifths of total length, remainder pointed and serves for anchorage in substratum; polyps up to 20mm long when fully expanded. Colour orange-yellow. Habitat sandy bottoms from about 20 to 40m. Similar species none.

Funiculina quadrangularis (Pallas) Long slender central filament reaching 400mm; region bearing polyps relatively flexible and a little more than half total length; lower region, part of which is embedded in the substrate, relatively strong and stiff and slightly wider except for lance-like tapering tip; two forms of polyps branch from upper region. Colour pink. Habitat on muddy bottoms from 40 to 400m. Similar species *Virgularia mirabilis* below.

Virgularia mirabilis (O. F. Müller) Slender feather-like delicate erect colony reaching 200mm high; central column slender and when in situ, polyp-bearing side branches pass right down to the substrate; side branches slender. Colour cream-yellow. Habitat on muddy bottoms from 30m. Similar species on account of its slender delicate nature it is unlikely to be confused with any other species of sea-pen.

Corynactis
viridis

Veretillum cynomorium

Virgularia mirabilis

Pteroeides spinosum

Pennatula phosphorea

Funiculina
quadrangularis

Order Alcyonacea Soft corals

Colonial anthozoans whose retractable polyps have 8 branching (pinnate) tentacles; polyps are embedded in the body mass which has a skeleton made of a great number of free calcareous ossicles making these colonies soft and flexible. Usually attached to rocks and stones. Mediterranean recorded species = about 5.

Alcyonium palmatum (Pallas) Erect stout branching colonies reaching 500mm high; colonies white, pink, brown or red with translucent white polyps; tentacles with 11 to 13 branchlets on either side. **Habitat** attached to stones or shells on muddy substrates or standing freely in shallow water down to 20m. **Similar species** *Alcyonium glomeratum* but this has opaque polyps.

Cornularia cornucopiae (Pallas) Small growths of polyps not embedded in a central mass but interconnected by stolons; polyps reaching up to 25mm high; lower part of polyps and stolon yellow-brown, upper part and tentacles terracotta to white. **Habitat** on stones and rocks from the shore down. **Similar species** none.

Parerythropodium coralloides (Pallas) Encrusting red colonies growing over shells and pebbles to form small lobe-like colonies or growing on dead gorgonian colonies (see below); size according to situation but not large. **Colour** varieties are found from red to yellow, pink or white; tentacles white. **Habitat** in sheltered localities on rocky substrates where suitable support can be found. **Similar species** might be confused with *Alcyonium* spp. but usually grows over another object.

Order Gorgonacea Sea-fans and their allies

Colonial anthozoans in which the polyps have 8 branching tentacles. The polyps are retractable, and are embedded in tissue supported by a branching central skeleton of calcium carbonate bound with a horn-like substance called gorgonin. They are anchored at the base by a holdfast. There are about 20 recorded species from the Mediterranean; further details are given by Carpine & Grasshoff 1975.

Corallium rubrum (Linnaeus) **Red 'Coral' or Precious 'Coral'** Colonial, hard growths branching in all planes, reaching 500mm in height; outer surface of main skeleton grooved; small polyps with white tentacles cover whole colony. **Colour** red, pink or white, occasionally brown or black. **Habitat** on hard substrates with poor illumination from 50 to 200m. **Similar species** may occasionally be confused with *Parerythropodium coralloides* above, or with calcified ectoprocts such as *Myriopora truncata* (see page 226).

The genus *Eunicella* is represented by 4 species in the Mediterranean. These are all superficially similar and all branch in one plane but attention should be paid to colour, branching pattern, distribution of polyps at the tips of the colonies and attachment to the substratum. May support the sea-anemone *Amphianthus dohrni*.

Eunicella verrucosa (Pallas) Attached by holdfast; colonies branch at all levels; may reach 300mm high. Higher polyp calyces on terminal branches with polyps arranged in two rows here (biserial). **Colour** usually white, may be pink. **Habitat** attached by base to rocks from 35–200m.

Eunicella cavolinii (Von Koch) Attached by holdfast. Colonies branch at all levels; may reach 300mm high; cylindrical terminal parts of branches have polyp calyces arranged all round. **Colour** dark yellow-orange to red. **Habitat** tends to grow on vertical rock faces between 10 and 30m; often with corals.

Eunicella singularis (Esper) *(= E. stricta)* Attached by holdfast; colonies branch mainly at the base so that the terminal limbs are long and upright; may reach 500mm high; the polyp calyces do not project much from the mass of the colony. **Colour** grey-white or greenish-white (in the presence of commensal zooxanthellae). **Habitat** horizontal rock platforms down to 50m, often with corals.

Paramuricea clavata (Risso) *(=P. chamaeleon)* Rigid colonies branching in one plane and with a bushy appearance, reaching up to 1m high. **Colour** carmine-red or violet, sometimes brown or yellow at the tips. **Habitat** often attached to vertical rocks and associated with *Eunicella cavolini* or in more horizontal positions with *E. singularis* from 15–100m. **Similar species** several.

Parerythropodium coralloides

Alcyonium palmatum

Eunicella cavolinii

Eunicella verrucosa

Eunicella singularis

Cornularia cornucopiae

Paramuricea clavata

Corallium rubrum

Phylum Ctenophora Comb jellies

These animals differ in several respects from the cnidarians proper and are thus often separately classified. The body may be variously shaped and is composed of two thin layers of cells separated by a volume of transparent, iridescent or luminous jelly which constitutes the animal's main bulk. The mouth is situated at the bottom of the body and leads into a series of digestive canals which open by one or two minute pores at the top. Radiating over the surface of the body from the top is a series of up to 8 swimming structures called comb-rows. Each comb-row is made up of a number of plates consisting of fused cilia. These beat up and down rhythmically, so driving the animal through the water.

Ctenophores are highly predatory and feed on other floating animals. Many have tentacles which can be protruded from pits on either side of the body and trailed along like fishing lines. These tentacles cannot sting the prey, but trap it with special lasso cells, thus securing it until it is passed to the mouth. Mediterranean species = about 13. For more details see Chun, C. 1880.

Class Tentaculata
Ctenophores with retractile tentacles

Pleurobrachia pileus (O. F. Müller) **Sea-gooseberry** Rounded oval body up to 30mm in length with conspicuous comb-rows running from the apex but terminating short of the bottom of the animal; relatively long branching tentacles. Colour transparent, white-orange gut. Habitat common in open water; rarely rock pools; may be in shoals. Similar species *Hormiphora plumosa*.

Bolinopsis infundibulum (O. F. Müller) (Not illustrated) Oval body reaching up to 150mm in length; 2 conspicuous comb-rows at either side of the mouth which may be half the length of the rest of the body; branched tentacles. Habitat open water sometimes in shoals. Similar species none.

Hormiphora plumosa Agassiz Similar to *Pleurobrachia pileus*. Pear-shaped body with conspicuous comb-rows. Colour gut brown, tentacles with brown and yellow branches. Habitat open water.

Callianira bialata Delle Chiaje Body rectangular when viewed end on; at the apex are two outward-twisted wing-like structures; the transparent body appears ribbed or veined; 8 comb rows; branching tentacles Colour tentacles pinkish. Habitat open water. Similar species none.

Deiopea kaloktenota Chun Compressed, almost transparent body reaching up to 40mm high; the beating elements of the comb-rows are relatively large; 6 comb-rows; two pointed projections arise from the side of the body. Habitat open water. Similar species *Eucharis multicornis*.

Eucharis multicornis Esch Scholtz Compressed, almost transparent body with 2 conspicuous lobes normally reaching up to 100mm high (large examples reaching 200mm have been recorded); in addition to the tentacles are 2 worm-like processes projecting from midway down the body; at the apex there are a number of small nipple-like processes set into the tissue; delicate. Colour sometimes with brownish-pink tints. Habitat open water. Similar species *Deiopea kaloktenota*.

Cestus veneris Lesueur **Venus' Girdle** Ribbon-like body 80mm long by 15mm high; 4 transparent comb-rows; main tentacles reduced but secondary tentacles lie in 2 grooves near the mouth. Colour transparent, sometimes green-violet. Habitat open water. Similar species *Vexillum parallelum* Fol.

Class Nuda Ctenophores lacking Tentacles

Beroë cucumis Fabricius Mitre-shaped body up to 100mm or more long; comb-rows run from the apex to the base; branched inner canals visible. Colour transparent. Similar species one.

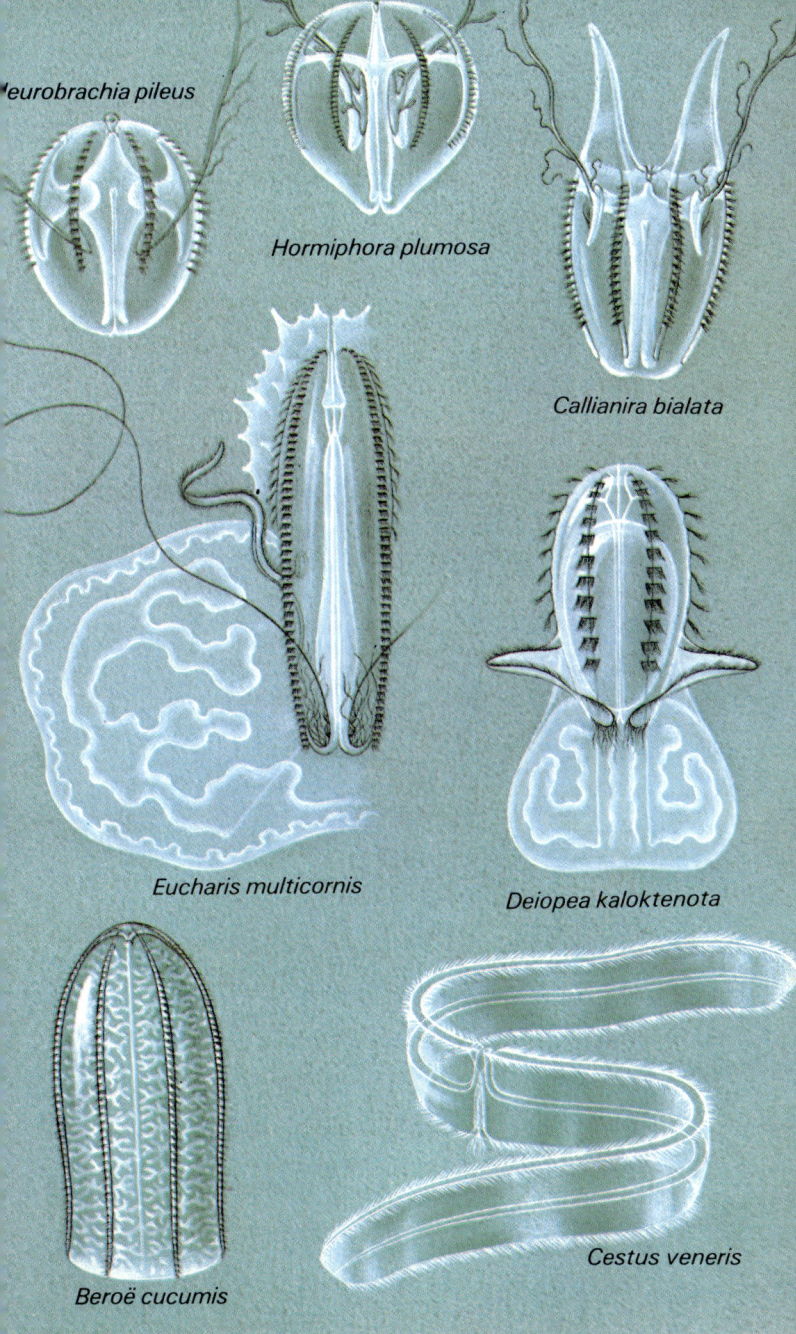

Pleurobrachia pileus

Hormiphora plumosa

Callianira bialata

Eucharis multicornis

Deiopea kaloktenota

Beroë cucumis

Cestus veneris

Marine worms

The animal kingdom includes many types of worms. The different types cover a range of body plans and life styles and are grouped as follows: Phylum Platyhelminthes (the flatworms); Phylum Nemertina (the ribbon worms); Phylum Nematoda (the round worms, outside the scope of this book); Phylum Annelida (the segmented worms), and various minor phyla including Echiura and Sipuncula which have no common names.

Apart from their bilateral symmetry, bodies composed of three cell layers and the fact that they all require a moist if not aquatic environment, these different groups have relatively little in common. Disposition of appendages, presence or absence of segmentation, number of body openings, habitat and the pattern of locomotion should all help in distinguishing between these various phyla.

Phylum Platyhelminthes Flatworms
Class Turbellaria

Generally free-living leaf-shaped worms. They lack a body cavity separating the gut from the remaining tissues. The mouth opens on the underside and the pharynx is often everted for feeding. The gut is simple or branching and sometimes visible through the skin; there is no anus. Rudimentary sense organs at the anterior end include eye-spots and tentacles. Locomotion is with a characteristic gliding movement effected by the combined action of thousands of cilia on the underside, although the animals can change their shape by muscle contraction. A fuller account is given by Ax 1956 and Westblad 1955 and 1956. Mediterranean species number unknown, at least 15.

Convoluta convoluta (Abildgaard) Length up to 6mm; head broader than tail and lacking distinct tentacles; body flattened and leaf-like, often turned up at the edges; chitinous mouth piece on underside but no gut. Colour bright green due to the presence of symbiotic algae. Habitat among seaweeds and on sand down to 15m. Similar species there are a number of flatworms of this size but none other than *C. convoluta* is bright green.

Stylochus pilidium Lang Length up to 35mm; head slightly broader than tail and with two tentacles; eyes at the edge of the body and on the tentacles; dappled appearance. Colour brown with black markings. Habitat among stones and shells in shallow water; often in swarms. Similar species several of similar size.

Leptoplana alcinoi Schmidt Length up to 14mm; body relatively long for width with edges thrown into folds; head rounded without tentacles but 2 conspicuous dark eyes near the midline; tail tapering; gut visible in the middle of the body; lively and active; body fragile. Colour translucent grey. Habitat often among seaweeds; may occur in swarms. Similar species several of similar size.

Thysanozoon brocchii Grube Length up to 50mm; strong body folded along edges and bearing many protuberances, giving it a hairy appearance; head with two tentacles; tail blunt. Colour brown to pinkish. Habitat among seaweeds and mussel banks from the shore down. Similar species none.

Monocelis lineata (O. F. Müller) Length up to 2mm; head not readily discernible from body and lacking distinct tentacles; body slightly pointed at the head end. Colour opaque white gut visible through the skin. Habitat among seaweeds, e.g. *Ulva* (see page 16) from the middle shore down.

Prosthederaeus vittatus (Montagu) Length up to 30mm or more; body striped cream-brown and often thrown into folds; head bears conspicuous tentacles; body flat and leaf-shaped with pointed tail. Colour cream-pink with brown stripes. Habitat under stones in mud. Similar species: *P. giesbrechtii* Lang (also illustrated) which reaches up about 30mm and is sometimes found feeding on sea-squirts.

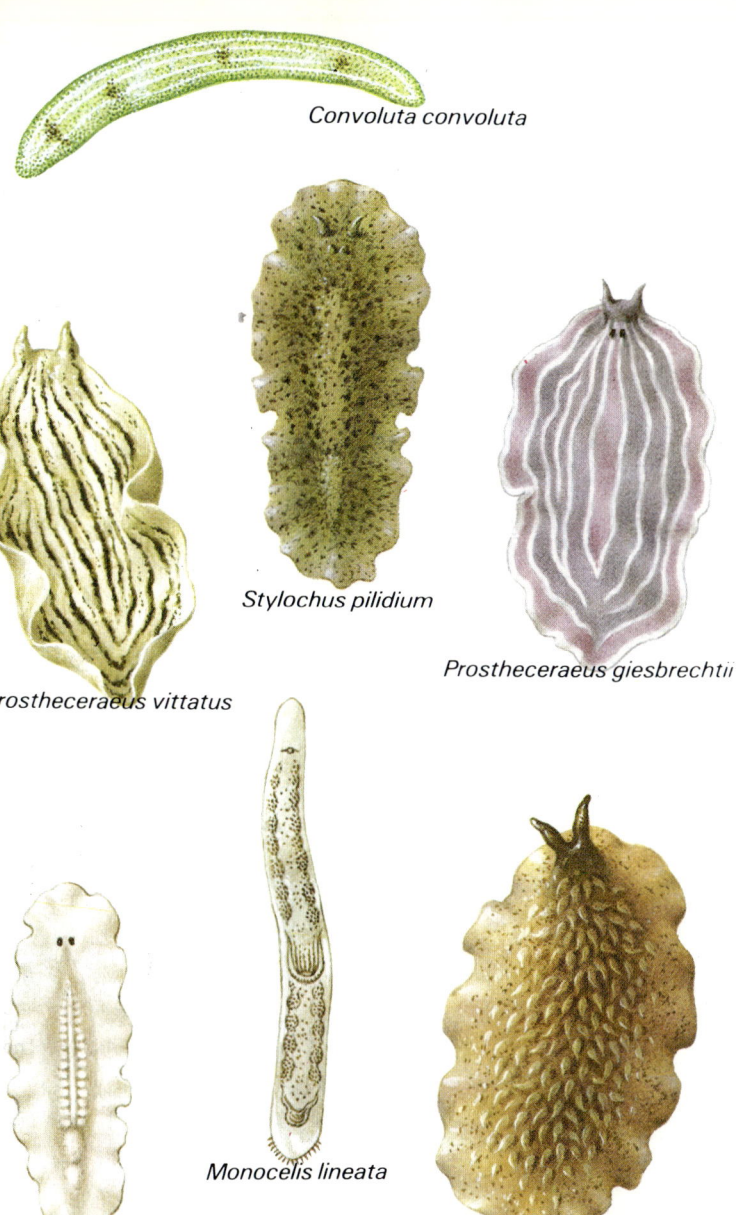

Convoluta convoluta

Stylochus pilidium

Prostheceraeus giesbrechtii

Prostheceraeus vittatus

Monocelis lineata

Leptoplana alcinoi

Thysanozoon brocchii

Phylum Nemertina Ribbon worms

These ribbon-shaped worms are often extremely long. The unsegmented body is composed of three cell layers and there is no body cavity separating the gut from the other tissues. The mouth is anterior and the anus posterior, with a characteristic proboscis opening via the mouth to capture and handle prey. Rudimentary sense organs including eyespots are present.

Nemertines are frequently abundant animals but because they are fragile and often burrow in sand and mud they are easily overlooked. The extension of the proboscis may greatly increase the apparent length of the animal. The form and disposition of the eyes together with the shape of the head are most helpful in assisting identification, and there is generally a slit along each side of the head which should be looked for with the assistance of a hand lens. For this reason the heads of the animals have been illustrated approximately one-and-a-half times larger than their bodies. See Bürger, O. 1895 and Gibson, R. 1972.

Tubulanus annulatus (Montagu) Length up to 250mm but may reach 1m. Body flat below, rounded above, narrowing behind the head and tapering towards the tail; no eyes visible but head slits open just behind the snout. Colour head usually paler than patterned body. Habitat in sand under stones, in rocky clefts or in disused annelid worm tubes from low water down to 10m or deeper. Similar species the pale annular rings and horizontal lines should prevent confusion with superficially similar species but *Tubulanus nothus* with a red-brown body has horizontal lines and more irregularly spaced paler annular rings, sometimes single, sometimes in groups of 3 or 4.

Tubulanus nothus (Bürger) (Not illustrated) Length up to 120mm; not as slender in proportion as *T. annulatus* (above); head rounded and wider than body. Colour see above. Habitat in mud, among shells and algae down to 20m.

Cephalothrix linearis (Rathke) Length to 100mm. Long slender body, narrow inconspicuous head. Habitat mud and under stones, etc. on the shore and shallow water.

Lineus bilineatus (Renier) Length up to 300mm. Head broad, lacks eyes but bears deep slits. Body tapering towards tail. Colour shades of brown with two fine dorsal white lines. Habitat deeper water among coralline seaweeds and shells, rare in shallow water. Similar species the patterning should prevent confusion

Micrura aurantiaca (Grube) Length up to 100mm. Head bears short white snout which lacks eyes; mouth has shallow slits. Body flat below and rounded above. Colour brick-red with a white proboscis. Habitat under stones in rock pools and in deeper water. Similar species several.

Cerebratulus species. These large and fragile nemertines are difficult to identify and there is some doubt as to the validity of the names. Microscopical investigations are necessary. *C. fuscus* (McIntosh) (illustrated) reaches 100mm in length; head with 4–8 eyes and deep side slits; body flat, tapering towards tail which bears a terminal filament. Colour skin colour to grey. Habitat among coralline algae, shells, pebbles and mud, and algal holdfasts to 100m. *C. marginatus* Renier may be a separate species. Length up to 800mm. Shape and habitat similar to *C. fuscus*. Colour grey with pale margins.

Lineus ruber (O. F. Müller) Red Ribbon Worm Length up to 160mm. Head spatulate, fractionally wider than the adjoining part of the body and with shallow slits; 3 or 4 eyes in a row on each side. Body flattened, the latter part tapering towards the tail. Colour red-brown, ventral surface paler than dorsal. Habitat stones, muddy gravel from low water down. Similar species several.

Lineus geniculatus (Delle Chiaje) Length up to 400mm. Head flattened and blunt but not much wider than the body; somewhat similar in shape to *L. ruber.* Body flat, may be twisted and coiled; no terminal thread-like tail appendage. Colour green-brown-black ground colour with white rings. Habitat under stones and on hard bottoms from the shore down. Similar species several.

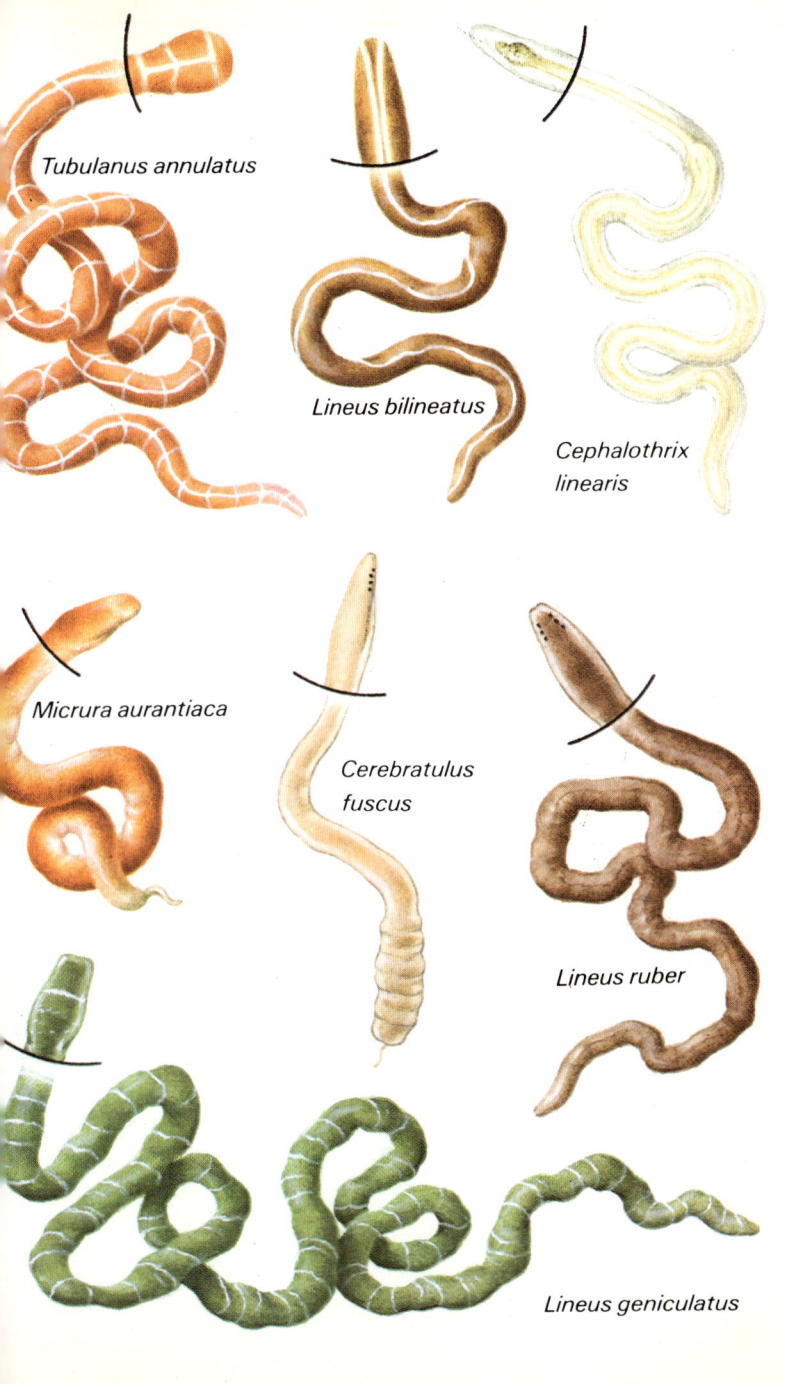

Tubulanus annulatus

Lineus bilineatus

Cephalothrix linearis

Micrura aurantiaca

Cerebratulus fuscus

Lineus ruber

Lineus geniculatus

Drepanophorus spectabilis (Quatrefages) **Length** up to 80mm. **Head** conical or diamond shaped; many eyes situated dorsally and laterally; oblique slits with accessory furrows. **Body** flattened, thickening suddenly behind the head. **Colour** red-yellow above with 5 conspicuous ivory longitudinal stripes; pale below. **Habitat** under stones, in rock crevices and among shells down to 40m; **Similar species** several superficially similar including *D. crassus* below.

Drepanophorus crassus (Quatrefages) (Not illustrated) Similar in shape to *D. spectabilis* but body tapers towards head and tail; body much longer, reaching 140mm and wider. **Colour** above red-brown-yellow-orange, not striped; below pale pink. **Habitat** as above but down to 200m.

Prosorhochmus claparèdii Keferstein **Length** up to 35mm. **Head** broad and spatulate, often slightly wider than the body and with a conspicuous central notch from which a pale streak leads beyond the grey ganglia; 4 eyes situated well back, the anterior pair larger than the posterior pair and the left eyes widely separated from the right; 1 pair of slits. **Body** flattened with a slight constriction behind the head, tapering in the region of the tail. **Colour** pale yellow but occasionally orange. **Habitat** in crevices and fissures in rocks on the shore. **Similar species** several superficially similar.

Amphiphorus lactifloreus (Johnston) **Length** up to 80mm but often less. **Head** flat and spatulate; several groups of eyes arranged in a marginal row on each side and in groups over or close to the conspicuous pink ganglia, giving an overall near triangle of eyes; 2 sets of slits run obliquely. **Body** flattened ventrally, rounded dorsally and not tapering towards the blunt tail. **Colour** various shades from pink to white but pink predominates; a translucent line along the dorsal surface indicates the position of the retracted proboscis. **Habitat** under stones and associated with algae from low water down to deep water. **Similar species** several superficially similar.

Tetrastemma melanocephalum (Johnston) **Length** up to 35mm. **Head** flattened and generally wider than the body, with a conspicuous frontal notch; 4 eyes, the first pair of which lie within the conspicuous patch of black skin so that they are not readily distinguished, while the second pair lie behind the pigment patch and are conspicuous; 1 pair of obliquely set slits. **Body** flattened when extended but more rounded when contracted; slightly constricted behind the head and not tapering until near the tail. **Colour** dull yellow-green with a squarish black patch on the head. **Habitat** on seaweeds and under stones on the lower shore and down to 60m. **Similar species** several superficially similar but the black head patch is a key character.

Oerstedia dorsalis (Abildgaard) **Length** up to 25mm but often less. **Head** slightly notched in front; 4 eyes arranged in a square; 1 pair of slits. **Body** almost circular in cross-section and slightly tapering at each end. **Colour** dorsal surface brown-red with either yellow granules or a dorsal stripe, and ventral surface paler; or green-brown with annular markings and a white stripe. **Habitat** among algae and other growths from low water down to 20m or more. **Similar species** several superficially similar.

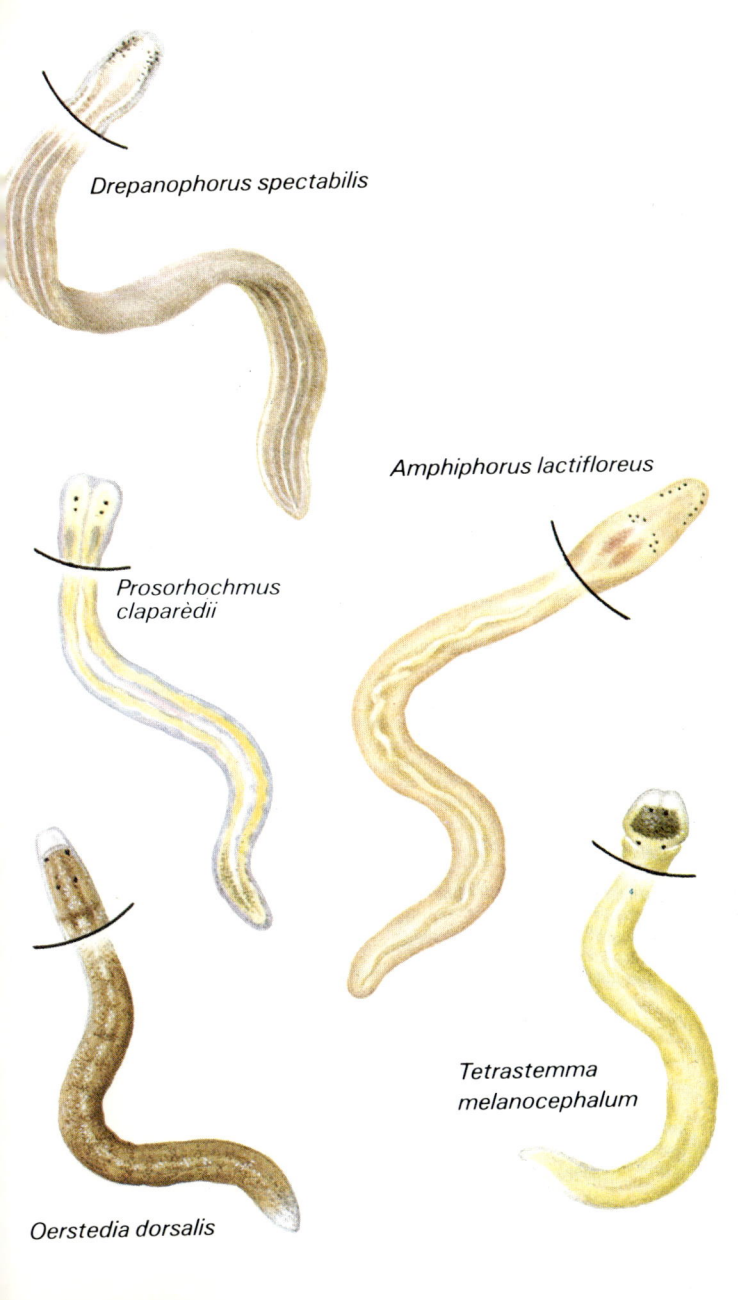

Drepanophorus spectabilis

Amphiphorus lactifloreus

*Prosorhochmus
claparèdii*

Oerstedia dorsalis

*Tetrastemma
melanocephalum*

Phylum Annelida Segmented worms

A very important group of worms of between seven and eight thousand species. The body is composed of three cell layers, with the middle layer divided into two by a true, fluid-filled body cavity (the coelom). The body is divided lengthwise into a number of recognizable segments, each one of which usually carries bristles known as *chaetae* as well as other structures. The head is often well developed, bearing sense organs and a simple brain, and the mouth opens by the second segment (the first segment is the prostomium or preoral segment). The anus opens on the terminal segment. Well-developed longitudinal and circular muscles of the body wall allow extension and contractions of the body, and are often associated with locomotory segmental appendages.

This phylum is divided into three large classes: the bristle worms (Polychaeta), the earthworms and their allies (Oligochaeta) and the leeches (Hirudinea). Of these, the polychaetes have representatives in almost all marine environments, but they rarely occur in fresh water and only very rarely in damp terrestrial habitats. The other classes have a few marine representatives, but are more significant in fresh water or on land. This brief accountant cannot do justice to the evolutionary and ecological significance of the annelids in general, but something of the importance of the polychaetes will be appreciated by readers exploring the sea and the shore because of their abundance, diversity and their exceptionally wide geographical distribution.

Throughout the accompanying line diagrams, the following key letters have been applied: ac=aciculum (rod which supports the parapodium); an = antenna; c=cirrus (small outgrowth of head or parapodium); ch=chaeta, bristle or seta; d=dorsal surface; e=eye; ft=free tooth; g=gill; lo=lateral organ (small structure on parapodium of some species); j=jaw; mi=marginal membrane of head of some species which is incised at the edge; mm=marginal membrane of head of some species which has a scalloped edge; p=palp; pr=prostomial process; pro=proboscis; s=scale; t=tooth; tc=tentacular cirrus of head; v=ventral surface.

Class Polychaeta Bristle worms

Annelids with a preoral segment or prostomium. The head is formed from several highly modified segments fused together and carries a variety of specialized structures such as antennae, eyes, palps, jaws and tentacular cirri (see fig. 5). The rest of the body is composed of a relatively large number of similar segments, most of which bear a pair of locomotory appendages called parapodia; these sometimes have a respiratory function also. The parapodia are composed of several structures which are important in the identification of the various species (see fig. 6). The sexes are usually separate, and fertilization takes place in the sea; a pelagic larva is often formed. Habits and habitats are very variable.

This diverse group of worms cannot be divided satisfactorily into orders, but falls into a number of families of which twenty-six are treated in this book. The basic body plan has been described, but it is modified in the various groups. The polychaetes may be described loosely as errant (free-living and predatory) or sedentary (burrowing or tube-dwelling); yet within these groupings many variations of form occur. Three characteristics in particular help to identify the worms correctly. The presence or absence of a tube and the form of the tube itself is a clue to a number of families. A hard, calcareous tube occurs in the Serpulidae; e.g. in the Amphictenidae and Terebellidae it consists of

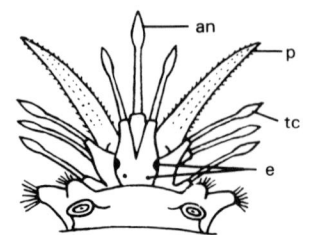

Fig. 5 Head of *Lepidonotus clava*

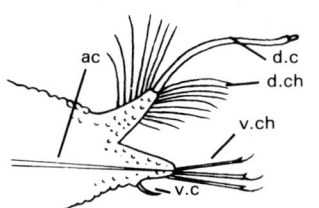 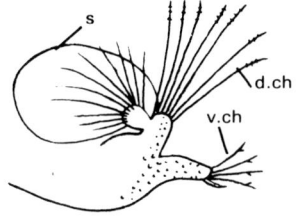

Fig. 6 Cirrus-bearing (left) and scale-bearing parapodia of *Hermione hystrix*

grains of sand cemented into an organic matrix; and in the Sabellidae the tube is built of mud. If the worm is a burrowing variety then the shape and form of the burrow should be considered. Next, the form of the head and the disposition of the appendages should be examined. These are shown in fig. 5 and in a number of other cases on the following pages. Finally, the form of the parapodia should be investigated. A hand lens may be essential, and mounting parapodia on a slide and examining them with a microscope will often assist further. Fig. 7 shows the arrangement of bristles or chaetae as well as the dorsal and ventral parapodial cirri and other structures in a typical polychaete.

When examining an unidentified worm it should be remembered that errant polychaetes generally have well-developed eyes and tentacles for receiving information from the environment and for seeking prey. They often have a protrusible proboscis which may be armed with powerful jaws, and their parapodia are powerfully built for crawling or swimming. Generally, the form of the proboscis may be discerned only when it has been everted. Pressing the pharyngeal region of such worms either when they are alive or narcotized may cause the proboscis to be everted. In other proboscis-bearing species the shape of the organ can sometimes be made out through the relatively transparent body wall of the anterior region.

Fig. 8 shows the general characters of an errant polychaete's anterior end. Sedentary polychaetes usually have reduced sensory structures and reduced parapodia, but their gills may be large and conspicuous, and are often also used to filter food from the water as well as to extract oxygen.

Fauvel, P. 1923 and 1927 provides a detailed identification for most of the Mediterranean polychaetes.

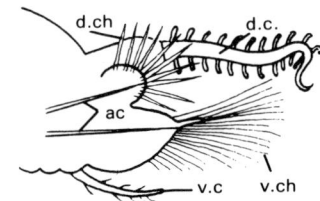

Fig. 7 Parapodium of *Harmothoë impar*

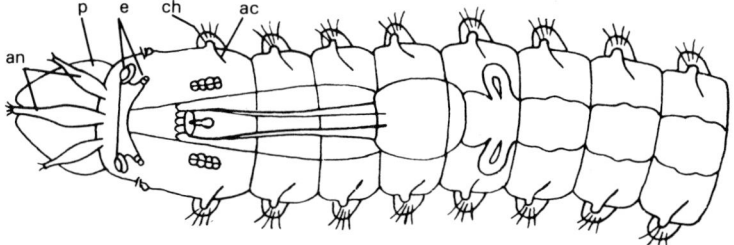

Fig. 8 Anterior end of male *Exogene gemmifera* (family *Syllidae*)

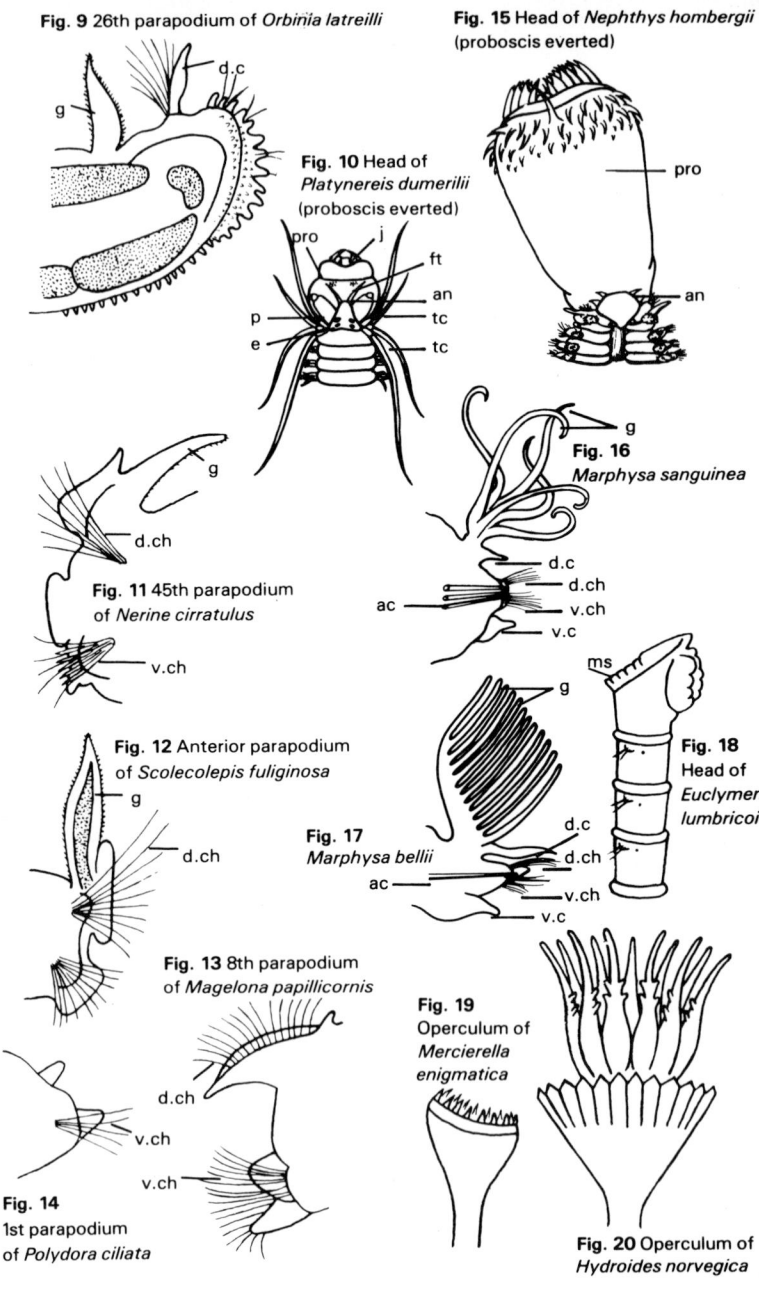

Fig. 9 26th parapodium of *Orbinia latreilli*

Fig. 15 Head of *Nephthys hombergii* (proboscis everted)

Fig. 10 Head of *Platynereis dumerilii* (proboscis everted)

Fig. 16 *Marphysa sanguinea*

Fig. 11 45th parapodium of *Nerine cirratulus*

Fig. 12 Anterior parapodium of *Scolecolepis fuliginosa*

Fig. 17 *Marphysa bellii*

Fig. 18 Head of *Euclymene lumbricoid*

Fig. 13 8th parapodium of *Magelona papillicornis*

Fig. 19 Operculum of *Mercierella enigmatica*

Fig. 14 1st parapodium of *Polydora ciliata*

Fig. 20 Operculum of *Hydroides norvegica*

96

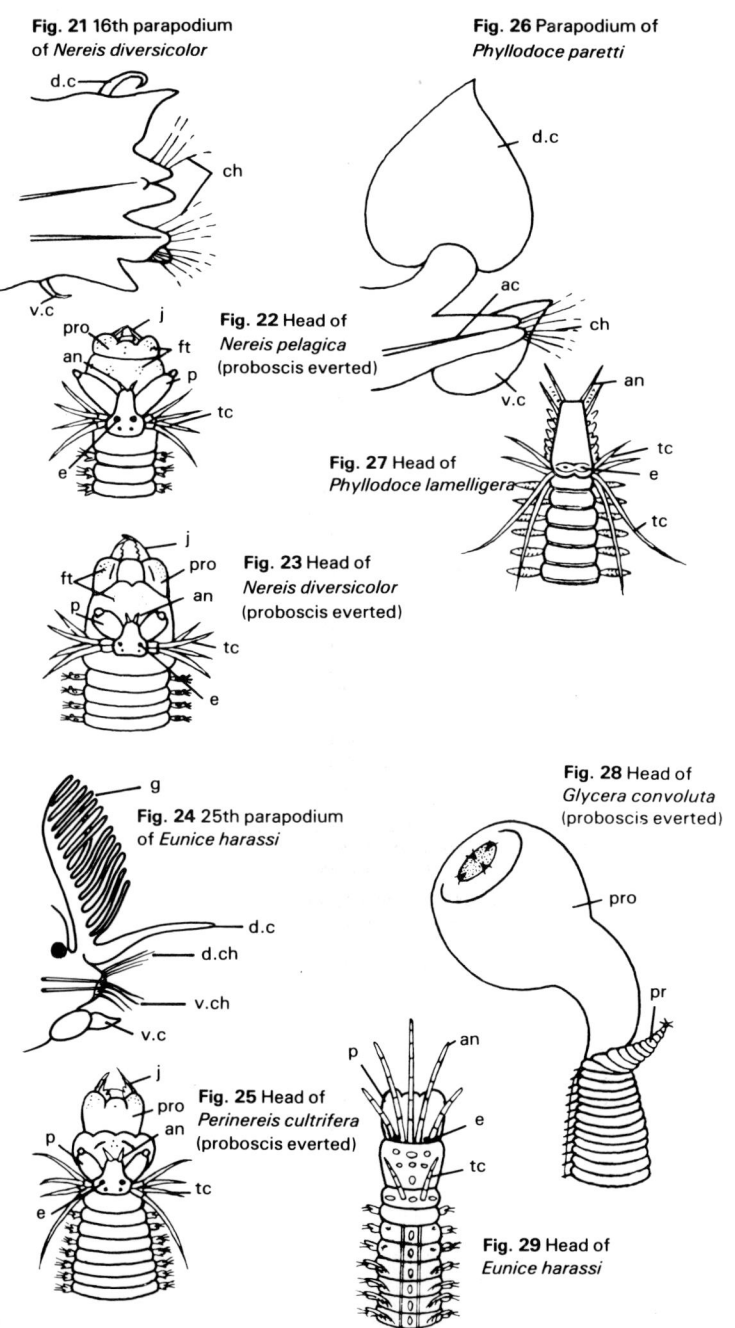

Fig. 21 16th parapodium of *Nereis diversicolor*

Fig. 22 Head of *Nereis pelagica* (proboscis everted)

Fig. 23 Head of *Nereis diversicolor* (proboscis everted)

Fig. 24 25th parapodium of *Eunice harassi*

Fig. 25 Head of *Perinereis cultrifera* (proboscis everted)

Fig. 26 Parapodium of *Phyllodoce paretti*

Fig. 27 Head of *Phyllodoce lamelligera*

Fig. 28 Head of *Glycera convoluta* (proboscis everted)

Fig. 29 Head of *Eunice harassi*

Family Aphroditidae: Scale worms

Free-living polychaetes; proboscis eversible, bearing papillae; dorsal surface wholly or partly covered by interfolded scales which mask the head when viewed from above; parapodia divided into dorsal and ventral lobes; handling may dislodge these giving a false idea of their numbers. Mediterranean species = about 30.

Aphrodite aculeata Linnaeus **Sea Mouse** Length up to 200mm but often less. Head largely obscured by mat-like covering of hairs and scales. Body bulky and ovoid, convex above and flat below; about 40 chaeta-bearing segments. Colour dirty brown, chaetae iridescent green, brown or yellow. Habitat on soft substrates

Hermione hystrix (Savigny) Length up to 60mm. Head largely hidden by body scales; has 2 long palps, 1 long antenna, 2 eyes, 2 pairs of tentacle cirri. Body ovoid, flat and bearing scales above, about 34 chaeta bearing segments; cirrus bearing parapodia alternate with scale-bearing ones. Colour reddish-brown above Habitat among shells and on soft substrates down to 100m.

Lepidonotus clava (Montagu) Length up to 30mm. Head somewhat hidden by body scales; has 2 palps, 1 median antenna, 2 lateral antennae about two-thirds as long, 1 pair of large and 1 pair of small eyes, 3 pairs of tentacle cirri. Body flat bearing rounded scales; about 24 chaeta-bearing segments; cirrus-bearing segments alternate with scale-bearing ones. Colour brownish. Habitat under rocks.

Harmothoë imbricata (Linnaeus) Length up to 50mm, often less. Head somewhat hidden from above by body scales; 2 quite long tapering palps; 1 long median and 2 much shorter lateral antennae; 2 pairs of eyes; 2 pairs of tentacle cirri. Body flat, tapering towards tail, covered by 15 pairs of scales with papillae on sides; 37 chaeta-bearing segments; cirrus-bearing segments alternate with scale bearing ones. Colour variable, bluish-grey to brownish-grey with a paler spot to centre of scales. Habitat shallow water, under stones and rocks.

Harmothoë impar (Johnston) (=**Evarne impar**) Length up to 25mm Head somewhat obscured by body scales; 2 fat palps; antennae, eyes and tentacle cirri similar to H. imbricata. Body very fragile, flat and bearing 15 pairs of overlapping scales the sides of which carry papillae; 30–40 chaeta-bearing segments; cirrus bearing segments alternate with scale-bearing ones. Colour brownish-green, scales have a yellowish central spot. Habitat under stones, shells and weeds on lower shore and shallow water.

Polynoë scolopendrina (Savigny) Length up to 120mm. Body has 15 pairs of overlapping scales which cover about half the body; 80–100 segments bear chaetae, alternate parapodia from segment 2 to 32 carry scales. Colour variable; often red with shiny scales. Habitat in other worm tubes, cracks in rock and in sand.

Sthenelais boa (Johnston) Length up to 200mm. Head partially obscured by body scales; 2 long slender palps; 1 long median antenna, 2 pairs of thinner shorter lateral antennae; 2 pairs of eyes; 1 pair of tentacle cirri. Body long, tapering, convex above bearing many pairs of interfolding kidney-shaped scales, the outer edges ornamented with papillae; about 150–200 chaeta-bearing segments. Colour very variable, red, yellow or brownish with dark transverse bands. Habitat in sandy and muddy places under stones and near weeds like Zostera.

Family Amphinomidae

Active free-living polychaetes, proboscis eversible, unarmed, generally with a small fleshy wattle on the head (caruncle); parapodia divided into dorsal and ventral lobes and bearing gills, 1–2 dorsal cirri, 1 ventral cirrus, 2 groups of chaetae per parapodium. Mediterranean recorded species = about 5.

Hermodice carunculata Pallas Length from 60 to 300mm. Head with oval caroncula reaching back to 4th chaeta-bearing segment, 2 small palps; 1 median antenna slightly larger than palps and 2 lateral ones about the same length as palps Body lacking scales, up to 150 chaeta-bearing segments; quite conspicuous gills borne from the first chaeta-bearing segment back and arranged dorsally on each parapodium in 2 groups. Colour brownish above with red gills and white spots Habitat on rocks and floating objects associated with Lepas (see page 184).

Lepidonotus clava

Hermione hystrix

Aphrodite aculeata

Harmothoë imbricata

Harmothoë impar

Sthenelais boa

Polynoë scolopendrina

Hermodice carunculata

Family Phyllodocidae Paddle worms

Active free-living polychaetes; proboscis eversible lacking jaws; numerous segments with parapodia bearing conspicuous leafy or paddle-like dorsal cirri; ventral cirri of parapodia less conspicuous but similar in form; normally one group of chaetae per parapodium; terminal segment bearing 2 anal cirri. Mediterranean recorded species = about 35.

Phyllodoce lamelligera (Gmelin) **Length** 60–600mm. **Head** tapering to front; 4 short antennae; 2 black eyes; 4 pairs of long tentacle cirri. **Body** between 300 and 400 segments; dorsal cirri of parapodia conspicuous, lance-shaped and greenish. **Colour** body bluish-brown. **Habitat** under large pebbles, rocks and weeds from the lower shore down.

Phyllodoce mucosa Oersted (Not illustrated) **Length** 50–100mm. **Head** tapering to front, 4 very short antennae, 2 black eyes, 4 pairs of tentacle cirri, the rearmost the longest and reaching back to the 8th or 10th chaeta-bearing segment. **Body** of 100 to 200 chaeta-bearing segments, thin at the rear; parapodia have dorsal cirri, ovoid at front, subrectangular in the middle of the body and lance-like at the rear; 2 cylindrical anal cirri. **Colour** dirty white with brownish spots. **Habitat** in sand and muddy places in shallow water.

Phyllodoce paretti (Blainville) **Paddle Worm** **Length** 150 to 300mm. **Head** rounded; 4 short antennae; 2 large eyes; 4 pairs of inconspicuous tentacle cirri. **Body** long and tapering at both ends, about 200 segments; parapodia with large leafy dorsal cirri at the front and lance-like at the rear. **Colour** variable, often bluish with black, green or yellow marks on parapodia. **Habitat** under rocks and stones in shallow water.

Eulalia viridis (O. F. Müller) **Green Leaf Worm** **Length** 50–150mm. **Head** small and rounded; 1 small median antenna and 2 pairs of lateral antennae; 2 eyes; 4 pairs of tentacle cirri, the last two pairs reaching back to the 10th or 12th chaeta-bearing segment; proboscis when fully extended very long. **Body** up to 200 segments; parapodia with quite conspicuous dorsal cirri. **Colour** grey-green. **Habitat** in rock crevices, often creeping about on rocks exposed to air at low water.

Eulalia sanguinea Oersted **Length** up to 60mm. **Head** antennae, eyes, tentacle cirri and proboscis similar to *E. viridis*. **Body** relatively short and inflated; 60 to 140 segments; parapodia with somewhat pointed dorsal cirri. **Colour** variable, white to pale green or brown, sometimes with paler dorsal line. **Habitat** among rocks and weeds from lower shore down.

Family Alciopidae

Active pelagic polychaetes; head with eversible unarmed proboscis; very large red eyes; 5 short simple antennae; body transparent and long; parapodia with leaf-like dorsal and ventral cirri. Mediterranean recorded species = about 11.

Alciopa cantrainii (Delle Chiaje) **Length** up to 110mm. **Head** small with very conspicuous eyes; other appendages typical of family. **Body** sharply tapering towards either end; parapodia with long chaetae adapted for swimming. **Colour** crystal clear; eyes red. **Habitat** pelagic; likely to be caught in plankton samples.

Family Tomopteridae

Active, free-living planktonic polychaetes; bodies transparent; parapodia large; chaetae and acicula wanting. Mediterranean recorded species = about 6.

Tomopteris helgolandica (Greeff) **Length** up to 17mm. **Head** has a pair of conspicuous long palps behind which is a pair of short chaeta-bearing appendages followed by another pair about two-thirds the length of the body. **Colour** transparent, colourless. **Habitat** pelagic; most likely to be found in plankton samples.

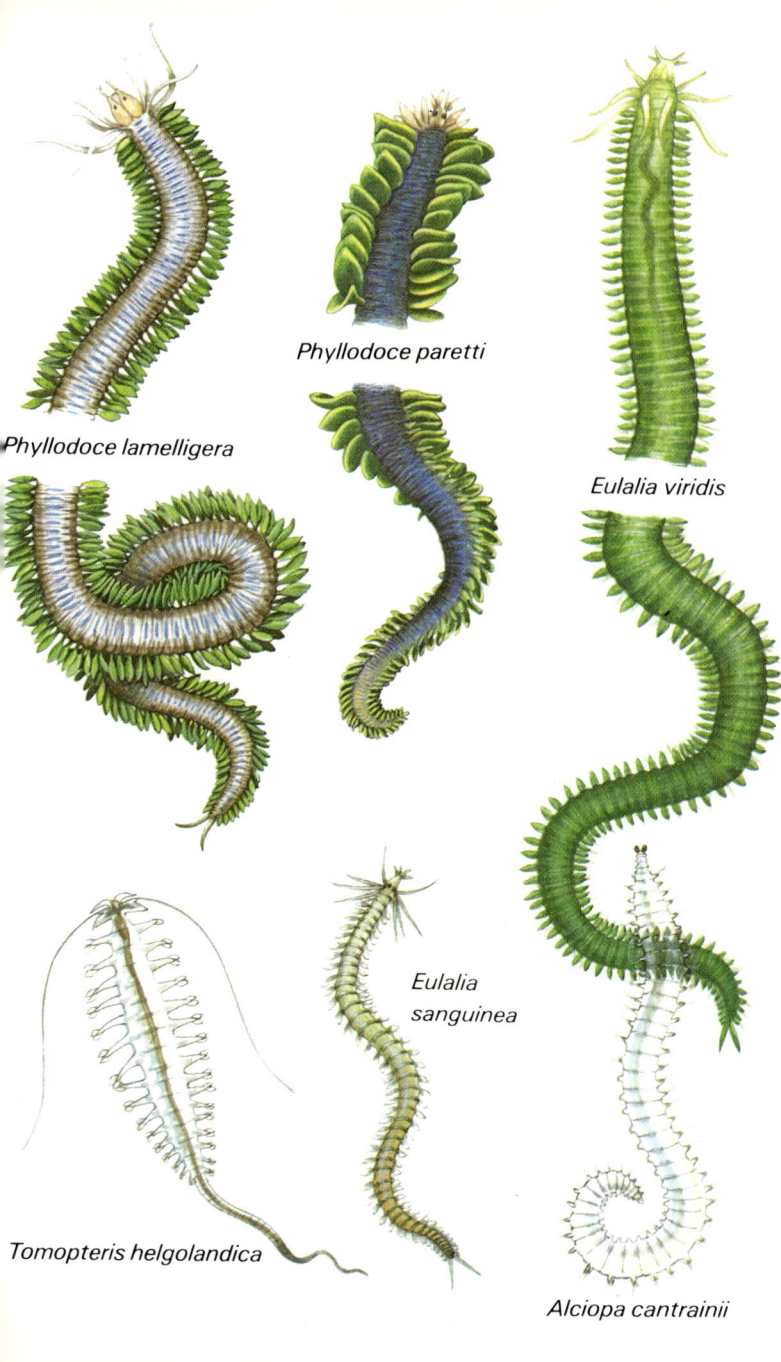

Phyllodoce paretti

Phyllodoce lamelligera

Eulalia viridis

Eulalia
sanguinea

Tomopteris helgolandica

Alciopa cantrainii

Family Hesionidae
Active free-living polychaetes; the head may be quite small, proboscis eversible and often short. Mediterranean species = about 10.

Hesione pantherina (Risso) Length up to 60mm. Head small; 2 small antennae, 2 large anterior eyes and 2 smaller ones behind; 8 pairs of tentacle cirri; proboscis cylindrical, lacking teeth and with a large opening. Body cylindrical, slightly tapering to rear; 16 chaeta-bearing segments, dorsal cirri of parapodia long, chaetae in 1 large bundle. Colour brownish with white marks. Habitat under stones and pebbles in shallow water, sometimes swimming.

Kefersteinia cirrata (Keferstein) Length up to 75mm, often less. Head has 2 small antennae lying between the 2 thicker, slightly larger palps; 2 pairs of eyes, the anterior ones larger; 8 pairs of tentacle cirri; illustration shows proboscis everted; proboscis lacks jaws, is wide and short and bears pappilae. Body fragile, 36–65 chaeta-bearing segments; dorsal cirri or parapodia long, chaetae in 1 large bundle. Colour green-brown-yellow according to sex and maturity. Habitat among weeds, shells and worm tubes, on the shore and in shallow water.

Family Syllidae
A very large family of small delicate free-living polychaetes, often beautifully coloured. The head usually carries 2 palps, 3 antennae, 4 eyes and 2 pairs of tentacle cirri. The proboscis is eversible and divided into two parts, the anterior being cylindrical, chitinous and armed with 1 or more teeth; the posterior is muscular. The body is generally quite thin, sometimes short; parapodia often bear very long dorsal cirri which under the microscope appear joined; there are usually 2 small anal cirri. Reproduction is sometimes by budding to give stolons of several individuals arranged in a chain, or in side branches. This is a large family of which only a few can be included here; identification is often difficult. Mediterranean recorded species = about 55.

Syllis prolifera Krohn Length 10–25mm. Head bears small triangular palps, 1 median antenna and 2 slightly shorter lateral antennae, 4 eyes and 2 minute eyespots. Body composed of many segments; parapodia with long 'segmented' dorsal cirri with 20 to 40 segments; chaetae arranged in one group. Colour very variable, grey-red, sometimes with brown, pink or orange markings. Habitat on the shore and in shallow water among weeds and stones.

Autolytus pictus (Ehlers) Length up to 25mm. Head palps difficult to see; 1 median antenna a little longer and wider than the 2 lateral antennae; 4 large eyes; tentacle cirri nearly as long as lateral antennae; eversible proboscis has 10 large teeth alternating with 10 smaller ones. Body rounded, 60 to 100 chaeta-bearing segments; form of parapodia varies: first chaeta-bearing segment has very long dorsal cirri, those of the second are short; 2 thick anal cirri. Colour variable; pinkish below, violet marks above.

Family Nephtydidae Catworms
Active medium to large-sized free-living polychaetes with small heads and flattened bodies; eversible proboscis has horny jaws; characteristic sinusoidal swimming motion. Mediterranean recorded species = about 6.

Nephtys hombergii Audouin & Milne Edwards Length up to 200mm. Head 4 small antennae, 2 small brown eyes; large cylindrical proboscis bearing several rows of papillae (visible when everted) and one conspicuous long dorsal papilla (lacking in the related species *N. caeca*, but this does not occur in the Mediterranean). Body flat and muscular; between 90 and 200 chaeta-bearing segments, chaetae densely arranged in 2 groups and quite short. Colour iridescent pink and blue-white. Habitat burrowing in sand and mud on the shore and in shallow water.

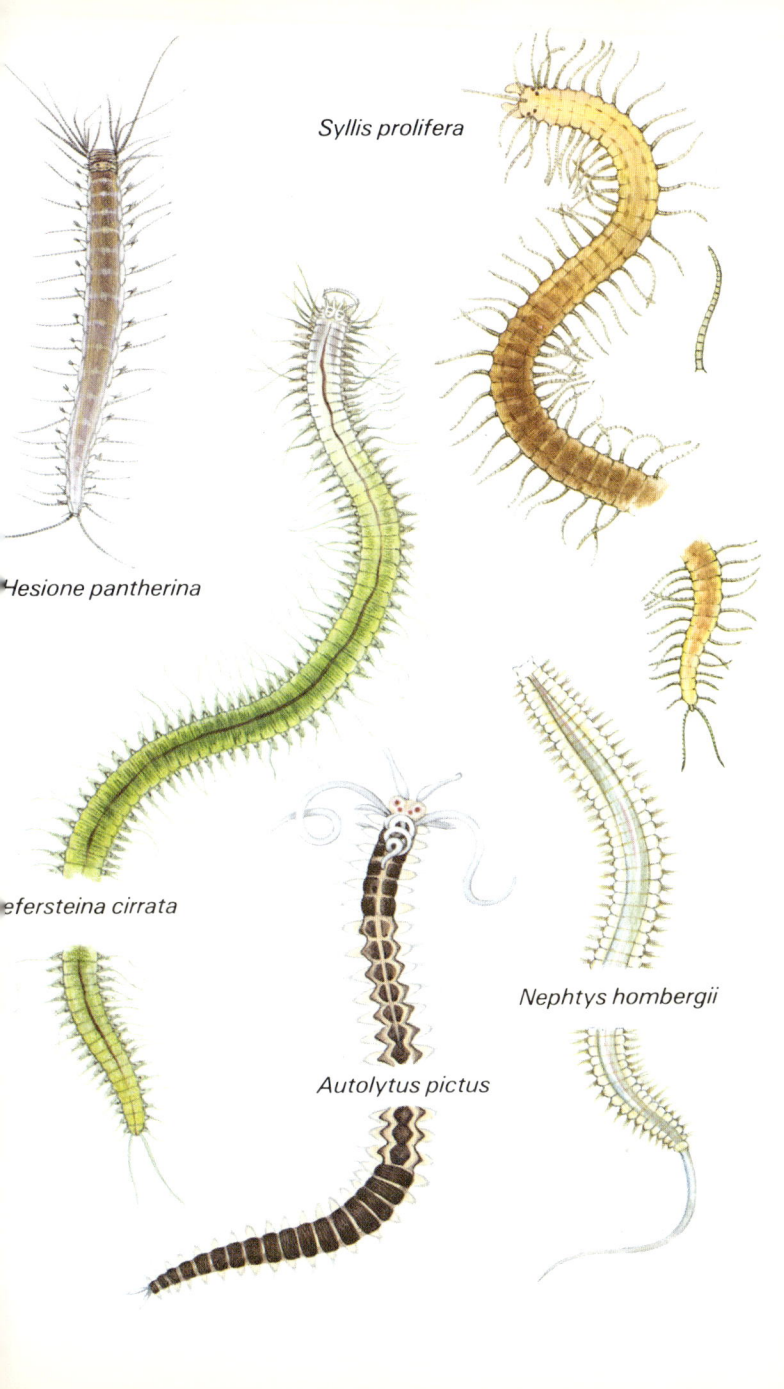

Syllis prolifera

Hesione pantherina

efersteina cirrata

Nephtys hombergii

Autolytus pictus

Family Nereidae Ragworms

Often large, active free-living polychaetes; head with 2 ovoid palps which have button-like tips, 2 short antennae, 4 eyes and 4 pairs of tentacle cirri; proboscis eversible, armed with large chitinous jaws and accessory or 'free' teeth; rounded muscular bodies with chaetae borne in 2 bundles on well developed parapodia; 2 anal cirri. Mediterranean species = about 21.

Nereis fucata (Savigny) Length up to 200mm. Head typical of the family; last tentacle cirri reach back to 3rd or 5th chaeta-bearing segment. Body composed of 90 to 120 chaeta-bearing segments. Colour usually brown-yellow with white markings in the middle of each segment. Habitat the adults generally live inside whelk shells inhabited by hermit crabs.

Nereis pelagica Linnaeus Length up to 120mm. Head typical of the family; proboscis when everted shows 2 powerful jaws with 5–7 teeth. Body rounded, tapering towards the tail; 80 to 100 chaeta-bearing segments; dorsal cirrus of parapodia fairly long and conspicuous. Colour adult normally red, brown or yellow with distinct dorsal red blood vessel. Habitat among rocks, shells and weeds on shore and in shallow water. N.B. a pelagic reproductive form, the epitoke, may be seen; in this the posterior parapodia are modified for swimming.

Nereis diversicolor (O. F. Müller) **Ragworm** Length up to 120mm. Head typical of family; proboscis when everted shows 2 powerful jaws each with 5–8 teeth as well as groups of free teeth; last tentacle cirri reach back to 5th or 7th chaeta-bearing segment. Body rounded, 90–120 chaeta-bearing segments. Colour variable: green-yellow with red to orange tints; distinct dorsal blood vessel runs all down the back. Habitat on the lower shore or in shallow water burrowing in sand.

Perinereis cultrifera (Grube) (=***Nereis cultrifera***) Length up to 250mm. Head quite typical of family; last tentacle cirri extend back to 5th or 6th chaeta-bearing segment. Body somewhat flattened and tapering towards tail; 100–125 chaeta-bearing segments. Colour adults brown-green with reddish tints on upper parts of parapodia. Habitat on gravel, sand or mud on the shore and in shallow water. N.B. there is a pelagic epitoke.

Platynereis dumerilii (Audouin & Milne-Edwards) Length up to 60mm. Head typical of family; everted proboscis shows jaws with 5–20 teeth and several groups of free teeth; very long tentacle cirri, the last reaching back to the 10th or 15th chaeta-bearing segment. Body with 70–90 chaeta-bearing segments. Colour variable: greenish-yellowish or pink-red. Habitat among weeds and rocks burrowing in a membranous tube. N.B. there is a pelagic epitoke.

Family Glyceridae

Free-living polychaetes of small to medium size; head tapering and carrying 4 tiny antennae; proboscis eversible and covered with papillae; body tapers to tail which bears 2 anal cirri. Mediterranean recorded species = 8.

Glycera convoluta Keferstein Length 60–100mm. Head small with 4 minute antennae; eversible proboscis carries 4 jaws and many papillae and appears bulbous when extended. Body 120–180 segments, each marked with two annular rings; rounded and somewhat earthworm-like; parapodia not large. Colour translucent pink-red. Habitat in sand or mud, often among weeds in shallow water.

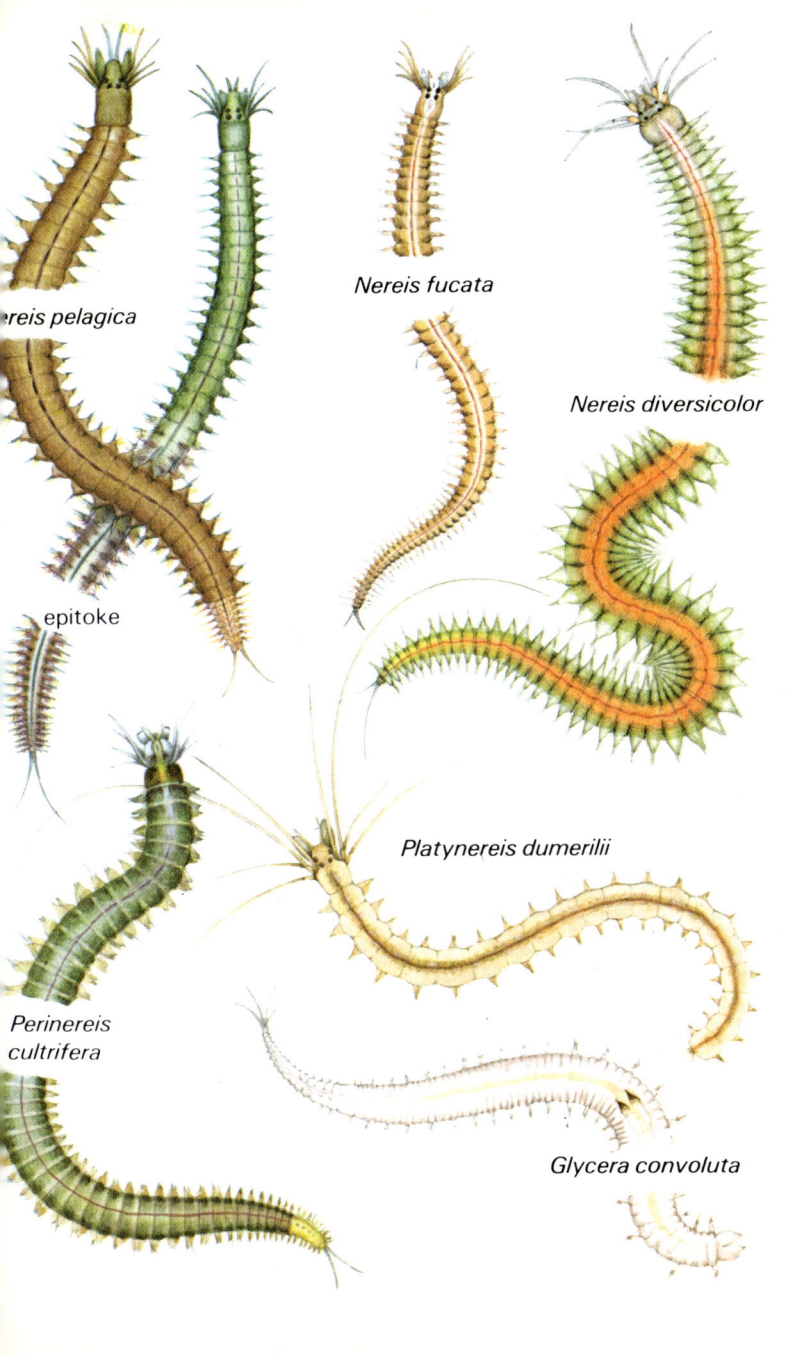

reis pelagica

Nereis fucata

Nereis diversicolor

epitoke

Perinereis cultrifera

Platynereis dumerilii

Glycera convoluta

Family Eunicidae

Active large free-living polychaetes; includes some of the largest Mediterranean species. The head bears 2 bilobed palps, 5 antennae, 2 eyes and 1 pair of tentacle cirri; proboscis contains a complex and powerful set of jaws; the first two apparent body segments lack parapodia and chaetae; large gills are present. Mediterranean recorded species = 34.

Eunice harassii Audouin & Milne-Edwards **Length** up to 250mm. **Head** typical of family, with the single pair of tentacle cirri set well up towards the mid-dorsal line; median antenna largest and about twice as long as the width of the head. **Body** rounded; parapodia with comb-like gills from about the 4th segment back; 2 long anal cirri. **Colour** violet, brown or red with paler marks on the head; gills blood-red; antennae pale yellow and ringed. **Habitat** under rocks and stones from the shore down. **Similar species** *E. rousseaui* which reaches 3m.

Eunice vittata (Delle Chiaje) (Not illustrated) **Length** up to 100mm. **Head** similar to *E. harassii* but the appendages are longer and more finely tapering; median antenna is about 3 times the width of the head. **Colour** ventrally the anterior is brownish, becoming paler; dorsally each segment has 3 transverse red stripes. **Habitat** among rocks and pebbles from the shore down.

Marphysa sanguinea (Montagu) **Length** 300 to 600mm. **Head** bilobed, 5 antennae, 2 eyes but no tentacle cirri. **Body** about 300 chaeta-bearing segments, fragile, flattened, with gills arranged in bunches of 4–7 filaments starting from between segment 16 to segment 30. **Colour** greyish, green or brown with red flashes and an iridescent green line down the back. **Habitat** in rock crevices and among weeds from the shore down.

Marphysa bellii (Audouin & Milne-Edwards) **Length** up to 200mm. **Head** rounded, 5 antennae weakly annulated, 2 eyes, no tentacle cirri. **Body** long and filiform; up to 300 chaeta-bearing segments bearing comb-like gills from segment 12 to segment 35; gills clearly missing at anterior and posterior which is not the case in *Eunice*. **Colour** red at anterior; iridescent brownish towards tail. **Habitat** in sand under stones, near weeds like *Zostera*.

Diopatra neapolitana Delle Chiaje **Length** large active polychaete from 150–500mm long. **Head** bearing 2 globular palps below, 2 small frontal antennae and 5 long tapering occipital antennae, one in the midline, the rearmost pair reaching back to the 5th to 8th chaeta-bearing segment; no tentacle cirri. **Body** rounded, fragile; 200–300 chaeta-bearing segments; gills borne from about the 5th chaeta-bearing segment back and comprising a central filament with many small side filaments arranged all round it. **Colour** yellowish with blue-green iridescence; parapodia with white dots; gills red with green spiral lines. **Habitat** living in a membranous tube buried in sand in shallow water.

Hyalinoecia tubicola (O. F. Müller) **Length** from 60 to 120mm. **Head** with 2 fat palps; 2 oval frontal antennae and 5 long tapering occipital antennae, the longest reaching back to about the 10th chaeta-bearing segment. **Body** not long; up to 130 chaeta-bearing segments, and housed in a free horny tube resembling the quill of a feather. **Colour** pinkish. **Habitat** in deep water on sand and gravel.

Halla parthenopeia (Delle Chiaje) **Length** from 500–800mm. **Head** short, rounded, with 3 small antennae lying between the 2 anterior eyes and 2 posterior eyes lying either side of the base of the median antenna; complex jaws. **Body** tapering to the front and flattened at the rear; 700–800 chaeta-bearing segments; parapodia lacking gills; dorsal cirri large and red. **Colour** orange with iridescent surface. **Habitat** in coastal waters.

Ophryotrocha puerilis Claparède & Mecznikow **Length** up to 10mm. **Head** with 2 small palps, 2 small antennae and 2 eyes. **Body** short and tapering to rear; 20–30 segments; parapodia bilobed, lacking gills; chaetae arranged in 2 bundles. **Colour** transparent, creamy-white, jaws can be made out within anterior. **Habitat** creeping on other invertebrates e.g. bryozoans, echinoids and ascidians.

Eunice harassii

Marphysa sanguinea

Marphysa bellii

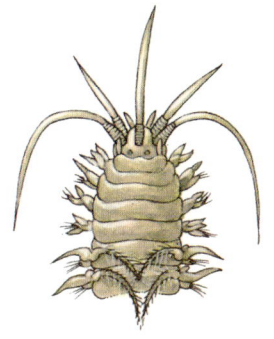

Diopatra neapolitana

Halla parthenopeia

tube

Hyalinoecia tubicola

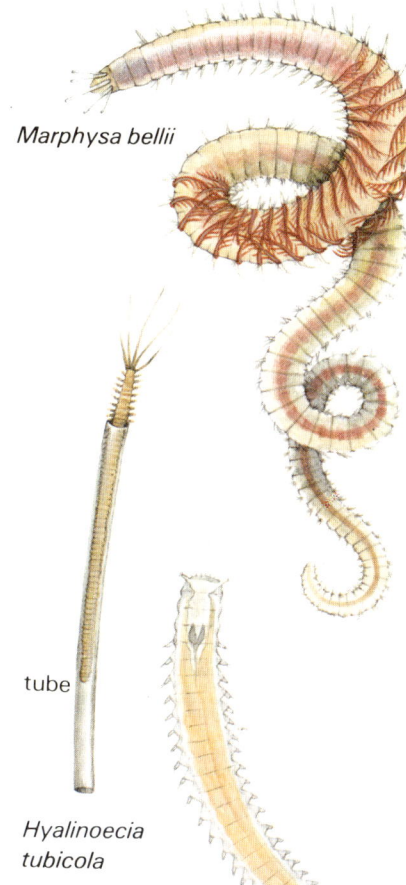

Ophryotrocha puerilis

Family Orbiniidae (=Ariciidae)

Sedentary polychaetes burrowing in mud or sand; the head bears 2 small eyes but no appendages; proboscis unarmed; parapodia with simple dorsal gills; many segments. The body is divided into a flattened thoracic region and an abdominal half-cylindrical region. Mediterranean recorded species = about 9.

Orbinia latreilli (Audouin & Milne-Edwards) **Length** up to 400mm. **Head** small, conical, lacking appendages and eyes. **Body** long, fragile, flat above, round below, 300–400 chaeta-bearing segments of which 30–34 may be in the thoracic region; gills borne from the 5th chaeta-bearing segment back; tail ends in 2 long anal filaments. **Colour** pink in front, yellowish towards rear. **Habitat** in sand and mud from the shore.

Family Spionidae

Sedentary polychaetes without distinct regions to the body; head with 2 long conspicuous palps, no antennae, 4 eyes, front of head is often developed into 2 swollen extensions (frontal horns). The body has parapodia with the dorsal and ventral cirri often lamella-like; gills are carried dorsally on a number of parapodia. Mediterranean recorded species = 16.

Scolecolepis fuliginosa (Claparède) **Length** up to 60mm. **Head** with 2 small frontal horns sticking out sideways and 2 long banded palps. **Body** delicate, thin; 100–150 segments, gills borne on the first chaeta-bearing segment and back; between 6 and 8 lobed processes on tail. **Colour** reddish, head darker than body. **Habitat** burrowing in sand on the shore and in shallow water.

Nerine cirratulus (Delle Chiaje) **Length** up to 80mm. **Head** pointed, with 4 small eyes arranged in a square; 2 long palps reach back as far as the 24th chaeta-bearing segment. **Body** long and slender; about 200 chaeta-bearing segments; gills borne from 2nd segment back but may be missing around 10th. **Colour** variable; blue-green. **Habitat** burrowing in sand or gravel.

Polydora ciliata (Johnston) **Length** up to 30mm. **Head** with 4 eyes in a square and 2 long thin palps. **Body** thin, up to 180 chaeta-bearing segments, slender gills carried from the 7th segment back until the 10th before the last; tail carries fan-shaped appendage. **Colour** brown-yellow. **Habitat** boring into oysters and shells from which only the thread-like palps may be visible under a lens.

Family Magelonidae

Sedentary burrowing polychaetes; the oval flattened 1st segment lacks eyes and antennae; large proboscis; 2 long palps covered by papillae; body is divided into 2 regions; gills lacking; parapodia bilobed. Mediterranean recorded species = 1.

Magelona papillicornis O.F. Müller **Length** up to 170mm. **Head** typical of family. **Body** 150 chaeta-bearing segments. **Colour** palps and anterior pink; posterior grey-white. **Habitat** burrowing in sand; in deep water.

Family Chaetopteridae

Sedentary polychaetes, tube-dwelling; body divided into 3 regions on basis of 'parapodial' form and function. Mediterranean recorded species = 6.

Chaetopterus variopedatus (Renier) **Length** up to 250mm. **Head** broad, 2 palps, large terminal mouth. **Body** segments 1–9 bear chaetae; middle region has a 'waist' and 3 flap-like parapodia; posterior part has bristled parapodia. **Colour** greyish-brown. **Habitat** living in a parchment-like U-shaped tube buried in the substrate down to deep water.

Orbinia latreilli

Scolecolepis fuliginosa

Nerine
cirratulus

Polydora ciliata

Chaetopterus
variopedatus

Magelona papillicornis

Family Chlorhaemidae

Sedentary burrowing polychaetes; chaetae of anterior segments surround the head which has a retractile mouth siphon, eyes, 2 large palps and retractable gills; parapodia are bilobed. Mediterranean recorded species = 5.

Flabelligera diplochaitos (Otto) Length up to 100mm. Head 4 eyespots, 2 groups of 40–50 green gills, 2 large palps a little longer than the gills; a 'cage' of chaetae protecting these appendages. Body 40–50 segments, tapering to tail. Colour violet to greenish with semi-transparent body. habitat among detrital sediments and corallines down to 50m.

Stylarioides eruca (Claperède) Length 40 to 60mm. Head 2 moderately long palps, a variable number of cylindrical green gills arranged in 3 dorsal and 2–4 ventral rows; chaetae of the first 3 or 4 segments form a protective 'cage' over these appendages. Body tapering, 60 to 80 segments. Habitat in sand in the roots of sea-grasses like *Posidonia*, in shallow water.

Family Capitellidae

Earthworm-like sedentary polychaetes with a conical retractile pre-oral segment, a large unarmed proboscis, mouth opens ventrally; 2 eyes are often present; the body has a short anterior swollen part and thinner longer posterior part, often with twisted gills. Number of Mediterranean recorded species=11.

Capitella capitata (Fabricius) Length up to 100mm. Head bears 2 small eyes on underside, small proboscis. Body 90 or more chaeta-bearing segments; not typical of family; fragile. Habitat often burrowing in dirty sand or under pebbles on the shore and in shallow water.

Family Arenicolidae

Sedentary tube-dwelling polychaetes. Head lacks palps and antennae, proboscis is unarmed; body is composed of 2 or 3 regions; bilobed parapodia, dorsal lobe conical, ventral lobe twisted; gills often conspicuous. Mediterranean recorded species = 3.

Arenicola marina (Linnaeus) **Lugworm** Length up to 200mm. Head small and typical; proboscis eversible, bearing papillae; prostomial lobes are more or less equal. Body rounded; 6 swollen anterior segments without gills, followed by 13 similar with conspicuous red tufted gills; posterior less swollen. Colour brown-green. Habitat in U-shaped burrows from shore downward (relatively rare).

Arenicola claparèdii Levinsen (Not illustrated) Very similar to *A. marina*. Length up to 150mm. Head typical, with an eversible proboscis bearing papillae; prostomium has 2 large lateral lobes and a smaller central lobe. Body rounded, strongly annulated. Colour reddish tinges on green; gills red. Habitat burrowing in U-shaped tubes from shore down to about 6m. N.B. quite difficult to distinguish from *A. marina*.

Flabelligera diplochaitos

Stylarioides eruca

Capitella capitata

Arenicola marina

Family Maldanidae Bamboo worms

Sedentary burrowing polychaetes. The head lacks appendages and terminates obliquely when viewed from side; the body is cylindrical with relatively few long segments, not divided into distinct regions. Mediterranean recorded species = about 10.

Euclymene lumbricoides (Quatrefages) **Length** up to 150mm. **Head** inconspicuous eyes; pre-oral segment conical. **Body** slender; 19 chaeta-bearing segments, tapering slightly after segment 15. **Colour** pink to brown with transverse reddish marks at the anterior. **Habitat** burrowing in sand and mud from the shore down.

Family Oweniidae

Sedentary tube-dwelling polychaetes. The head lacks appendages and is capped by a small folded membrane; the body is cylindrical with few segments, the anterior ones longer than the posterior ones; the worms build tubes using small shelly fragments or shell gravel. Mediterranean recorded species = 1.

Owenia fusiformis Delle Chiaje **Length** up to 100mm. **Head** typical; has 6 branching gills. **Body** 20–30 segments, 3 short ones follow head, then 5–7 longer ones, followed by the remainder which are shorter. **Colour** green-yellow. **Habitat** in a membranous tube with grains of sand or shell particles, the tube usually showing above the substrate; from the shore down.

Family Sternaspisidae

Sedentary polychaetes of aberrant form: head reduced, body short with short segments, anterior chaetae short and stout; posterior chaetae with long gills. Mediterranean recorded species = 1.

Sternaspis scutata (Ranzani) Strangely-shaped worm up to 30mm long. **Head** typical; mouth ventral. **Body** 20–22 segments, anterior 3 bear chaetae arranged in arcs each side; 7th segment carries 2 conspicuous tubuler genital papillae. **Colour** white-grey-yellow. **Habitat** in sand and mud down to quite deep water.

Family Sabellariidae

Sedentary tube-dwelling polychaetes with bodies divided into 3 parts. The head has a complex arrangement of chaetae in 3 concentric rings to act as a stopper for the tube; thoracic region consists of 2 segments with reduced chaetae and 3 or 4 other segments; abdominal region of about 30 segments with parapodia followed by a narrow tail. The tubes are of large sand grains, often forming colonies and reef-like structures. Mediterranean recorded species = 2.

Sabellaria alveolata (Linnaeus) Typical of family. **Length** up to 40mm. **Body** 32–37 segments. **Colour** whitish-red-violet. **Habitat** living in tubes arranged in colonies overlying rocks or other objects, from the shore down.

Family Amphictenidae

Sedentary tube-dwelling polychaetes with short bodies divided into 3 parts: the head/thorax has gills and includes the first 3 chaeta-bearing segments; the abdomen carries bilobed parapodia; the short tail region is concave on dorsal side. The tubes are conical, rigid, open at both ends, made of sand grains embedded in matrix. Mediterranean recorded species = 3.

Pectinaria koreni (Malmgren) *(=**Lagis koreni**)* Typical of family. **Length** up to 50mm. **Head** carries many chaetae in a dorsal shield and bears club-like papillae. **Colour** white-pink; gills red. **Habitat** in tubes of medium-sized sand grains set free in sand with the worm standing on its head.

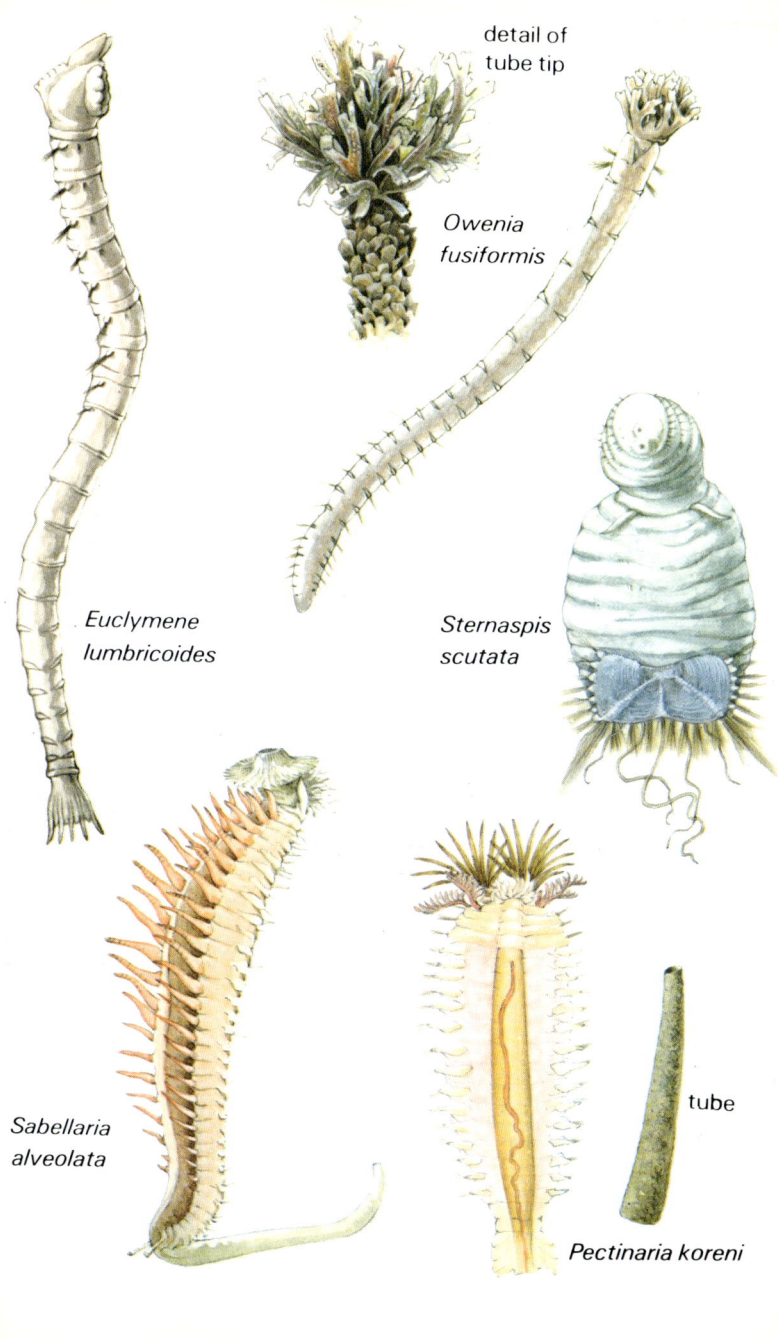

detail of
tube tip

Owenia
fusiformis

Euclymene
lumbricoides

Sternaspis
scutata

Sabellaria
alveolata

tube

Pectinaria koreni

Family Terebellidae

Sedentary tube-dwelling or burrowing polychaetes with bodies divided into two regions: the swollen thoracic region carries the reduced lobe-like head which has eyes and modified segments bearing many feeding tentacles and pairs of blood-red branching gills; the tapering abdominal region has reduced appendages. The membranous tubes are mud or sand covered. Mediterranean species = 23.

Amphitrite gracilis (Grube) Length 60–120mm. **Head** lobe without lateral processes; 2 pairs of branching gills, the first pair bigger than the second. **Body** 100–200 segments of which 17 to 19 chaeta-bearing segments form the thorax; long, gelatinous and fragile. **Colour** pale red-yellow; tentacles pale, gills blood-red. **Habitat** burrowing in twisted galeries in mud on the shore and in shallow water.

Amphitrite rubra (Risso) (Not illustrated) Length 50–100mm. **Head** lobe with marked lateral processes on each side; 3 pairs of well developed gills. **Body** 80–100 segments of which 23 chaeta-bearing segments form the thorax. **Colour** variable: white to pink; tentacles white, gills red. **Tube** of mud or sand. **Habitat** in shallow water under old shells and pebbles.

Lanice conchilega (Pallas) **Sand Mason** Length 250–300mm. **Head** bears 3 pairs of gills; eyes; a triangular lobe either side of the mouth; 2nd segment short, 3rd segment has 2 leafy lobes. **Body** 150–300 segments, swollen thorax of 17 chaeta-bearing segments; thin fragile abdomen. **Tube** typical; of moderate-sized sand grains cemented on to membranous matrix; frayed appearance at top; projects from the sand. **Colour** pinkish-yellow-brown; tentacles white, gills blood-red. **Habitat** tubes set in sand on the shore and in shallow water.

Polymnia nebulosa (Montagu) Length 50–150mm. **Head** lobe elongated with many small eyes and numerous tentacles; 3 pairs of irregularly branching gills. **Body** tapering to tail; round and fragile; about 100 segments; 17 chaeta-bearing segments form thorax. **Colour** orange, pink or brown with white spots; tentacles orange, pink or white; gills blood-red, often with white spots. **Tube** slimy, with shell fragments. **Habitat** tube fixed to pebbles and old shells in shallow water.

Family Sabellidae

Sedentary tube-dwelling polychaetes with bodies divided into two regions; short thoracic part carries reduced head with eyes and a crown of gills carried on 1st segment surrounding the mouth; long abdomen. The tube is of mucoid substance with particles of sand and mud usually embedded in it; there is no operculum to close tube. Mediterranean recorded species=27.

Sabella pavonina Savigny **Peacock Worm** Length up to 250mm. **Head** reduced; 1 pair of palps and 2 semi-circular equal groups of gills, each consisting of 8–45 filaments which join to give the appearance of a crown. **Body** rounded above, flattened below; 100–600 segments; 6–12 chaeta-bearing segments form thorax. **Tube** of fine particles, cylindrical, not vertical in the substrate, sticking out about 80mm; worm can withdraw into tube rapidly, the aperture then partly closes; bottom of tube attached to larger pebbles. **Colour** body yellow-orange or violet; gills variable, banded and often beautifully coloured. **Habitat** lower shore and shallow water in mud.

Spirographis spallanzanii Viviani Length 200–300mm. **Head** reduced, bearing 2 unequal groups of many gill filaments twisted into a spiral crown; no eyes. **Body** 100–300 chaeta-bearing segments of which 8 are thoracic; cylindrical in shape, tapering suddenly at the posterior. **Colour** yellow, maroon or brown; gills of variable colour: white, violet, yellow or brown, sometimes patterned. **Tube** cylindrical, encrusted. **Habitat** from shallow water down, usually in rocky crevices and silty places.

Bispira volutacornis (Montagu) Length 50–150mm. **Head** reduced, 2 small palps, 2 twisted gill clusters of 8–45 filaments and 2–3 pairs of eyes on each gill filament. **Body** about 100 chaeta-bearing segments of which 8 are thoracic;

Lanice conchilega

Amphitrite gracilis

tube

Polymnia
nebulosa

Bispira volutacornis

Sabella
pavonina

Spirographis spallanzanii

rounded but slightly flattened ventrally. **Colour** green to brown-violet; gills often white, rarely violet. **Tube** short, membranous and supple, colourless except at opening where it is grey. **Habitat** often in colonies growing under rocks from shallow water down.

Potamilla reniformis (O. F. Müller) **Length** up to 100mm. **Head** reduced, 2 tapering palps, 2 gill clusters of 10–15 gill filaments with each filament bearing up to 8 eyes; 2 leaf-like lobes at base of gills on dorsal side. **Body** 60–200 segments of which 9–12 are thoracic. **Colour** orange to brick-red; gills white-pink with brownish-violet marks. **Tube** transparent, horny and usually covered in mud. **Habitat** from shallow water down.

Branchiomma vesiculosum (Montagu) **Length** 100–150mm. **Head** reduced and bearing 2 equal semicircular groups of gill filaments forming a corolla-like circular tentacle crown; about 40 filaments in all; dark eyespot on tip of each filament. **Body** 100–200 chaeta-bearing segments of which 8–9 are thoracic. **Colour** variable: yellow-brown or red, sometimes spotted white; gills brown, green, yellow or violet. **Tube** leathery, encrusted with gravel and debris; quite large pieces except for top; protrudes about 25mm from substrate; animal can quickly contract into tube. **Habitat** muddy places; on the shore and in shallow water.

Myxicola infundibulum (Renier) **Length** up to 200mm. **Head** reduced; 2 dark crescentic palps; 2 semicircular clusters of 20–40 lance-like gill filaments giving a crown effect. **Body** flattened and capable of quick contractions into tube; about 130 double-ringed chaeta-bearing segments of which 7–8 form the thoracic region. **Tube** transparent thick-walled mucous structure. **Colour** gills violet, body brownish-yellowish. **Habitat** lower shore and shallow water; tubes set in mud. It superficially resembles small specimens of *Cerianthus* see page 72.

Dasychone lucullana (Delle Chiaje) (Not illustrated) **Length** 10–30mm. **Head** reduced, with 2 equal groups of gill filaments numbering up to 18, not spiral; each filament carries 7–15 pairs of violet eyespots either side of the midline. **Body** 40–60 chaeta-bearing segments of which 8 form the thorax; short, thick and tapering. **Colour** greyish-brown, gills violet or patterned white and violet. **Tube** cylindrical, elastic and encrusted with mud and grains of sand etc. **Habitat** among algae etc. down to quite deep water.

Family Serpulidae

Sedentary polychaetes living in a calcareous tube into which they can fully withdraw; bodies cylindrical and divided into a thoracic region of few segments which carries the reduced head, and an abdominal region with many segments. The head carries two groups of filamentous gills which unite to form a crown; gill filaments carry eyes; palps poorly developed. Body carries a conspicuous collar-like membrane (the collar) which sheaths most of the thorax but does not quite meet on either side of the dorsal surface. Operculum is pulled down by special muscles to close the tube when the worm has withdrawn. Mediterranean recorded species = about 27.

Serpula vermicularis Linnaeus **Length** 50–70mm. **Head** reduced, bearing 2 lobes of gills of 30–40 filaments united at their bases, which are fuller on the ventral face than on the dorsal; operculum trumpet-shaped and edged with fine teeth. **Body** 200 chaeta-bearing segments of which 7 form the thorax. **Colour** variable: pale yellow, yellow-red, brick-red etc.; gills blood-red to pink. **Tube** fixed to rocks and shells at base, remainder free.

Hydroides norvegica (Gunnerus) **Length** 15–30mm. **Head** reduced, carrying two lobes of gills, each comprising 15–20 filaments, united at their bases; operculum of complicated shape and slightly like a thistle crown. **Body** about 100 chaeta-bearing segments of which 7 form the thorax. **Colour** red; gills patterned in zones red and white; operculum stalk red. **Tube** white, cylindrical and twisted, sometimes with growth lines, sometimes keeled. **Habitat** fixed on stones, shells, boat bottoms, piers etc. in shallow water.

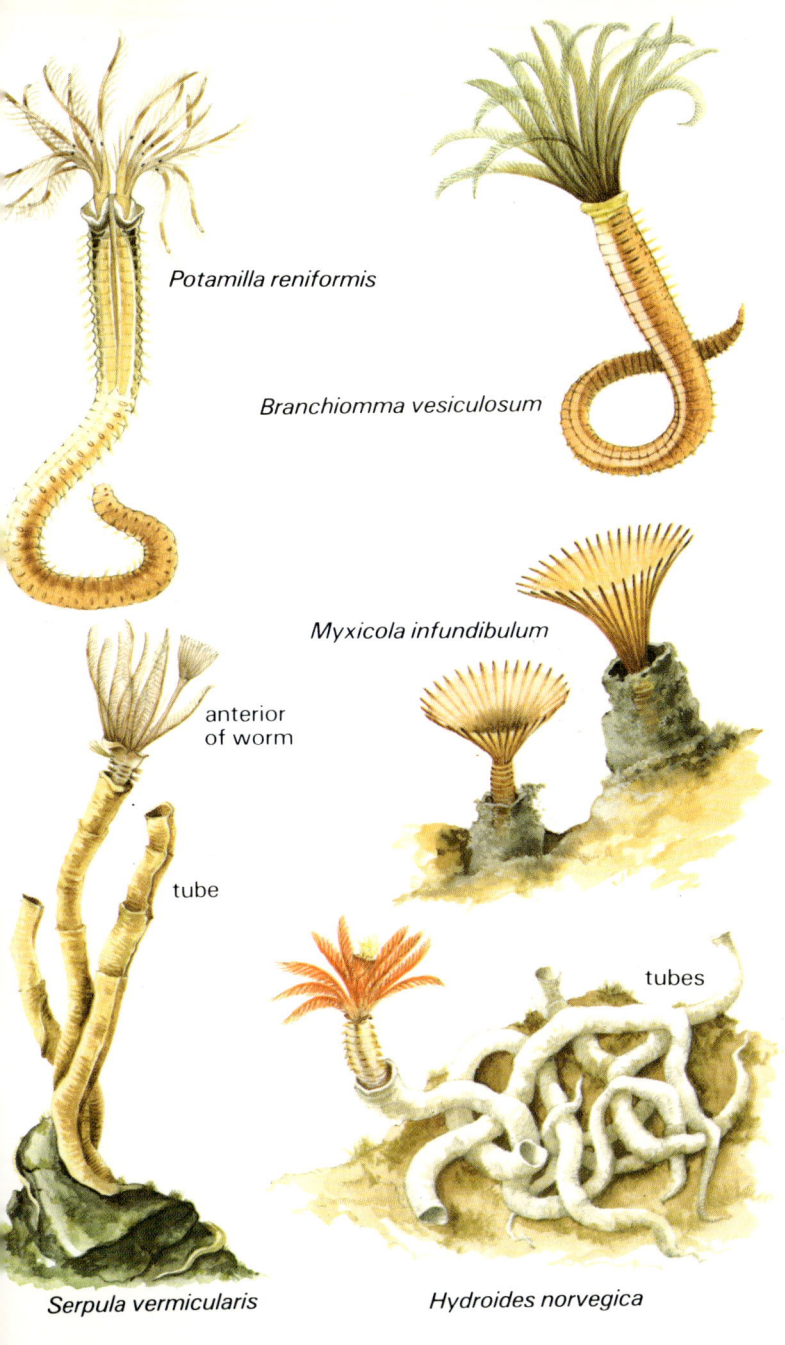

Potamilla reniformis

Branchiomma vesiculosum

Myxicola infundibulum

anterior
of worm

tube

tubes

Serpula vermicularis

Hydroides norvegica

Mercierella enigmatica Fauvel Length 6–25mm. **Head** reduced, bearing 2 groups of gills of 6–10 thick short filaments not apparently united at their bases by a membrane; operculum ornamented by small spines. **Body** 70–120 chaeta-bearing segments of which 7 form the thorax. **Colour** male greenish, female red-orange; gills greenish and striped brown. **Tube** white when new but becoming dirty; thin; ringed by successive collars; only terminal part occupied by the animal. **Habitat** growing in enormous numbers to form a reef, often encrusting rocks, reeds or jetsam; in shallow water, often in brackish places.

Pomatoceros triqueter (Linnaeus) Length 15–25mm. **Head** reduced, bearing 2 groups of gills of 18–20 thick short filaments bound at the base by a membrane; operculum of variable form, often with a cone mounted like the mute of a trumpet and sometimes ornamented by a few small spines. **Body** 80–100 chaeta-bearing segments of which 7 form the thorax. **Colour** very variable: body red, brown, yellow or greenish; gills variously coloured blue and white, or brown, red, yellow and white. **Tube** white, chalky, triangular in section, with a small pointed protrusion over the opening, tapering posteriorly and of irregular shape. **Habitat** tube encrusting stones or rocks for all its length; on the shore and in shallow water.

Filograna implexa Berkeley Length 3–5mm; **Head** reduced, bearing 2 gill lobes of 4 filaments each; operculum transparent and scoop-shaped. **Body** 25–35 segments bear chaetae, sharply divided into thoracic and abdominal regions separated by a waisted segment; 6–9 chaeta-bearing segments form thorax. **Colour** transparent, pale yellow intestine may be visible. **Tube** chalky-white, very fine, cylindrical; often many grow together to form chains which may get washed up. **Habitat** shore and shallow water.

Protula tubularia (Montagu) Length 2–5mm. **Head** reduced, bearing 2 gill lobes of 20–45 filaments each, slightly spiralled; no operculum. **Body** 100–125 chaeta-bearing segments of which 7 are thoracic. **Colour** orange to red; thorax greenish; gills white to pink with bands of red or orange. **Tube** white, chalky, cylindrical, fixed at the base, often free-standing for rest of length. **Habitat** from shallow water down to about 100m, usually in stony or rocky places.

Note on the genus *Spirorbis* The sub-family of Spirorbinae contains many species of small serpulid worm. These mostly belong to the genus *Spirorbis* and are characterised by a minute spiral tube encrusting rocks, stones, weeds, piers etc. The precise identification of spirorbids is a specialist task. About 7 species are known from the Mediterranean.

Spirorbis borealis Daudin Length 3–3.5mm. **Head** reduced, bearing 2 gill lobes each with 4–5 filaments; operculum with a calcareous tip. **Body** coiled in a left-handed spiral; 21–35 chaeta-bearing segments of which 3 form the thorax. **Colour** tints of blue or indigo; brown eggs may be held in tube. **Tube** chalky, left-handed coil with 2–4 turns. **Habitat** on weeds and stones, on the shore and in shallow water.

Spirorbis pagenstecheri Quatrefages Length up to 2mm. **Head** reduced, bearing 2 gill lobes each with about 4 filaments, operculum terminating in a calcareous piece. **Body** covered in a right-handed spiral; 11–15 chaeta-bearing segments of which 3 form the thorax. **Colour** thorax uncoloured; abdomen red. **Tube** chalky-white, somewhat triangular in section; keeled. **Habitat** tubes fixed to rocks, pebbles and weed, on the shore and in shallow water.

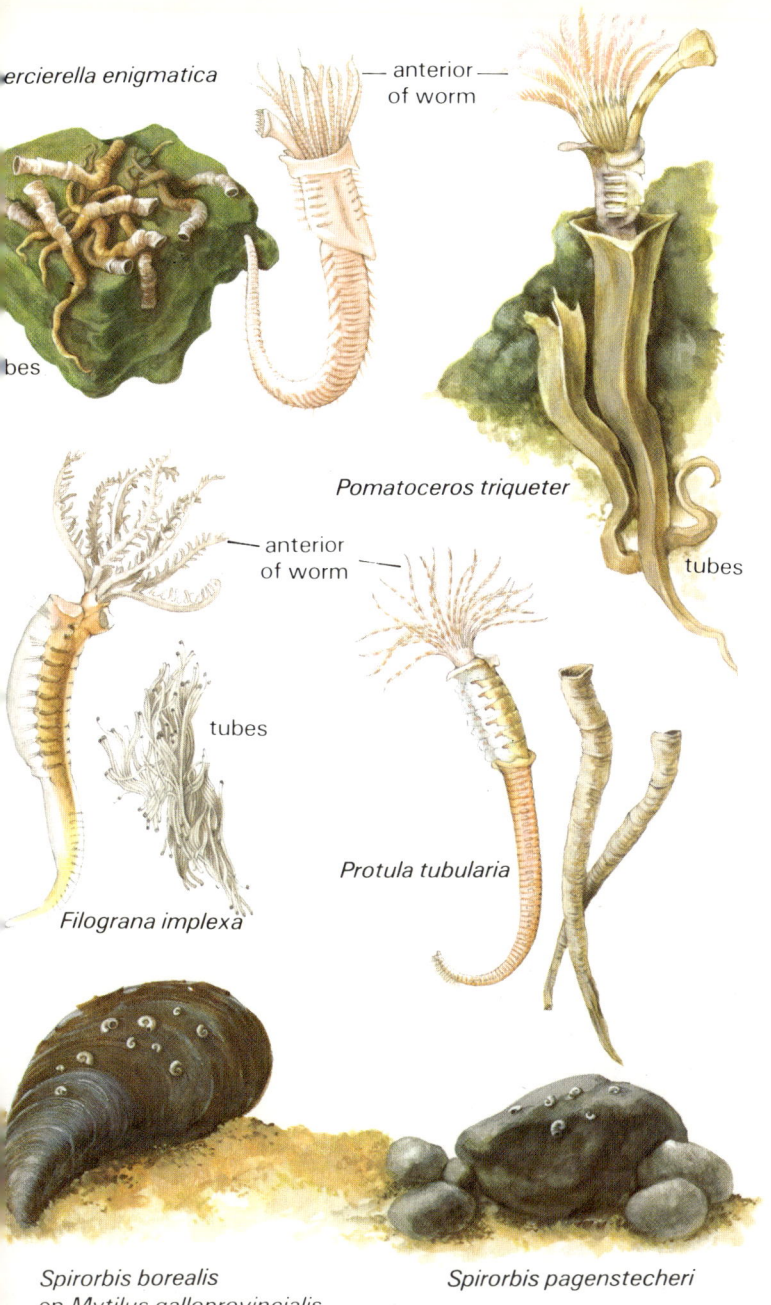

ercierella enigmatica

tubes

anterior of worm

Pomatoceros triqueter

tubes

anterior of worm

Filograna implexa

tubes

Protula tubularia

Spirorbis borealis
on *Mytilus galloprovincialis*

Spirorbis pagenstecheri

Phylum Echiura

The echiurans constitute a small and little known group of marine worms. The pear-shaped or carrot-shaped female bears a long retractile proboscis. The mouth is ventral and the anus terminal. The body is composed of three layers of cells and has a coelom. The female feeds on small animals and organic deposits; the male is small and parasitic on the female. The life-cycle includes a pelagic larval phase whose sex is determined by the substratum it settles on. Should it land on an existing female it becomes male and a dwarf parasite; if it lands elsewhere it becomes female. Cuenot, L. 1922 gives a fuller account of the Echiura. Mediterranean recorded species uncertain, probably less than five.

Thalassema gigas M. Müller Length parasitic male is minute. Female up to 120mm. **Body** slimy, sausage-shaped, tapering at both ends, without bristles and covered with conspicuous papillae; strap-like unbranched proboscis usually folded into a groove near where it joins the body; proboscis and body somewhat contractile. **Colour** greenish. **Habitat** under rocks and stones among mud and sand from the shore down to about 50m. **Similar species** probably none.

Bonellia viridis Rolando Length parasitic male up to 2mm. Female up to 150mm. **Body** plump, pear-shaped, bearing a very long proboscis which in some large examples may reach 1m when fully extended; proboscis protrudes from holes or crevices; divides into 2 parts at tip and has a characteristic gutter leading to the mouth. **Colour** green. **Habitat** in holes in rocks, from which the proboscis protrudes; 1 to 100m. **Similar species** none.

Phylum Sipuncula

The sipunculans comprise another relatively small and little-known phylum. These worms have cylindrical bodies with an anterior mouth borne on a protrusible proboscis and surrounded by small frilly tentacles. The anus is situated on the upper side of the body. The sexes are separated in nearly all cases, but alike in appearance. Gibbs, P. E. 1977 and Cuenot, L. 1922 give further details on Mediterranean Sipuncula. Mediterranean recorded species = about 5.

Golfingia elongata (Keferstein) Length up to 150mm. **Body** highly contractile, slender, cylindrical and tapering at both ends; skin smooth and lacking prominent papillae; proboscis at anterior end may reach 50mm when fully extended and carries the mouth at its tip surrounded by a simple whorl of very small tentacles numbering 24–28 in most adults. **Colour** pale and straw-like. **Habitat** burrowing under stones in mud from low water down. **Similar species** probably only *G. vulgaris* (de Blainville) **Length** up to 200mm. **Body** skin more wrinkled and with papillae quite noticeable on trunk; tentacle crown of proboscis complex, juveniles having about 20 very small tentacles, while the average adult has about 50–60.

Sipunculus nudus Linnaeus Length up to 350mm. **Body** cylindrical: large folded lobe around mouth not divided into tentacles: skin patterned with small rectangles. **Colour** grey-yellow-brown. **Habitat** in soft substrates down to deep water.

Aspidosiphon mülleri Diesing Length up to 80mm. **Body** with shield-like plates at both ends of trunk. **Colour** grey-brown-black. **Habitat** in discarded gastropod shells and crevices of rocks, calcareous algae and corals.

Phascolosoma granulatum Leuckart Length up to 100mm. **Body** broad and tapering: from 12 to 60 tentacles above mouth according to size: covered by rounded warts of varying size tipped with clear spot in a dark ring. **Habitat** mud and gravel down to 90m.

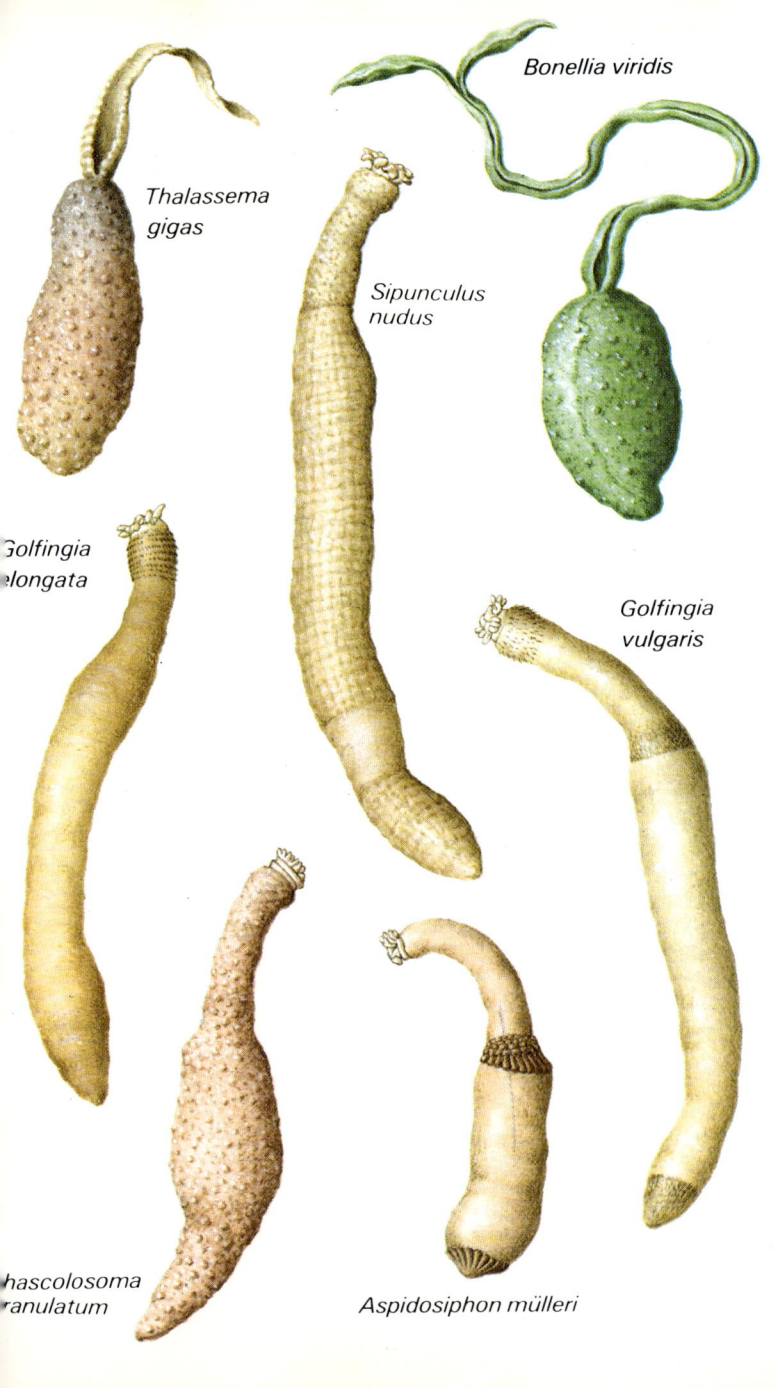

Thalassema gigas

Bonellia viridis

Sipunculus nudus

Golfingia elongata

Golfingia vulgaris

Phascolosoma granulatum

Aspidosiphon mülleri

Phylum Mollusca

The molluscs are usually bilaterally symmetrical animals with an unsegmented body composed of three layers of cells, and possessing a body cavity (coelom) which is very much reduced. The head is often well developed, and a chitinous, toothed ribbon called the *radula* is frequently found in the pharynx (this is used for scraping algae from rocks or for drilling into the shells of prey, etc.). The muscular foot is used for many different functions. The dorsal viceral hump is covered by a mantle which secretes the shell (when present), and which also encloses the mantle cavity where the gills lie and into which the anus and other ducts discharge. Sexes are sometimes separate, and there is a pelagic larva in marine species.

This important group of invertebrates is probably the second largest phylum in the animal kingdom, the arthropods being the largest. Most molluscs are aquatic, and many of them are marine, but a few species (the terrestrial snails and slugs), have conquered land. This phylum shows a very characteristic type of body plan, but each of its seven constituent classes has pursued a particular variation of this, so that when viewed overall the group appears diverse. Of the seven classes, five are discussed in this book. These are the *Polyplacophora* (chitons), *Gastropoda* (snails and slugs), *Scaphopoda* (tusk shells), *Bivalvia* (bivalves) and *Cephalopoda* (squids and octopuses). The remaining two classes are of great evolutionary importance, but their members are relatively small and obscure, one class is very restricted in its distribution.

The molluscs have led with great success various life styles in the sea. In many cases they show a remarkable level of evolutionary development, and this is often evident in the range of sensory and locomotory structures they display. Consequently in a number of species, notably within the Cephalopoda, complex and sophisticated behaviour patterns occur. Because of their hard shells many molluscs have fossilized well, therefore more is known of their evolutionary history than is the case with many other types of invertebrate.

The most characteristic feature of the molluscs is their shell, although in some groups it is internal or lacking. It provides support and protection for the soft parts of the body and often performs other functions such as regulating buoyancy in floating or swimming species. The symmetry of the shell is dictated by that of the animal which occupies it. Most molluscs are bilateral, but many of the gastropods are asymetrical as adults. This is due to a peculiar phenomenon known as *torsion*, which results from the unequal development of the left and right sides of the body and leads to an apparent twisting of the visceral hump. The effects of this are that the anus and gills, which originally developed behind the hump, now come to lie in front of it. Torsion first appeared in the evolution of the gastropods, and these evolutionary origins are reflected in the larval development of most gastropods, where it occurs at about the time of metamorphosis. Torsion should not be confused with spiralization of the visceral hump – that part of the body which encloses most of the internal organs – and the shell, which is a different phenomenon allowing an increase in the volume of the viscera. The shell is essentially arranged in several layers, of which the outermost is normally the horny periostracum. In some cases this is rubbed away. Beneath this lies the ostracum or prismatic layer, made up of layers of calcium carbonate laid down on an organic framework; sometimes this is followed by another layer (the nacreous layer), which resembles mother-of-pearl. The secretion and maintenance of the shell is the function of the mantle. Unlike the arthropod exoskeleton, which must be moulted to permit growth, the molluscan shell grows by the addition of new shell material. The newest part of the shell is that adjacent to the actively secreting part of the mantle. The gastropod shell can be regarded as a coiled tube. If it is held upright so that the mouth faces the observer and the tip points up, the coil is right-handed (dextral) if the mouth is on the right; and left-handed (sinistral) if the mouth is on the left. As the animal grows, the tube is made longer and wider so that the newest part is that nearest the mouth. The mantle is generally active at secreting the shell at one point along its length, that being near to the opening or periphery of the shell. In bivalves there are two valves hinged together. Sometimes these valves are equal, and sometimes they are not. The bivalve shell grows peripherally due to mantle activity, and the point of mantle attachment to the shell is marked by the pallial line. The form and symmetry of the shells are indicated in figs. 30 and 31.

The forms and habits of the various classes of molluscs vary greatly. The chitons (Polyplacophora) are relatively inconspicuous and creep slowly over rocks and shells in search of their algal food. They number about 1000 species.

The Gastropoda is the largest class, comprising roughly 90 000 known species. It is divided into three sub-classes, all of which move on a flat-tened foot. The first of these is the Prosobranchia (limpets, winkles, whelks, etc.). These are familiar sea-shore animals and, although not swiftly moving, they search actively for their food which is taken with the help of the radula. The subclass Opisthobranchia includes the sea-slugs. The adults have undergone detorsion of their bodies and are frequently colourful and attractive, unlike their terrestrial namesakes. They are usually carnivorous. In the third sub-class (the Pulmonata) the mantle cavity has developed into a lung which breathes air. Very few pulmonates are marine, but some freshwater species are found where the water is brackish.

The Scaphopoda is a small class of about 350 species. The head is reduced and these animals live partly buried in sand and mud.

The class Bivalvia numbers about 15 000 species. Most of these are marine, and many are burrowers. They all filter sea water with their gills to collect particles of food. The foot is frequently developed to form an efficient digging organ.

In the class Cephalopoda the foot has been greatly modified to form eight or ten suckered tentacles. The head merges with these and houses the highly developed brain and sensory organs. These animals are active predators and rapid movers. There are about 750 species.

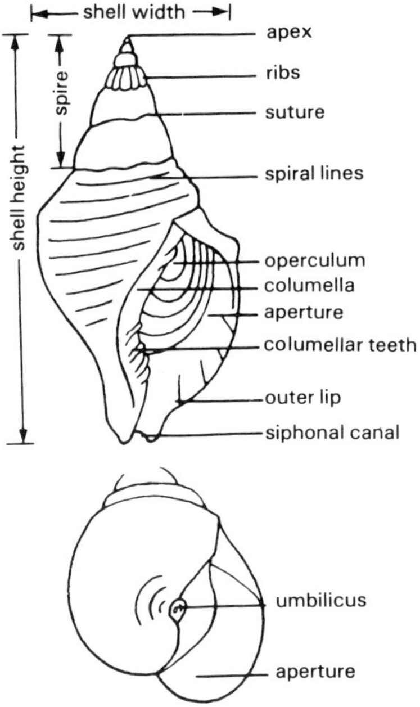

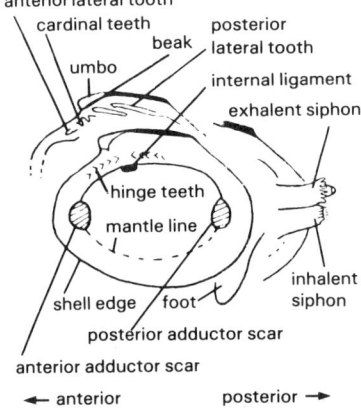

Fig. 30 External features of a gastropod shell

Fig. 31 Internal features of a bivalve (one shell removed)

123

Class **Polyplacophora** Chitons or coat-of-mail shells

Bilaterally symmetrical molluscs in which the mantle forms 8 transverse, calcareous plates containing spicules; these plates are surrounded by a fleshy girdle. The ventral surface is formed entirely by the foot. The body is ringed by a row of gills lying in the mantle groove on the underside. N.B. the precise classification of chitons is often a difficult process, and may involve the removal and detailed examination of one or more dorsal plates. The values given below are an *approximate* ratio for apparent shell width to total width. Further descriptions of many of the European chitons are given by Matthews, G., 1953.

Chiton olivaceus Spengler **Chiton** Length up to 40mm. **Dorsal plates** lightly etched, yellow-brown to olive; ratio about 5:6. **Fleshy edge** with alternating pale and darker bands, not spiny. **Gills** in the mantle groove on the underside. **Habitat** on rocks and stones on the shore from the splash zone and in shallow water. **Similar species** all chitons are quite difficult to identify without a lens.

Acanthochiton communis (Risso) Length up to 50mm. **Dorsal plates** with fine spots; greyish to yellowish colour; ratio about 5:7; conspicuous clusters of spines set in 7 groups at the edge of the shell plates. **Fleshy edge** finely spotted and fringed. **Habitat** on hard substrates in shallow water. **Similar species** several.

Callochiton achatinus (Brown) Length up to 20mm. **Dorsal plates** smooth and shiny; ratio 3:5. **Fleshy edge** broad with rounded granules, characteristic red-brown spots on first and last plates and also occasionally on other plates; 20–25 pairs of gills. **Habitat** on rocks and under stones on the shore. **Similar species** few.

Callochiton laevis (Montagu) Length up to 20mm. **Dorsal plates** smooth and shiny and spotted black; background reddish to olive; ratio about 3:5. **Fleshy edge** relatively broad with about 18 groups of bristles arranged round the edge. **Habitat** down to about 10m, especially among calcareous red algae. **Similar species** few.

Lepidopleurus cancellatus (Sowerby) Length up to 10mm. **Dorsal plates** ornamented with fine spots, brown-grey to yellow-white; ratio about 4:5. **Fleshy edge** quite narrow. **Habitat** on hard bottoms down to about 10m. **Similar species** few.

Lepidopleurus cajetanus (Poli) Length up to 30mm. **Dorsal plates** first and last ornamented with concentric lines, remaining plates bear sculptured markings rising up from the fleshy edge and narrowing off to the posterior; ratio 4:5.5; colour brownish. **Fleshy edge** smooth. **Habitat** on stony substates and shells down to quite deep water.

Chitona squamosus Linnaeus Length up to 25mm. **Dorsal plates** slightly ridged, grey with black or brown spots and uneven brown lines; ratio about 7:9. **Fleshy edge** with pale and dark stripes. **Habitat** on rocks on the shore and in shallow water. **Similar species** few.

Ischnochiton albus (Linnaeus) Length about 10mm. **Dorsal plates** often shiny with conspicuous central crest and very fine ridges; ratio about 2:3. **Fleshy edge** covered by large smooth granules and edged with small spines; slightly separated from the shell plates; 12–16 pairs gills. **Habitat** among rocks on the shore and in shallow water. **Similar species** few.

Lepidochiton cinereus (Linnaeus) Length up to 20mm; less flattened than many other species. **Dorsal plates** slightly granular and variously coloured olve-grey to dull red; ratio about 3:4. **Fleshy edge** red-brown-green with minute granules; 16–19 pairs gills. **Habitat** on rocks and under stones on the shore. **Similar species** several.

Chiton olivaceus

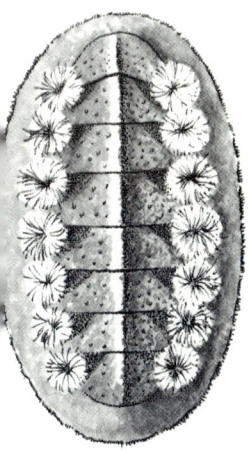

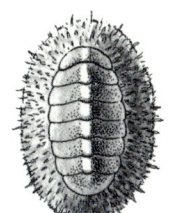
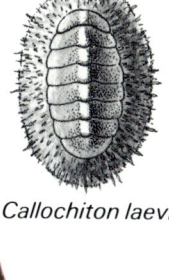

Lepidopleurus cancellatus

Callochiton laevis

Acanthochiton communis

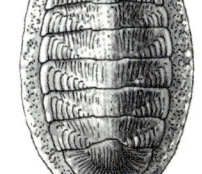
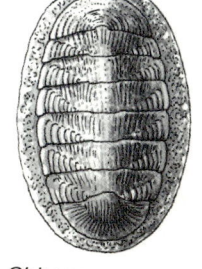

Lepidopleurus cajetanus

Chitona squamosus

Ischnochiton albus

allochiton achatinus

Lepidochiton cinereus

Class Gastropoda Snails and slugs

Asymmetrical molluscs with a well developed head and broad flattened foot. The shell is normally single and coiled in a helix; this coiling is a separate feature from the twisting of the visceral mass known as torsion. The larval phase is sometimes passed in the egg case.

Subclass Prosobranchia Sea-snails

Adults show torsion. Prosobranchs are generally marine and breathe by gills. The shell is made up of a series of whorls joined by sutures. The first is the apical whorl and the last is the body whorl; the whorls appear wrapped around a central column or columella which may be solid or hollow; if the former it opens via a pore at the base known as the umbilicus. Prosobranch larvae are often free-living and planktonic, serving to distribute the species. The sexes are separate. Graham, A. 1971, and Fretter, V. and Graham, A. give a detailed account of many European prosobranchs.

Order Archaeogastropoda Limpets, ormers, top-shells etc.

Algal browsers, usually lacking an operculum for closing the shell. The shell is usually lined with mother-of-pearl.

Haliotis tuberculata Linnaeus **Green Ormer** Flattened shell up to 80mm long, with a very large opening on the underside and an obvious row of about 6 respiratory openings on the upper side; evidence of further such openings now sealed off may be seen. **Colour** outside of shell green, brown or red; inside with a well developed mother-of-pearl lining. **Habitat** adhering strongly to rocks and stones, often near weeds on the shore and in shallow water. **Similar species** *H. lamellosa*.

Haliotis lamellosa Lamarck **Common Ormer** Flattened shell up to 70mm long, like *H. tuberculata* but upper surface has a more creased or folded appearance and may be encrusted with calcareous algae.

Emarginula reticulata Sowerby **Slit-limpet** Conical shell up to 20mm long having a conspicuous slit in the leading edge; when viewed sideways the shell apex will be seen to be turned over slightly. **Colour** outside of shell white, grey, green or yellow. **Habitat** firmly adhering to rocks and stones on the shore and down to 60m. **Similar species** three, including *E. elongata* and *E. cancellata*.

Emarginula elongata da Costa **Slit-limpet** Conical shell up to 8mm long having a conspicuous slit in the leading edge; when viewed sideways the shell apex will be seen to be turned over. **Colour** outside of shell yellow to white. **Habitat** on rocks and stones on the shore. **Similar species** three.

Emarginula cancellata Philippi **Slit-limpet** (Not illustrated) Conical shell up to 12mm long, with high vaulted appearance; delicate and thin-walled; relatively small but still conspicuous slit in leading edge; apex set well back. **Colour** outside of shell yellowish-white with close-set ridges of small tubercles radiating from apex. **Habitat** under stones and pebbles from the shore down. **Similar species** three.

Diodora apertura (Montagu) *(=D. graeca* and ***Fisurella reticulata)*** **Keyhole Limpet** Conical shell up to 40mm long with apex towards the rear and slightly turned over; characteristic small respiratory opening or 'keyhole' in apex; outside of shell ribbed. **Colour** usually greyish but sometimes yellow, orange or red. **Habitat** down to deep water among rocks. **Similar species** two.

Diodora italica (Defrance) **Keyhole Limpet** Conical shell up to 45mm long; apex towards the rear; conspicuous sculptured ribs. **Colour** outside of shell white, grey or violet. **Habitat** adhering to rocks in shallow water. **Similar species** two.

Diodora gibberula (Lamarck) **Keyhole Limpet** Bowl-shaped or basin-

Haliotis lamellosa

Haliotis tuberculata

Emarginula reticulata

Emarginula elongata

Diodora gibberula

Diodora apertura

Diodora italica

shaped shell up to 12mm long; shell more smooth than *D. italica*. **Colour** white to yellow with greyish radiating marks. **Habitat** in shallow water adhering to stones. **Similar species** two.

Note on the limpets About 5 species of *Patella* have been recorded from the Mediterranean but the exact status of these is not fully understood. Limpet identification is often very difficult and may require expert knowledge.

Patella coerulea Linnaeus **Mediterranean Limpet** Low conical shell up to 45mm long; exterior of shell with coarse radial corrugations leading to a wavy margin and many fine unequal radial ribs. **Colour** outside grey-brown-red with white spots and radial marks; concentric growth rings may show as contrasting colour bands; inside coloration of fresh shells with dark stripes leading to the apex which is blue mother-of-pearl. N.B. the appearance of this limpet is very variable: colours fade with age and after dark. **Habitat** firmly attached to rocks, usually in the horizontal plane, on the shore. **Similar species** *P. lusitanica*.

Patella lusitanica (Gmelin) Conical shell up to 40mm long and narrower than *P. coerulea* which it resembles. **Colour** outside of shell spotted black; interior with dark radial rays running in a little way from the lip; apical scar pale. **Habitat** adhering to rock, usually on vertical surfaces on the shore.

Calliostoma zizyphinum (Linnaeus) *(=C. conulus)* **Common Topshell** or **Painted Topshell** Sharply conical shell about 25mm high and 25mm wide, consisting of about 9 whorls with sutures so inconspicuous as to give nearly straight and not concave sides to the conical outline; whorls with spiral ridges. **Colour** outside of shell yellow to pink with dark brown or red stripes; a variety exists with a nearly white shell. **Habitat** on rocks and pebbles on the shore and down to about 100m. **Similar species** *C. papillosum* and one other.

Calliostoma papillosum (da Costa) *(=C. granulatus* or *Trochus granulatus)* Sharply conical shell up to 35mm high and 35mm wide; very similar to *C. zizyphinum* but with granular spiral ridges on the whorls, and sides of spire concave at the sutures. **Habitat** never on the shore, usually in sandy and muddy bottoms down to 300m.

Cantharidus striatus (Linnaeus) *(=Jujubinus striatus* or *Trochus striatus)* **Grooved Topshell.** Tall conical shell about 10mm high but not so wide. Apex not sharply pointed; sides steep but not quite straight; sutures not conspicuous; the body whorl having about 6 spiral ridges, the lowest one being most strongly developed and slightly like a keel; no umbilicus. **Colour** outside grey to white with brownish-red markings. **Habitat** on soft substrates and weeds, e.g. sea-grasses from the shore down to 100m. **Similar species** three.

Cantharidus exasperatus (Pennant) *(=Jujubinus exasperatus* or *Trochus exasperatus)* Tall shell, up to 8mm high and about 7mm wide; similar to *C. striatus* but more sharply conical; the shell has a rough texture with coarse ornamentation of ridges, up to 7 on each body whorl; sutures not conspicuous; 6–8 whorls. **Colour** outside pale brown to deep red to green, with patches of darker pigment, and white on the spiral notches. **Habitat** on soft substrates and sea-grasses from the shore down to 200m. **Similar species** *C. striatus* (above), *C. montagui* and *C. clelandi* (below).

Cantharidus montagui (Wood) and ***Cantharidus clelandi*** (Wood) (not illustrated) are sublittoral topshells found down to 200m in the Mediterranean but not on the shore (see Fretter, V. and Graham, A. 1977).

Patella coerulea

Patella lusitanica

Calliostoma papillosum

Calliostoma zizyphinum

Cantharidus striatus

Cantharidus exasperatus

Cantharidus montagui

Monodonta turbinata (Born) *(=Gibbula lineata)* **Toothed Winkle** or **Thick Topshell** Conical shell about 25mm high and the same width; about 6 whorls; sutures not conspicuous; a weak 'tooth' is visible on the inside of the mouth opening. **Colour** outside white to yellow with darker oblong marks arranged in parallel rings around spine. **Habitat** on rocks on the shore. **Similar species** several including *M. articulata* below.

Monodonta articulata Lamarck (Not illustrated) Conical shell about 25mm high and similar width; generally similar to *M. turbinata* above. Shell quite thick and with a weak 'tooth' visible on the inside of the mouth. **Colour** outside grey to light red with red-white longitudinal bands. **Habitat** on rocks on the shore and in shallow water.

Gibbula magus (Linnaeus) Low conical shell about 25mm wide and less tall; up to 8 whorls. The whole upper surface is distinctly and regularly bumpy; sutures are conspicuous; large umbilicus. **Colour** outside yellow-white background with red to purple marks which are often faded in empty shells and those occupied by hermit crabs. **Habitat** in sand and mud down to 10m. **Similar species** none.

Gibbula cineraria (Linnaeus) **Grey Topshell** Flat conical shell up to 12.5mm high and about the same width; about 7 compressed whorls with relatively inconspicuous sutures; umbilicus clearly visible. **Colour** outside of shell greyish background with dark reddish-grey marks arranged as narrow bands, often faded. **Habitat** on rocks, stones and weeds from the shore down to about 20m. **Similar species** none.

Gibbula divaricata (Linnaeus) Tall shell up to 23mm high but less wide, with 6 uncompressed whorls; sutures conspicuous so that shell looks somewhat like a coiled tube; no umbilicus; mouth opening big. **Colour** outside grey-green background with reddish spots. **Habitat** on stones and among weeds from the shore down. **Similar species** none.

Gibbula adansoni (Payraudeau) Conical tall shell, about 13mm high and less wide, with about 6 whorls and conspicuous sutures. **Colour** outside reddish-brown background sometimes with blue marks. **Habitat** on rocks and stones on the shore. **Similar species** none.

Gibbula varia (Linnaeus) Conical shell, up to 12mm high and about the same width; up to 6 whorls. **Colour** background red, black, brown or grey to yellow, with variable brown radial bands and spots. **Habitat** among rocks, weeds and stones down to quite deep water. **Similar species** several; see Riedl, R. 1963.

Clanculus corallinus (Gmelin) Low conical shell about 10mm high and almost as wide, with about 6 whorls covered in small tubercles which are arranged in 6 spiral lines per whorl; 2 teeth on the inside of the mouth opening. **Colour** outside of shell pink, red or brown, sometimes patterned. **Habitat** on rocks on the shore down to about 10m. **Similar species** *C. cruciatus* (see page 132).

Clanculus cruciatus (Linnaeus) (Not illustrated) Low conical shell, similar in shape and texture to *C. corallinus* (page 130); about 10mm high and with 5–6 whorls, each bearing 5–6 rows of tubercles. **Colour** outside of shell red-brown with whitish flecks and stripes. **Habitat** under stones and among rocks on the shore and in shallow water.

Monodonta turbinata

Gibbula magus

Gibbula cineraria   *Gibbula divaricata*

Gibbula adansoni

Clanculus corallinus   *Gibbula varia*

Astraea rugosa (Linnaeus) **Rough Star-shell** Shell about 50mm high and of similar width, with about 7 whorls, their upper surfaces being ornamented with bumps and their sides with thorny spirals; sutures not highly conspicuous. A strong operculum may be present. **Colour** outside of shell usually reddish-brown. **Habitat** on rocks from the shore down. **Similar species** none.

Leptothyra sanguinea (Linnaeus) Shell about 7mm high, usually wider and with about 5 whorls cut by clear spiral grooves; sutures fairly conspicuous. **Colour** outside of shell blood-red but sometimes encrusted. **Habitat** on rocks and among weeds on the shore. **Similar species** none.

Tricolia pullus (Linnaeus) **Pheasant Shell** Shell about 8mm high and less wide; usually 4 whorls, the lowermost (body whorl) making up more than half the total height. **Colour** outside of shell shiny with irregular red and brown marks; conspicuous white operculum present when animal is alive. **Habitat** among rocks and weeds on the shore. **Similar species** one.

Order Mesogastropoda Periwinkles, tower shells, worm shells, etc.

Algae eaters, deposit feeders and predators. The inner shell lacks mother-of-pearl; a horny operculum is often present.

Littorina neritoides (Linnaeus) **Small Periwinkle** the only periwinkle in the Mediterranean. Shell about 5mm high, occasionally larger, with a strongly pointed spire; about 5 whorls; sutures conspicuous; surface smooth; outer lip of aperture almost parallel to spire where the two join. **Colour** blue-black. **Habitat** on rock from the shore down. **Similar species** none.

Hydrobia ventrosa (Montagu) Shell up to 6mm high with about 6 swollen whorls; outer lip of aperture meets spire at right-angles. **Colour** outside of shell brownish with V-shaped markings on head of snail itself between the eyes. **Habitat** among weeds, rocks and pebbles, often in lagoons and brackish places. **Similar species** several.

Truncatella subcylindrica (Linnaeus) **Looping Snail** Shell about 5mm high. Adults easily identified by the loss of the apical whorls (present in the juveniles) which give the shell a truncated appearance; adults usually have about 3 whorls left. **Colour** outside of shell yellowish-brown. **Habitat** on mud, among stones and weeds on the shore. **Similar species** none.

Astraea rugosa

Leptothyra sanguinea

Littorina neritoides

Hydrobia ventrosa

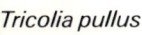

Truncatella subcylindrica

Tricolia pullus

Note on the small snails The family Rissoidae is a large group of small snails with possibly as many as 1500 species worldwide in marine and freshwater habitats. Rissoids are small snails and consequently they are not as much studied as some groups. Taxonomically and anatomically a good deal of work has still to be done on them, and their ecological significance is little understood. About ten species of rissoid including the genera *Rissoa, Alvania, Cingula* and *Barleeia* are commonly encountered in the Mediterranean sea; a detailed account of this group is beyond the scope of this book. A lens may be necessary to make out details.

Rissoa parva (da Costa) Small narrow shell about 7mm high with about 8 slightly inflated whorls; apical whorl pointed; 8–12 ribs per whorl running at right-angles to the fairly conspicuous sutures but dying out when they reach the periphery of the shell. **Colour** outside of shell generally cream, grey or brownish; dark comma-shaped mark on body whorl. **Habitat** on the shore and in shallow water, mainly living on weeds with branching fronds, e.g. *Lomentaria, Plumaria, Ceramium* and *Corallina*, but occasionally on *Fucus, Ulva*, etc.

Rissoa ventricosa (Desmarest) Similar to *R. parva* but shell about 8mm high; about 8 whorls, but these not so swollen; sutures not set deeply into spire; ridges not distinct, and spiral lines and grooves lacking. **Colour** outside of shell yellowish with greenish-grey marks. **Habitat** among weeds in shallow water.

Alvania crassa (Kanmacher) *(=Rissoa crassa)* Small narrow shell about 2mm high with about 6 slightly inflated whorls separated by deep sutures; each whorl carries about 10 ribs. These ribs are pronounced on the lower whorls but are lacking at the apex; the ribs themselves are considerably narrower than the intervening spaces; apical whorl is blunt-tipped; lip of aperture quite thick. **Colour** outside of shell white; shell itself has a transparent glassy appearance. **Habitat** under pebbles and amongst weed in shallow water and down to 50m, usually in sandy places.

Alvania cancellata (da costa) Small, somewhat conical shell about 4mm high; 6–7 moderately inflated whorls separated by distinct sutures, but these are not very conspicuous because they are not deep; whorls ornamented by lattice-like pattern of intersecting rib and spiral ridges; apical whorl not sharply pointed; no umbilicus; lip of aperture moderate. **Colour** outside of shell white, cream or orange. **Habitat** under pebbles and rocks, usually from shallow water down to 90m.

Alvania lactaea (Michaud) *(=Rissoa lactaea)* Shell with oval profile, about 6mm high with 5–6 moderately inflated whorls separated by sutures lying in deep grooves; whorls ornamented by intersecting ribs and spiral ridges to give a pattern of oblongs; apical whorl somewhat blunt at the tip; lip of aperture moderate; no umbilicus. **Colour** outside white to cream. **Habitat** under stones and pebbles and among weed on the shore and in shallow water.

Turritella communis Risso **Tower Shell** Tall narrow shell up to 60mm high, with many inflated whorls separated by conspicuous sutures and carrying spiral whorls. **Colour** outside of shell red-brown-yellow or white. **Habitat** in sand and mud, generally down to about 80m. **Similar species** *T. triplicata.*

Turritella triplicata (Brocchi) **Tower shell** (Not illustrated) Similar to *T. communis* but about 50mm high and straight-sided so that the whorls are not conspicuous; 3 main spiral stripes on each whorl. **Colour** outside of shell pale with reddish markings. **Habitat** in sand and mud.

Vermetus gigas Bivone **Giant Worm Shell** Long tubular shell with only a suggestion of coiling, reaching up to 200mm; irregular in shape. **Colour** outside of shell chalky-white. **Habitat** attached to rocks, stones and shells. **Similar species** two, including *V. triqueter.*

Vermetus triqueter Bivone (Not illustrated) Similar to *V. gigas* but usually reaching about 40mm with a conspicuous ridge running along one side of the tube.

Rissoa parva

Rissoa ventricosa

ania cancellata

Alvania crassa

Alvania lactaea

Turritella communis

Vermetus gigas

Aporrhais pes-pelecani (Linnaeus) **Pelican's Foot Shell** Assymetrical shell about 45mm high; about 9 whorls separated by distinct sutures, each whorl carrying distinct knobs or tubercles; body whorl of adults carries a raised outer lip which is drawn out into several conspicuous points; aperture narrow. These points and the shell walls between them shield the animal's head. **Colour** outside of shell black to brownish-yellow. **Habitat** burrowing in mud and gravel down to about 80m. **Similar species** none.

Bittium reticulatum (da Costa) **Needle Shell** Small narrow shell about 15mm high; about 11 whorls; sutures distinct; each whorl ornamented by sculpturing of fine ridges and spiral grooves giving a latticed effect. **Colour** outside of shell brownish. **Habitat** in small groups, on the shore and in shallow water. **Similar species** several superficially alike.

Cerithium vulgatum Bruguière **Common Cerith** Strong narrow shell up to 60mm tall; about 11 whorls; sutures quite conspicuous; whorls ornamented by knobs and tubercles; conspicuous notch in anterior of aperture. **Colour** outside of shell grey, green, red or black. **Habitat** among stones and on mud down to 10m. **Similar species** *C. rupestre* (below).

Cerithium rupestre Risso **Rock Cerith** Similar in shape to *C. vulgatum* but less high, reaching about 25mm; sutures not so noticeably sunk into recesses. **Colour** outside of shell whitish-yellow, grey or greenish with brownish bands and spots. **Habitat** on rocks and stones on the shore and in shallow water.

Clathrus clathrus (Linnaeus) *(=Scalaria communis)* **Common Wentletrap** Narrow pointed shell reaching up to 40mm high with about 10 whorls; these are traversed by conspicuous ribs which pass across the conspicuous sutures; umbilicus present; mouth rounded. **Colour** outside of shell white, brown or red. **Habitat** on rocks, often near sand and mud, from the shore (in breeding season) down to about 80m. **Similar species** *Epitonium lamellosum* and *Clathrus turtonis* (Not illustrated).

Dolium galea (Linnaeus) **Giant Tun Shell** (Not illustrated) Large rounded shell up to 150mm high and almost as wide; a low spire of about 6 whorls, the lowest being very inflated; characteristic spiral ridges. **Colour** white-brown-yellow. **Habitat** in deep water. N.B. often sold in souvenir shops.

Epitonium lamellosum (Lamarck) *(=Scalaria lamellosum)* Somewhat like *Clathrus clathrus* above; about 30mm high with about 10 whorls; no umbilicus. **Colour** outside of shell whitish with brown or purple patches. **Habitat** on sand and mud down to quite deep water.

Ianthina communis Lamarck **Violet Sea Snail** Delicate conical shell up to about 15mm high and with about 5 whorls. **Colour** outside of shell is a striking violet colour. **Habitat** a pelagic snail, floating by means of a raft of bubbles trapped in mucus, and feeding on surface-dwelling organisms.

Aporrhais pes-pelecani

Bittium reticulatum

Cerithium vulgatum

Cerithium rupestre

Ianthina communis

Epitonium lamellosum

Clathrus clathrus

Natica hebraea Montagu **Necklace Shell** Globular shell, about 45mm high with about 4 whorls; large aperture, and large umbilicus partly concealed by swelling of inner lip; outside of whorls patterned with dark spiral bands and fine dots. **Colour** background coloration grey .or yellow to red-brown. **Habitat** burrowing in sand and mud down to quite deep water. **Similar species** 3, including *N. alderi* and *Neverita josephina*.

Natica alderi (Forbes) *(=**Natica pulchella** or **Polynices alderi**)* **Common Necklace Shell** Globular shell up to 15mm high with about 4 whorls; umbilicus half closed off by swelling of the inner lip; outer lip slopes up where it meets the spire. **Colour** outside white to yellow background, with light red to brown flashing. **Habitat** on extreme lower shore and in shallow water, burrowing in sand.

Neverita josephina Risso *(=**Natica josephina**)* Globular shell up to about 30mm high and with about 4 whorls; umbilicus occluded with brownish to violet button-like swelling on inner lip of aperture; outer lip slopes up where it meets the spire. **Colour** outside background colour yellow-white to grey; aperture yellowish. **Habitat** burrowing in sand down to quite deep water.

Calyptraea chinensis Linnaeus **Chinaman's Hat Shell** Low conical shell about 5mm high and twice as wide or more; evidence of spire at apex; aperture rounded with an internal shelf: viewed in profile the lower edge of the shell may be moulded to fit exactly over the substrate. **Colour** outside of shell white, yellow or brownish; smooth. **Habitat** on pebbles and shells, on the shore and in shallow water. **Similar species** none.

Crepidula fornicata (Linnaeus) **Slipper Limpet** Low oval shell with some signs of spiral coiling and with growth lines; about 16mm long; aperture large with extensive internal shelf; often forms chains of individuals with old females at the bottom and males at the top, which as they age become females. **Colour** yellow, white, green-brown; sometimes with red marks; underside usually white. **Habitat** on rocks and shells; *Crepidula* may become a serious pest in oyster beds where it smothers the oysters, and being a filter feeder, competes with them for food. **Similar species** none.

Capulus ungaricus (Linnaeus) **Bonnet Limpet** or **Hungarian Cap Shell** Low oval shell up to 25mm long, tapering towards one end where there is slight evidence of coiling in the recurved apex. **Colour** outer surface of shell brownish and covered with a fringe of periostracum; inside white. **Habitat** attached to other shells from which it steals food by means of its long proboscis; down to 100m. **Similar species** none.

Trivia monacha (da Costa) *(=**Cypraea europaea**)* **European Cowrie Shell** with compressed spire and slit-like aperture, about 13mm high and with outer surface polished (due to the protection of the mantle which when fully extended covers this) and sculptured into about 20 delicate ribs. **Colour** outer surface of shell delicate pink to grey-brown with 3–6 dark brown spots. **Habitat** on stones and encrusting invertebrates like sea-squirts on the shore and in shallow water. **Similar species** *T. adriatica*.

Trivia adriatica (Monterosato) *(=**T. mediterranea**)* Similar to *T. monacha* but about 9mm high. **Colour** outside of shell greyish-white to pink. **Habitat** in shallow water on hard bottoms.

Cypraea lacrimalis (Monterosato) Shell with compressed spire and slit-like mouth, reaching 50mm high. Outside of shell is polished and shiny due to protection of the mantle, and has fine ribs which mark the aperture. **Colour** outside grey-violet-brown. **Habitat** under rocks and pebbles. **Similar species** *C. spurca*.

Natica hebraea

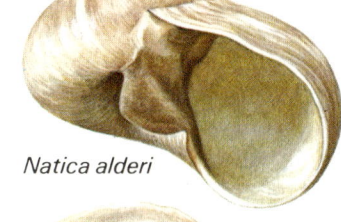

Natica alderi

Neverita josephina

Calyptraea chinensis

Crepidula fornicata

Capulus ungaricus

Trivia monacha

Trivia adriatica

Cypraea lacrimalis

Cassidaria echinophora (Linnaeus) **Knobbed Helmet Shell** Large rounded shell about 150mm high with about 6 whorls; sutures not very distinct; whorls ornamented with spiral lines and coarse nodules; horny operculum; large ear-shaped aperture with inner lip reflexed over spire; conspicuous siphonal notch. **Colour** outside of shell reddish to brownish-yellow; interior white. **Habitat** on sandy and hard bottoms from shallow water down. **Similar species** three, including *Cassidaria rugosa* and *Cassis sulcosa*.

Cassidaria rugosa (Linnaeus) *(=C. tyrrhena)* **Ribbed Helmet Shell** Similar in general form to *C. echinophora* and reaching up to 140mm high but lacking knobs. Shell ornamented by spiral grooves and ridges which make sutures less conspicuous; large ear-shaped aperture with very fine ribs inside. **Colour** outside of shell brownish-yellow; inside white. **Habitat** on soft substrates from shallow to deep water. **Similar species** three.

Cassis sulcosa Bruguière **Channelled Helmet Shell** (Not illustrated) Large rounded shell about 120mm high; whorls ornamented with broad spiral ribs separated by clearly defined channels; sutures not very conspicuous; large ear-shaped aperture with inner lip reflexed over spire. Conspicuous ribs on inside of outer lip; inside of shell shows spiral grooves. **Colour** outside of shell yellow to red-brown, with diagonal markings; furrows between ribs red-brown. **Habitat** on sandy bottoms. **Similar species** three.

Order Neogastropoda Whelks etc.

Deposit feeders and predators. The inner shell lacks mother-of-pearl; an operculum is present. The snails have well developed siphons which are often supported by a siphonal groove or canal in the shell.

Charonia nodifera (Lamarck) *(=C. lampas)* **Triton Shell** Large pointed shell about 400mm high; about 9 whorls with bumps or nodules on them arranged in spirals; periodic occurrence down the spire of previous apertures now merging with more recent whorls; conspicuous siphonal groove; large aperture. **Colour** grey-yellow background with brownish-red markings; sometimes encrusted. **Habitat** on soft and hard bottoms down to 50m. **Similar species** *Ranella gigantea*, *Cymatium corrigatum*, *C. cutaceum*.

Charonia variegata (Lamarck) Shell reaching up to 380mm high; about 8 whorls some of which are uneven; spire high, outer lip of aperture has 10 pairs of white rib-like teeth set in a brown patch. **Habitat** on rocky substrates in shallow water. **Similar species** see *Nodifera* which has brown teeth on the outer lip.

Murex brandaris Linnaeus **Murex** Ornamental shell about 80mm high; long, fairly straight siphonal canal which may be half length of whole shell; spire of about 6 whorls carries conspicuous spires; snail bears horny operculum on its foot. **Colour** outside of shell yellow to grey; snail possesses a gland in the mantle cavity secreting a substance which turns purple with exposure to light. This was used extensively in former times as a source of dye for costly garments. **Habitat** on soft substrates, often near sea-grasses down to 20m. **Similar species** *M. trunculus*.

Murex trunculus Linnaeus **Murex** Shell about 70mm high with about 6 whorls; whorls ornamented by knobs and very short spines arranged in spiral rows; siphonal canal takes up about one-quarter of total shell height. **Colour** grey-white with violet-brown bands. **Habitat** on soft and hard substrates from the shore down to about 10m. Also used formerly as a source of purple dye. **Similar species** *M. brandaris*.

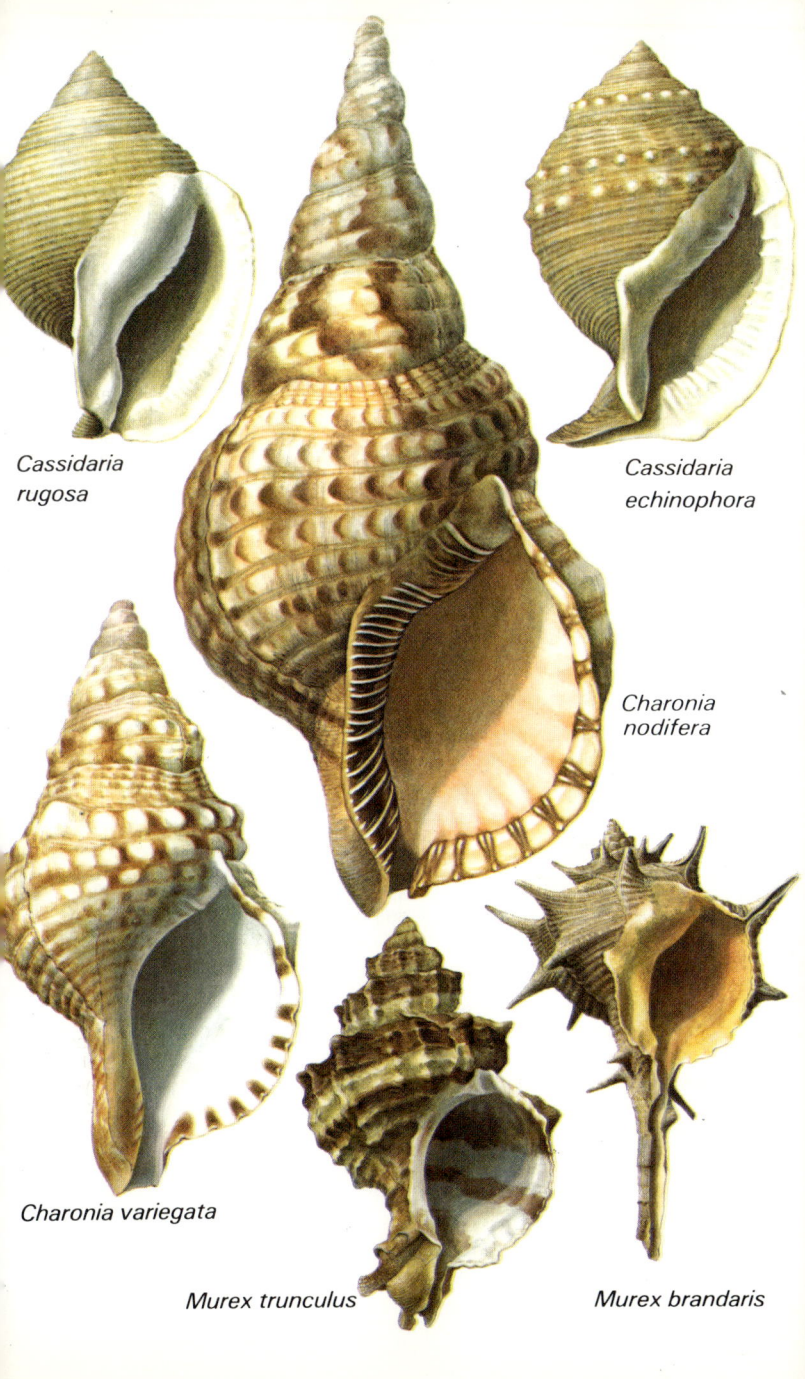

Cassidaria rugosa

Cassidaria echinophora

Charonia nodifera

Charonia variegata

Murex trunculus

Murex brandaris

Cymatium corrigatum Linnaeus Shell about 100mm high, with about 8 whorls ornamented with spiral rows of tubercles and small bumps; inner lip of aperture is strongly notched; long siphonal canal. Colour grey-yellow with brownish periostracum. Habitat on hard substrates and sometimes sand down to quite deep water. Similar species *C. cutaceum* Linnaeus (not illustrated) the whorls of which are more strongly sculptured.

Thais haemastoma Lamarck **Rock Shell** Shell about 80mm high with about 5 whorls; whorls ornamented by low bumps arranged in spiral whorls; large aperture; outer lip fully ribbed; short siphonal canal. Colour outside of shell brownish-grey; inside of shell orange near lip, turning pale pink further inside. Habitat on rocks and among shells. Similar species several small whelks.

Ocenebra erinacea (Linnaeus) **Sting Winkle** or **Oyster Drill** Thick ornamented shell up to 60mm high with about 5 ribbed whorls; siphonal canal open in juvenile species and closed in adult ones. Colour outside of shell unevenly sculptured and coloured yellow-white with dark brown marks. Similar species several small whelks.

Columbella rustica (Linnaeus) **Rustic Dove Shell** Small shell up to 20mm high; about 6 whorls of which the last is larger than all the others put together; sutures distinct; aperture elongated with teeth on inside of outer lip; very short siphonal canal. Colour background coloration whitish with reddish spots and stripes. Habitat on rocky bottoms. Similar species *Pyrene scripta*.

Pyrene scripta (Linnaeus) Narrow shell about 20mm high with about 6 whorls; sutures indistinct; nearly straight sides; long toothed aperture; short siphonal canal. Colour background whitish with orange-brown markings. Habitat on rocky bottoms. Similar species *C. rustica*.

Euthria cornea (Linnaeus) Long pointed narrow shell about 60mm high; about 7 whorls with conspicuous sutures and spiral lines; aperture large with curved open siphonal canal. Colour pale brownish with darker brown marks. Habitat on soft bottoms. Similar species several, including *Pisania maculosa*.

Pisania maculosa (Lamarck) **Mottled Triton** Long narrow shell about 25mm high, not sharply pointed; about 5 whorls; elongated aperture with toothed outer lip; sutures not deeply indented. Colour background green to grey, mottled brown to white violet patches inside aperture. Habitat on rocky bottoms. Similar species *Euthria cornea*.

Nassarius incrassatus (Ström) **Thick-lipped Dog Whelk** Conical shell up to 15mm high; about 7 convex whorls sculptured with ribs and spiral lines; aperture oval with thick outer lip, serrated on inside. Colour white-yellow-grey with brown stripes. Habitat on soft substrates in deep water. Similar species five, including *N. mutabilis*.

Nassarius mutabilis Linnaeus Shell up to 30mm high with about 6 whorls each of which is conspicuously broader than its predecessor; whorls marked by shallow ribs; outer lip of aperture finely toothed. Colour pale brown with red-brown on sutures, and brown markings. Habitat on soft substrates down to quite deep water. Similar species five including *N. incrassatus*

Tritonalia aciculata (Lamarck) **Shell** up to 20mm high; about 6 ribbed whorls; opening of shell toothed on the outer lip. Colour brownish; interior of opening brown. Habitat on rocks and sand from about 10m down.

Cymatium corrigatum

Thais haemastoma

Ocenebra erinacea

Columbella rustica

Pyrene scripta

Pisania maculosa

Nassarius incrassatus

Tritonalia
aciculata

Euthria cornea

Nassarius mutabilis

Fusus rostratus (Olivi) *(=Fusinus rostratus)* **Rostral Spindle Shell** Long narrow shell about 40mm high; about 8 convex whorls ornamented with spiral grooves and ribs; sutures conspicuous; lowermost whorl carries conspicuously elongated siphonal canal. **Colour** outside red-brown; inside paler. **Habitat** on soft substrates. **Similar species** *F. syracusanus* and *Fasciolaria tarentina*.

Fusus syracusanus (Linnaeus) **Syracusan Spindle Shell** Shell about 75mm high, broader than *F. rostratus*; about 8 whorls ornamented with spiral and vertical grooves to give a fairly coarse-ribbed appearance; sutures conspicuous. **Habitat** on soft substrates. **Similar species** *F. rostratus* and *Fasciolaria tarentina*.

Fasciolaria tarentina Lamarck **Tarentine Tulip Shell** About 50mm high with about 8 whorls which are ornamented with spiral rows of nodules giving the shell a moderately jagged appearance; sutures conspicuous; aperture large; large inner lip somewhat reflexed over base of spire. **Colour** white, green or brown. **Habitat** on soft substrates down to quite deep water. **Similar species** *Fusus rostratus* and *Fusus syracusanus*.

Mitra cornicula Linnaeus **Horny Mitre-shell** Shell about 20mm high with about 8 smooth whorls, each very slightly convex; aperture with smooth outer lip; several ridges on columellar edge. **Colour** brownish, resembling horn. **Habitat** on hard substrates, often under stones. **Similar species** two including *M. ebenus*.

Mitra ebenus Lamarck **Ebony Mitre-shell** Pointed narrow shell about 20mm high with about 9 sculptured whorls; aperture narrow with smooth outer lip and several ridges on the columellar edge plumper than *M. cornicula*. **Colour** dark brown-blackish with a narrow white spiral line on the whorls. **Habitat** on hard substrates, often under stones. **Similar species** two.

Cancellaria cancellata Linnaeus **Nutmeg Shell** Rough rounded shell up to 40mm high with about 5 whorls strongly ornamented with nodules and ribs. **Colour** white to brownish with brown spiral stripes. **Habitat** on soft substrates down to deep water. **Similar species** none.

Persicula miliaris (Linnaeus) **Millet-grain Shell** Small shell up to 10mm high with compressed spire and long narrow aperture. **Colour** white to yellow with red marks. **Habitat** on hard substrates. **Similar species** *P. clandestina* which is white and only up to 2mm high (not illustrated).

Philbertia purpurea Montagu **Purple Magnolia** Narrow shell about 20mm high and with about 6 whorls ornamented by spiral ridges and vertical grooves and ribs; edge of aperture toothed; no operculum. **Colour** brown to purple, sometimes with a pale spiral stripe. **Habitat** on soft substrates in deep water. **Similar species** none.

Conus mediterraneus Bruguière **Mediterranean Cone Shell** Shell with compressed spire and shallow conical tip up to 50mm high; short siphonal canal and long slit-like smooth-edged aperture. **Colour** yellow, brown, green; mottled. **Habitat** on soft substrates, often quite shallow. **Similar species** none although sometimes divided into 3 sub-species. N.B. can inject venom (like its tropical relatives) and cause pain.

Fusus rostratus

Fusus syracusanus

Cancellaria cancellata

Fasciolaria tarentina

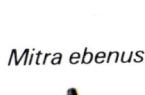

Mitra cornicula

Persicula miliaris

Mitra ebenus

Conus mediterraneus

Philbertia purpurea

Subclass Opisthobranchia

Gastropods which undergo torsion followed by de-torsion. The shell is reduced or absent. These often have conspicuous external gills and are brightly coloured. Further information is given in Pruvot-Fol, A. 1954, and Thompson, T. E. and Brown, G. H. 1976.

Order Bullomorpha Bubble shells

Shell fragile, external, internal or sometimes absent; mantle may be extensive, enfolding the shell, or reduced; gills internal; foot sole-like for creeping or extended into flaps for swimming.

Acteon tornatilis (**Linnaeus**) Resembling a prosobranch; animal can withdraw completely into shell. **Shell** about 20mm high and half as wide; about 7 whorls; sutures conspicuous. **Colour** shell pink-grey-yellow with yellow-white banding; animal has a cream-coloured body, an extensive mantle covering the shell and head when creeping. **Habitat** crawling on and burrowing into sand on the shore and in shallow water. **Similar species** none.

Bulla striata Bruguière *(=**Bullaria striata**)* Not closely resembling a prosobranch; animal not able to withdraw completely into its shell. **Shell** up to 60mm long and slightly more than half as wide; spire reduced; aperture long and slit-like towards apex. **Colour** shell brownish with patterning. **Habitat** on sand and mud, often among weeds on the shore and in shallow water. **Similar species** none.

Haminea hydatis (**Linnaeus**) Not resembling a prosobranch; animal not able to withdraw completely into shell. **Shell** up to 15mm high, not quite so wide, fragile; aperture height greater than spire height; aperture large and eel-like. **Animal** up to 30mm long; discordal head carries flat tentaculate processes which hide the front of the shell, the rear being covered by the hind flap of the mantle; comb-like sense organs with up to 12 pairs of lamellae lie on either side of the head. **Colour** shell translucent yellow-white; animal brownish. **Habitat** creeping on muddy sand and weeds from the shore down, sometimes swimming. **Similar species** several, including *H. navicula* (da Costa). **Shell** up to 32mm and body up to 70mm; comb-like sense organs with about 20 pairs of lamellae; never swimming.

Philine aperta (**Linnaeus**) Shell reduced, thin and white, enclosed in the body; body up to 25mm long and about half as wide, flat and without signs of coiling. **Colour** grey-white and translucent. **Habitat** on sand on the shore and down to deep water. **Similar species** several.

Order Pleurobranchomorpha

Shell usually internal, sometimes external or lacking; head with 2 pairs of tentacles; mantle cavity open on the right side and housing a comb-like gill; body covered dorsally by a membranous shield; foot well developed, often adapted for swimming.

Umbraculum mediterranea (**Lamarck**) Shell present as an umbrella-like structure borne on the apex of the body; a low cone about half total length of animal. **Animal** unable to withdraw into shell, reaching 150mm long and somewhat slug-like. **Colour** shell white and spotted with concentric marks; animal warty and red-brown or yellowish. **Habitat** on mud and sand down to 200m. **Similar species** *Tylodina citrina* (not illustrated) which reaches 35mm long and is yellowish with brownish stripes.

Pleurobranchus membranaceus (**Montagu**) *(=**Oscanius tuberculatus**)* Shell internal, up to 50mm long, delicate and transparent with a wide aperture. **Animal** up to 120 mm long with a dorsal white rounded shield-like mantle bearing conical warts; head with 2 pairs of tentacles, the upper pair (rhinophores) enrolled or tube-like; broad foot; gill on right side. **Colour** animal orange-yellow, foot red. **Habitat** on mud, sand and gravel down to 70m. **Similar species** about 4.

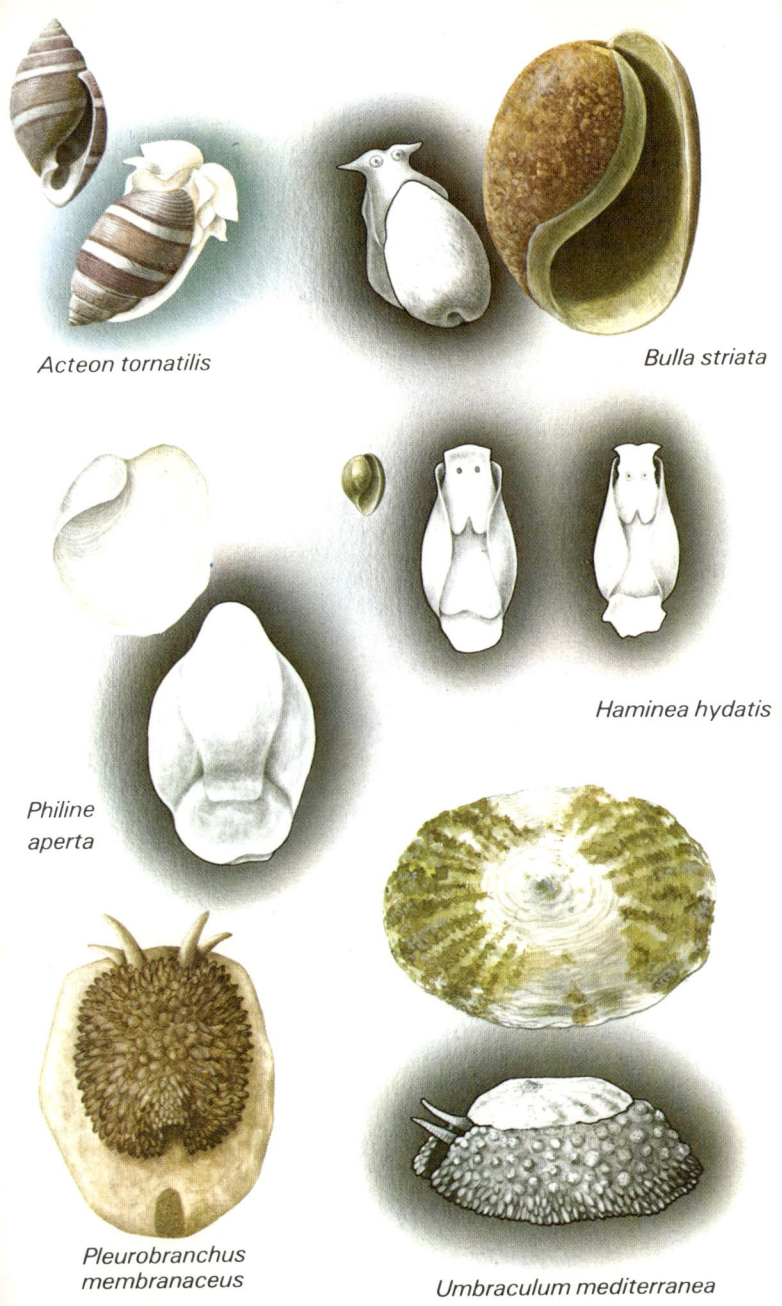

Acteon tornatilis

Bulla striata

Haminea hydatis

Philine aperta

Pleurobranchus membranaceus

Umbraculum mediterranea

Pleurobranchaea meckeli Leue (Not illustrated). Somewhat like *P. membranaceus* but up to 100mm long overall; mantle shield more ovoid and a conspicuous gill on right-hand side; a terminal papilla on the tip of the tail. **Colour** whitish with darker patterns. **Habitat** on sandy and muddy bottoms. **Similar species** about 4.

Berthellina citrina (Rüppell & Leuckart) **Shell** hidden by mantle; reaching up to 7mm, delicate and translucent with a wide aperture, small in proportion to animal. **Animal** up to 30mm long and roughly half as wide; mantle surface shield-like and smooth; a long gill lies under the right edge of the mantle; head with two pairs of tentacles, the lower pair somewhat flat, the upper pair enrolled or tube-like. **Colour** yellowish. **Habitat** often in shallow water, crawling on compound sea-squirts etc. on which it feeds. **Similar species** *Pleurobranchaea membranaceus* (above) and *Berthella plumula*.

Berthella plumula (Montagu) Similar to *Berthellina citrina* above; shell up to 30mm and animal up to 60mm; shell is much larger in relation to body. **Colour** body pale yellow or orange, sometimes with patterns on the shield-like mantle. **Habitat** on the shore in pools and in shallow water down to 10m, feeding on compound sea-squirts etc.

Order Aplysiomorpha

Shell internal or absent; animal slug-like; head long with anterior oral tentacles and posterior tentacles (rhinophores); body with reduced mantle, mantle cavity open; parapodial flaps and large foot. The precise identification may be difficult: see Thompson T.E. 1976 or Thompson T.E. & Brown G.H. 1976.

Aplysia punctata Cuvier **Sea Hare** **Shell** internal, covered by the mantle and up to 40mm long; very delicate; aperture wide. **Animal** body up to 200mm long and generally slug-like; at the sides extended up into two flaps or parapodia which join together high up at the rear; head with two pairs of tentacles, the second pair (rhinophores) somewhat enrolled or tube-like, the first pair with leaf-like edges. **Colour** shell yellowish-amber; animal purple-brownish or olive-green. **Habitat** in shallow water, especially in spring and summer. N.B. when disturbed can release purple dye into the water. **Similar species** *A. depilans* and *A. fasciata*.

Aplysia depilans Gmelin **Sea Hare** Very similar to *A. punctata* (above); the body flaps or parapodia are joined together high up; sole of creeping foot is quite wide so that the foot may appear sucker-like; animal up to 300mm long. **Colour** brown to green with darker marks. **Habitat** in shallow water, sometimes swimming. N.B. Colour removed to highlight structure.

Aplysia fasciata Poiret **Sea Hare** Like the two preceding species; however the body flaps or parapodia are not joined together high up at the rear, but low down, and appear well separated both at the front and the rear. Body up to 40mm. **Colour** dark brown-black; parapodia sometimes have red borders. **Habitat** as above, often swimming. N.B. Colour removed to highlight structure.

Order Sacoglossa

The shell is lacking in many species or may be present externally or internally; head carries very small oral tentacles, lacking in some species as are rhinophores; sides of body often developed into flap-like parapodia.

Elysia viridis (Montagu) No shell. Body up to 45mm long and not nearly so wide; delicate somewhat lance-shaped and flattened; edges raised to form parapodial folds which may be stretched out or folded over the rest of the body; the main head tentacles are enrolled and tube-like. **Colour** variable from green to red. **Habitat** usually associated with green weeds like *Codium* on which it feeds. **Similar species** two, including *E. hopei*.

Elysia hopei (Vérany) *(=Thuridilla hopei)* Body up to 15mm long; similar to *E. viridis* but the main head tentacles are much longer and slightly club-like in outline as well as enrolled; parapodial folds large. **Colour** upper side of the body dark violet with yellowish, bluish and whitish striations.

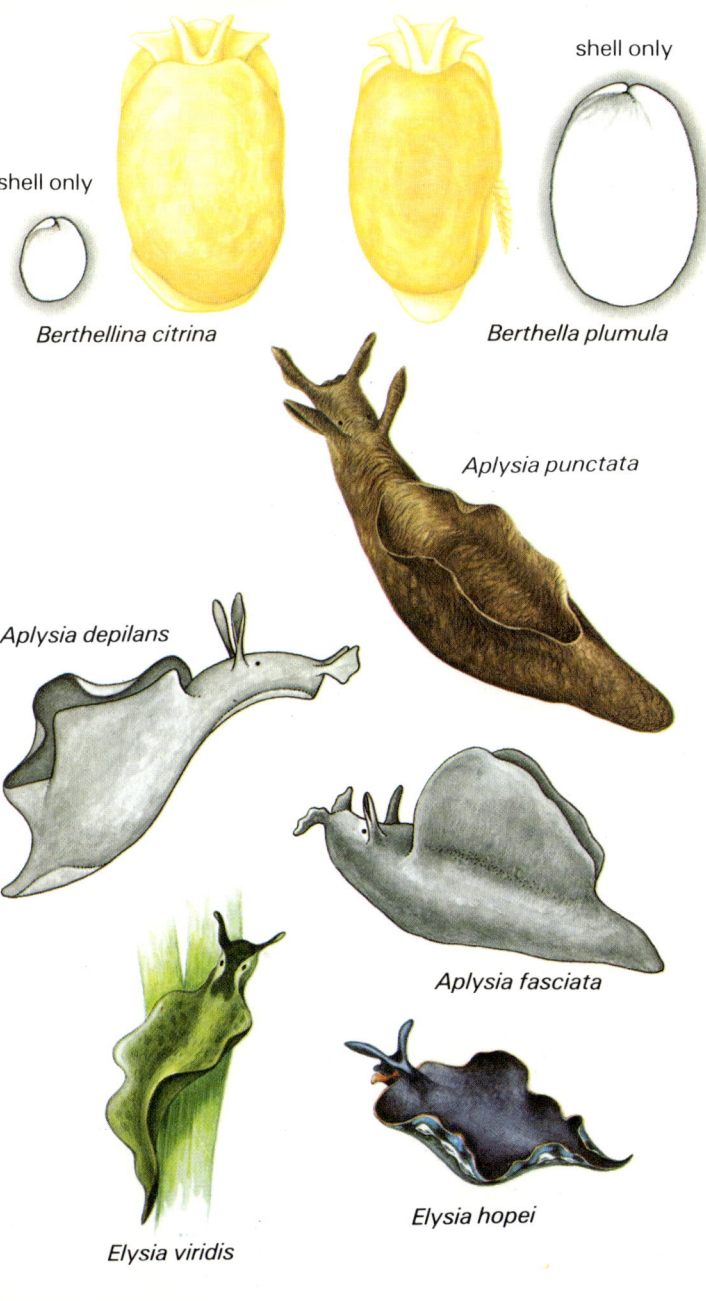

shell only

shell only

Berthellina citrina

Berthella plumula

Aplysia punctata

Aplysia depilans

Aplysia fasciata

Elysia viridis

Elysia hopei

Order Nudibranchia

Slug-like opisthobranchs lacking a shell; often highly coloured; external gills of feathery form arranged on the back near to or surrounding the anus; flanks sometimes carry groups of tentaculate defence organs called cerata; head may be clearly visible, carrying smooth oral tentacles and prominent rhinophores divided towards their tips into feathery plates; the front of the foot may have parapodial tentacles.

Doto splendida Trinchese *(=Doto pinnatifida)* Body up to 30mm long; quite narrow; head low and flat with slight frontal flaps; conspicuous pair of rhinophores surrounded by sheaths which carry dark brownish dots. **Colour** body light to dark brown, carrying up to 9 pairs of cerata; these carry branches arranged in rings; the tips of the branches are spotted dark brown or black; there are also black-tipped warts along the sides of the body. **Habitat** in shallow water, often among hydroids. **Similar species** four, including *D. coronata* (Gmelin), which is up to 15mm long, often less; with conspicuous side flaps on the front of the head; rhinophore sheaths open; white to yellow or pink with reddish or purple spots or marks at the base of each ceras; 8 pairs of cerata which bear branches arranged in rings; branch tips red.

Acanthodoris pilosa (Müller) Body may reach up to 60mm long but often less; nearly half as wide; front end rounded; oral tentacles usually concealed when viewed from above; a pair of rhinophores with feathery tips; mantle surface covered with soft tubercles; a circlet of 9 large branching gills surrounds the anus. **Colour** whitish-grey-brownish, even to purple. **Habitat** on the lower shore and in shallow water down to 80m. **Similar species** several.

Crimora papillata Alder & Hancock Body up to 35mm long, a little less than half as wide; front end rounded; tapering; most of the anterior and lateral margins carry elaborately divided yellow-orange tubercles which are better developed at the front; orange rhinophores; between 3–5 branching gills surround the anus. **Colour** body white to pale yellow. **Habitat** in shallow water down to 80m feeding on leafy bryozoa such as *Flustra*. **Similar species** none.

Polycera quadrilineata (Müller) Body up to 30mm long, quite narrow; head having 2 pairs of smooth narrow tentacles with orange tips and 1 pair of thicker rhinophores with orange-yellowish tips; up to 11 branching gills with yellow tips surround the anus and these lie between a further pair of smooth orange-tipped tentacles. **Colour** body white with patches of yellow and sometimes black marks. **Habitat** shallow water down to 30m. **Similar species** none.

Palio dubia (Sars) Body up to 30mm long, often less, quite narrow; a pair of slender tentacles occurs below the head; relatively short rhinophores; front edge of mantle carries many rounded yellow tubercles and the sides of the body are warty; more tubercles occur behind on either side of the gill circlet; 3–5 branching gills present. **Colour** yellow to green. **Habitat** in rock pools and down to 100m, usually feeding on bryozoans. **Similar species** none.

Greilada elegans Bergh Body up to 40mm long, quite narrow; conspicuous rhinophores; front edge of the mantle carries numerous small projecting processes, these lacking towards the rear; 5–7 branching gills just anterior to the anus. **Colour** unmistakeable: orange-yellow with conspicuous blue markings on the mantle and sides. **Habitat** in shallow water down to 25m, often associated with bryozoans such as *Bugula* on which it feeds. **Similar species** none.

Thecacera pennigera (Montagu) Body up to 30mm long, quite narrow; anterior end of foot bearing a pair of small flat parapodial tentacles; head carries rhinophores borne in sheaths, the lips of which are drawn out into a tubercle behind; 3–5 branching gills lie around the anus and posterior to these lie a pair of conspicuous club-like tentacles. **Colour** white with many small black spots interspersed with larger orange ones. **Habitat** in shallow water down to 20m, often associated with the bryozoan *Bugula* on which it feeds. **Similar species** none.

Limacia clavigera (Müller) Body up to 20mm long; a pair of quite thick

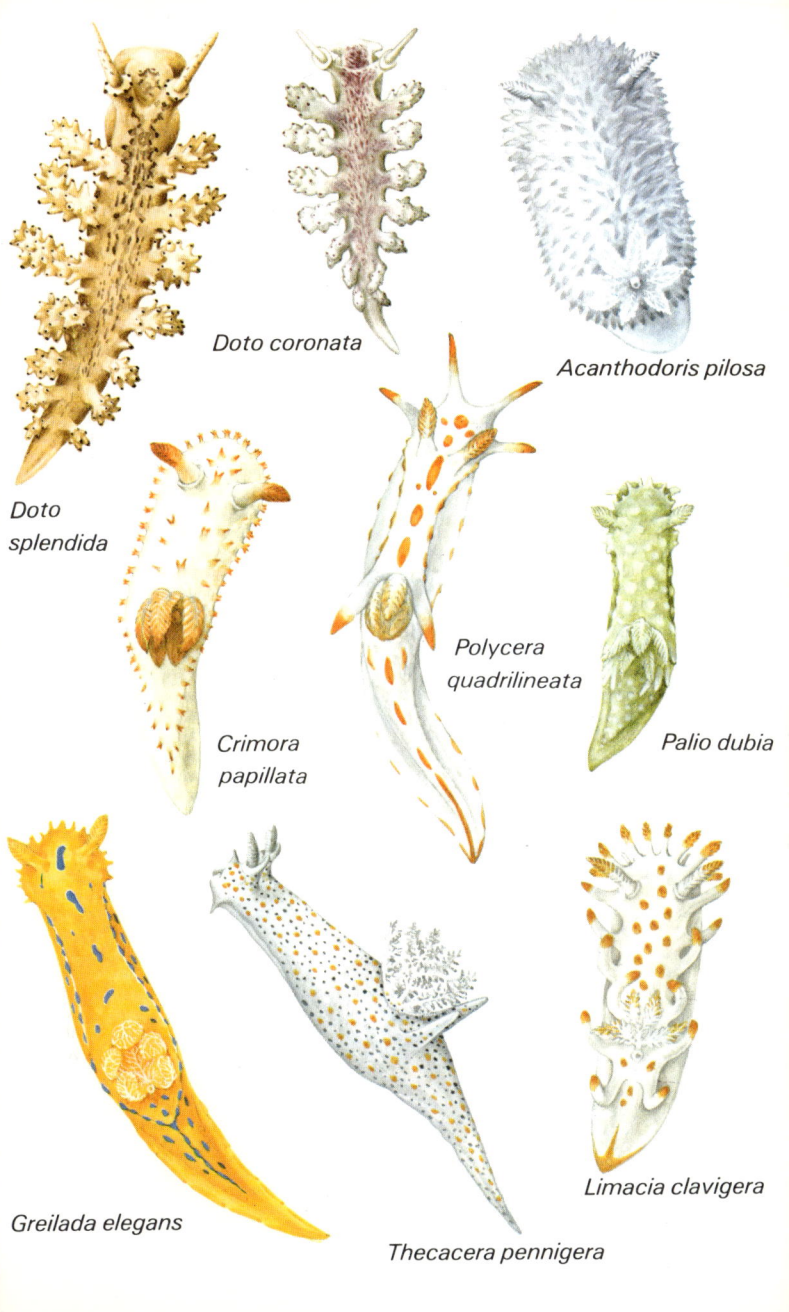

Doto coronata

Acanthodoris pilosa

Doto splendida

Crimora papillata

Polycera quadrilineata

Palio dubia

Greilada elegans

Thecacera pennigera

Limacia clavigera

rhinophores together with a number of paired yellow-tipped processes lying along the frontal margin of the body and terminating in small papillae; similar structures with smooth ends lie along both sides of the mantle; other much shorter structures lie on the back; 3–4 branching gills lie immediately in front of the anus. **Colour** background white; the yellow-tipped processes make identification easy. **Habitat** in shallow water feeding on bryozoans such as *Bugula*. **Similar species** none.

Rostanga rubra (Risso) *(=R. rufescens)* Body up to 15mm long, a little less than half as wide; front end blunt; head and finger-like oral tentacles not visible from above; yellowish rhinophores each with about 12 pairs of plate-like branches; mantle bearing short stubby tubercles; 10 branching gills surround the anus. **Colour** scarlet, occasionally yellowish, with a scattering of black spots. **Habitat** among sponges the colour of which it often matches in shallow water. **Similar species** few.

Archidoris pseudoargus (Rapp) *(=A. britannica* or *A. tuberculata)* Body up to 120mm but often less, not quite half as broad; head hidden by front of mantle which is round and blunt; oral tentacles reduced; conspicuous rhinophores; mantle covered by small warts; 8–9 branching gills surround the anus. **Colour** mantle a variety of colours but often white-yellow with pink, reddish or darker blotches. **Habitat** generally among sponges on the lower shore or in shallow water. **Similar species** several.

Peltodoris atromaculata (Alder & Hancock) Body to 60mm, less than half as wide; head hidden by front of mantle which is round and blunt; conspicuous rhinophores; 9 branching gills surround anus. **Colour** mantle whitish with dark brown markings. **Habitat** among rocks on lower shore in summer; in deeper water at other times. **Similar species** none.

Jorunna tomentosa (Cuvier) Body up to 55mm long, less than half as wide; tapering towards tail; head overlain by mantle and has narrow digit-like oral tentacles; rhinophores conspicuous, swollen before the tapering tip; mantle soft and bearing small tubercles which have a retractile core. **Colour** pale brown with darker marks and patches. **Habitat** down to 400m. **Similar species** few.

Arminia neapolitana (Chiaje) Body up to 50mm long, tapering towards pointed tail and less than 25mm at broader part; head is quite large and bears a pair of large smooth tapering oral tentacles which are just visible from above; rhinophores stubby and grouped close together at the front of a recess of the mantle; mantle surface thrown into numerous longitudinal furrows. **Colour** yellowish-brown with whitish ridges between the furrows. **Habitat** on sandy bottoms down to 40m, usually associated with sea-pens. **Similar species** none.

Antiopella cristata (Chiaje) Body up to 75mm long, quite narrow, gradually tapering to tail. The head is not easily made out from above and carries two oral tentacles; rhinophores quite well developed and almost joined at their bases; mantle bearing many conspicuous cerata, the tips of which have small white dots. **Colour** yellow, pale brown or creamy. **Habitat** in shallow water, usually associated with bryozoans like *Bugula*. **Similar species** few.

Coryphella lineata (Loven) Body up to 40mm long, much narrower; head clearly visible, bearing conspicuous tapering oral tentacles and tapering rhinophores; either side of the mantle extended into numerous cerata arranged in clusters of up to five. **Colour** background translucent-whitish with moe opaque patches and a conspicuous midline which runs up to the bases of the oral tentacles, where it branches to ascend each; such a line also passes down each side and joins the midline about half-way along the body; there is a similar one on the back of the rhinophores; each of the cerata contains a bright red mark and an opaque white tip. **Habitat** among hydroids down to quite deep water. **Similar species** not easily confused with others like *C. pedata* on account of the colour.

Coryphella pedata (Montagu) Body up to 40mm long, quite narrow and tapering towards the rear; head clearly visible, with conspicuous oral tentacles and rhinophores; the mantle bears cerata grouped along the dorsal aspects of the flanks

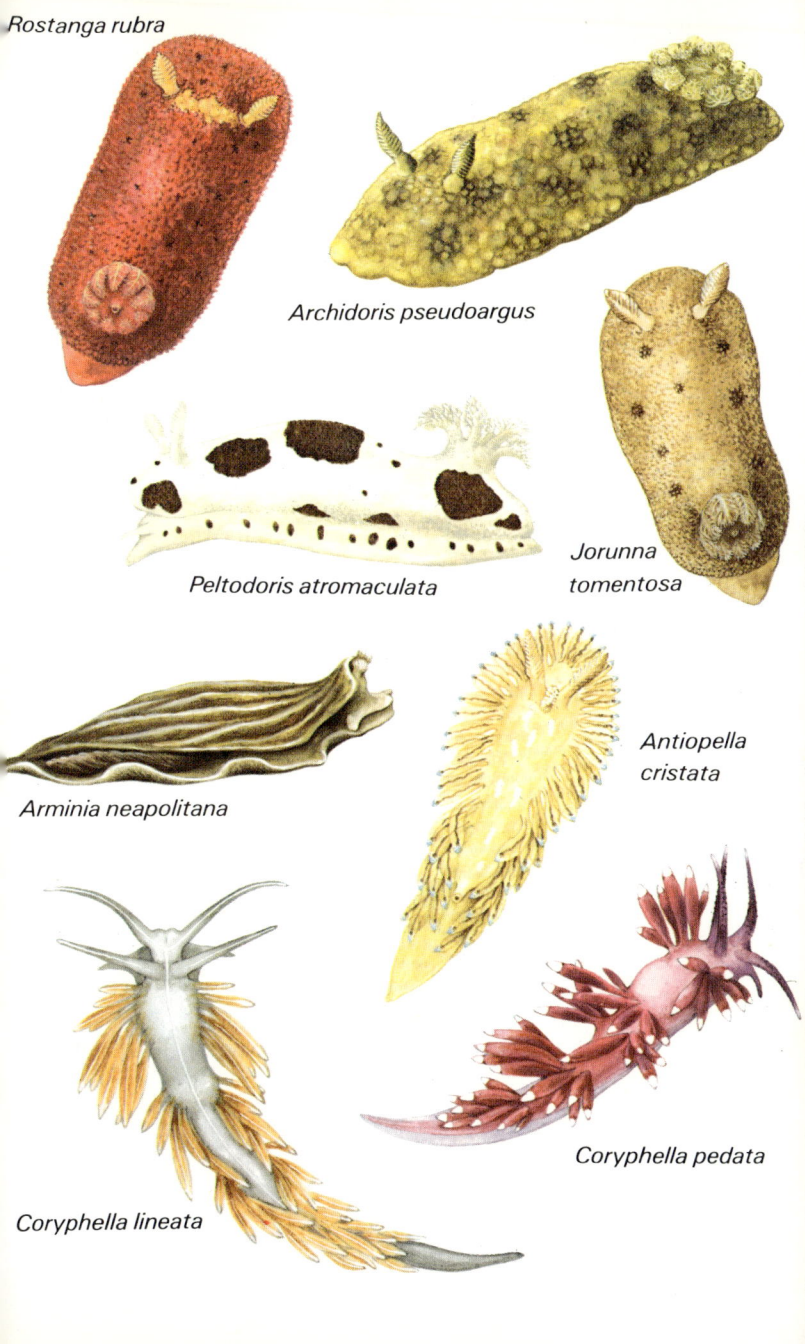

Rostanga rubra

Archidoris pseudoargus

Jorunna tomentosa

Peltodoris atromaculata

Arminia neapolitana

Antiopella cristata

Coryphella lineata

Coryphella pedata

in clusters of about 6. **Colour** each of the cerata contains a bright orange spot and has a white tip; body violet. **Habitat** among hydroids down to about 40m. **Similar species** *Flabellina affinis*.

Facelina bostonensis (Couthouy) (formerly included in *F. auriculata*) **Body** up to 50mm long, may be less, fairly narrow; head with long narrow oral tentacles, slightly shorter rhinophores and flattish propodial tentacles; many cerata arranged on either side of the body in clusters of 8, each contains a brownish or greenish pigment strip and a white band near the tip; no iridescent sheen. **Colour** body whitish with traces of pink. **Habitat** among rocks on the shore and in shallow water; often associated with hydroids. **Similar species** *F. coronata*.

Facelina coronata (Forbes and Goodsir) **Body** up to 38mm long; very similar to *F. bostonensis* but even narrower. **Colour** white with pinkish tinge and normally a bluish iridescence in the cerata, which have an underlying pinkish-red strip. **Habitat** as for *F. bostonensis* (above).

Caloria elegans (Alder and Hancock) **Body** about 35mm long, fairly narrow; head with conspicuous oral tentacles, each bearing a fine white stripe, and a pair of rhinophores; body bearing many cerata with an orange-red core, a dark subterminal band and a white tip. **Colour** background colour greyish-brown. **Habitat** in shallow water. **Similar species** few.

Spurilla neapolitana (Delle Chiaje) **Body** up to 60mm long, less than half as broad, tapering to tail; conspicuous head carries oral tentacles and rhinophores; branching gills arranged in pairs along the body. **Colour** overall brownish. **Habitat** in shallow water. **Similar species** none.

Flabellina affinis (Gmelin) **Body** up to 50mm long, narrow; conspicuous head with long slender oral tentacles and long rhinophores bearing fine ring-like ridges; small propodial tentacles; cerata in groups along either side of body. **Colour** pinkish-purple. **Habitat** in shallow water, often near weeds. **Similar species** few.

Cuthona caerulea (Montagu) **Body** up to 18mm long; head clearly visible and bearing long narrow smooth tentacles and rhinophores; the cerata are arranged in up to 10 rows across the body. **Colour** body white, sometimes slightly greenish; cerata characteristically pigmented with a subterminal blue band and a terminal orange tip; sometimes there is an inner orange band on the body side of the blue one. **Habitat** in shallow water among hydroids. **Similar species** none coloured like this.

Calmella cavolini Vérany **Body** up to 10mm long; head clearly visible and bearing long narrow oral tentacles and even longer smooth tapering rhinophores; two spots of pigment lie on the head immediately anterior to the rhinophore bases; groups of rather short stubby cerata are arranged along the flanks. **Colour** whitish with reddish markings. **Habitat** in shallow water, often among weeds. **Similar species** few.

Facelina bostonensis

Facelina coronata

Caloria elegans

Spurilla neapolitana

Flabellina affinis

Cuthona caerulea

Calmella cavolini

Fimbria fimbria (Bohadsch) *(=Tethys leporina)* Body up to 300mm long but often less; a large sea-slug of unusual shape with a very flat wide head, the anterior margin of which is furnished with many fine filamentous processes; the body carries well developed cerata arranged along either side; these are easily lost when handled. **Colour** whitish with purple marks. **Habitat** usually on soft bottoms down to 200m, but sometimes on the surface feeding on plankton. **Similar species** none.

Subclass Pulmonata

Gastropods which undergo torsion and bear a shell. There is no operculum. The mantle cavity is modified to form a lung for breathing air, and may be further modifed to permit respiration under water. Pulmonate snails are generally freshwater or terrestrial but a few species live on the shore.

Gadinia garnoti Payraudeau Shell about 5mm high and about twice as wide, being conical and somewhat limpet-like in appearance; apex curves towards posterior; solid; marked with rays and concentric spiral lines. **Colour** white-yellow. **Habitat** on rocks on the shore. **Similar species** none.

Ovatella myosotis (Draparnaud) *(=Phytia myosotis)* Shell about 9mm high and half as wide; about 7 whorls; sutures moderately conspicuous; aperture may show up to 3 ridges on the inner lip and one on the outer. **Colour** grey-brown. **Habitat** in estuaries, salt marshes and under stones on the shore. **Similar species** several prosobranchs may be confused but aperture tooth should prevent this.

Class Scaphopoda

Bilaterally symmetrical molluscs. The three-lobed foot is somewhat reduced and projects from the wider end of the tapering tubular shell. The foot is used for burrowing.

Dentalium vulgare da Costa **Tusk Shell** Shell up to 60mm long with longitudinal fine ribs and cross ridges; curvature of shell is even. **Habitat** usually buried in sand and mud from shallow water down. **Similar species** several, including *D. corneum*.

Dentalium corneum Linnaeus Shell up to 20mm long, greyish-white; curvature not even. **Habitat** burrowing in sand and mud down to quite deep water. **Similar species** several.

Class Bivalvia (=Lamellibranchia or Pelecypoda)

Bivalves are bilaterally symmetrical molluscs with the body compressed sideways and enclosed in a shell made up of two parts known as valves. These are linked dorsally by a ligament and held in correct alignment by a hinge. The form of the hinge and distribution of its teeth are sometimes important in identification. The head is reduced and the tentacles and radula are lacking. The foot is ventral and not used for crawling but is important, either in attachment to hard surfaces using a sticky secreted thread called a byssus, or in digging into mud, sand or gravel to facilitate burrowing. Bivalves are usually filter or deposit feeders. The sexes are separate but a few are hermaphrodite. A free-living planktonic larva develops after fertilization of the eggs. They inhabit almost all benthic marine habitats and some species live in fresh and brackish water.

There are some important bivalve characters which are used in identification. The two valves can vary in shape and strength; the elastic ligament may be inside the shell or outside it, or both. It forces the shell open and in life it is opposed by the contractions of the adductor or closing muscles, the attachment scars of which can usually be seen inside the shell. The hinge prevents the valves slipping out of alignment and allows them to open and close. It may be smooth, crenulate or toothed. Each valve develops from a beak, above which lies the convex 'umbo'

Fimbria fimbria

Gadinia garnoti

Ovatella myosotis

Dentalium corneum

Dentalium vulgare

Arca tetragona (Entry on page 158)

(plural umbones). In some genera, e.g. *Nucula*, the two valves may be similar (equal) but in others like *Pecten* they are not similar (unequal). The margin or edge of the valves may be smooth, crenulate or toothed, and the outer surface of the valves can be of many textures and is sometimes overlain by part of the periostracum. In many burrowing species the mantle is extended posteriorly through a gap in the valves as two siphons (sometimes united for most of their length) which enable the animal to draw in fresh seawater bearing food and oxygen, even though it may be buried below the substrate surface. In the following illustrations all the bivalves are shown with their anterior ends pointing towards the right-hand side and their dorsal surfaces uppermost, unless otherwise indicated. A more detailed account of many bivalves will be found in the following: Nordsieck F. 1969; Riedl, R. 1963; Tebble, N. 1976; or Yonge, C.M. and Thompson, T.E. 1976. Mediterranean species = not less than 130.

Nucula nucleus (Linnaeus) **Common Nut-shell** Shell up to 12.5mm long, less deep; valves equal; ligament internal; dark brown hinge with more anterior teeth than posterior; edges crenulate; adductor scars similar; siphons lacking. Colour periostracum brown-green; outside of shell brownish, inside pale. Habitat in soft substrates down to 150m. Similar species none.

Leda fragilis (Chemintz) *(=Nuculana fragilis)* **Thin Nut Clam** Shell up to 10mm long, less deep; outline drawn out posteriorly to form a sharpish point; valves equal; ligament external, dark. Colour periostracum brown; outside of shell white, inside slight mother-of-pearl. Habitat burrowing in sand and mud. Similar species several.

Arca noae Linnaeus **Noah's Ark Shell** Shell up to 80mm long and half as deep, plump; valves equal; ligament external; hinge straight with many small equal teeth; beaks and umbones well forward of midline and widely spaced when viewed from above; outside ribbed; edge smooth but crenulate towards rear; adductor scars equal. Colour periostracum brownish and hairy; outside of shell brownish, inside paler. Habitat attached to rocks and stones by byssus, down to quite deep water. Similar species four, including *A. tetragona* and *A. barbata*.

Arca tetragona Poli **Cornered Ark Shell** (Illustrated on page 157) Shell up to 50mm long; like *A. noae* but more rectangular in outline and white-yellow with brownish markings outside.

Arca barbata Linnaeus Shell up to 50mm long; like *A. noae* but with bristly black periostracum and light brown shell marked by red-brown rays.

Glycymeris glycymeris Linnaeus **Dog Cockle** Shell up to 80mm long and about as deep, round in outline; valves equal; quite heavy; ligament external; hinge with two rows of up to 12 teeth in each valve; beaks and umbones about in midline, slightly separated in dorsal view; outer surface slightly sculptured with concentric lines; edges crenulate. Colour outside yellowish with brown patterns, inside white. Habitat burrowing just below the surface of mud, sand and gravel down to about 80m. Similar species four, including *G. violascens*.

Glycymeris violascens Lamarck *(=Petunculus violascens)* Shell up to 70mm long, similar to *G. glycymeris* above; periostracum dark. Colour outside violet with lighter patterns.

Nucula nucleus

Leda fragilis

Arca noae

Glycymeris violascens

Glycymeris glycymeris

Arca barbata

Modiolus barbatus (Linnaeus) **Bearded Horse Mussel** Shell up to 60mm long, but often less, and half as deep; outline wedge-shaped, a typical mussel; valves equal; ligament external; hinge lacking teeth; beaks at anterior but not quite terminal; edge smooth; anterior adductor scar small, posterior scar larger; mantle margin not frilled; periostracum persists over posterior of shell and is formed from many semicircular rows of serrated bristles. **Colour** outside brownish-purple, inside paler. **Habitat** attached to rocks and shells by byssus, down to 100m. **Similar species** *M. adriatica*.

Modiolus adriatica Lamarck Shell up to 60mm long but often less; like *M. barbatus* but periostracum lacks hairs.

Mytilus edulis Linnaeus **Common Mussel** Shell up to 100mm long but often less; about half as deep; outline wedge-shaped; valves equal; ligament external; hinge lacking conspicuous teeth but with up to 12 crenulations near the umbones; beaks right at the anterior of the shell, not subterminal; edge smooth; anterior adductor scar small; posterior larger; mantle margin frilled, white or yellow-brown; periostracum very thin, brownish. **Colour** outside of shell brown, blue or black with brown markings; inside slight mother-of-pearl with dark border. **Habitat** attached by byssus threads to rocks and stones on the shore and in shallow water. **Similar species** about 7.

Mytilus galloprovincialis Lamarck **Mediterranean Mussel** This has been regarded as a separate species for a long time, however it is probably a race of *M. edulis* differing only in that the umbones of the shell are more pointed and turn down. The shell is generally deeper and less angular. Colour of mantle edge and outside of shell is dark; not found in estuaries.

Lithophaga lithophaga (Linnaeus) **Date Mussel** Shell up to 70mm long, not deep; outline elongated with rounded ends, upper and lower edges not parallel; rounded and plump; valves similar; outside sculptured by fine lines; edges smooth. **Colour** periostracum and outside of shell brownish. **Habitat** boring into coral colonies and limestone rocks, usually in shallow water. **Similar species** none.

Pinna nobilis Linnaeus **Fan Mussel** Shell up to 450mm long, fan-shaped; valves similar; ligament external; hinge lacks teeth; outside of shell bears prominent overlapping scales; edge smooth; anterior adductor scar small, posterior larger. **Colour** outside red-brown, inside grey to mother-of-pearl. **Habitat** standing upright in muddy sand or gravel, attached to sunken objects by byssus threads; usually in quite deep water. **Similar species** *P. squamosa* and *P. rudis*.

Pinna squamosa Gmelin **Fan Mussel** Like *P. nobilis* but up to 900mm long; outside of shell red-brown and sculptured with growth lines.

Pinna rudis Linnaeus **Fan Mussel** Like *P. nobilis* but up to 250mm long; outside of shell yellow to brown and covered in strong U-shaped teeth arranged in diverging rows along the shell.

Pteria hirundo (Linnaeus) *(=Avicula hirundo)* Shell up to 75mm long, unusual shape with each valve bearing a long drawn-out posterior process and a short anterior one; valves unequal; edge smooth. **Colour** periostracum brownish; outside of shell grey-brown, inside pearly white. **Habitat** attached to shells and stones in mud, sand or gravel. **Similar species** none.

Modiolus adriatica

Modiolus barbatus

Mytilus galloprovincialis

Mytilus edulis

Pinna squamosa

Pinna nobilis

Lithophaga lithophaga

Pinna rudis

Pteria hirundo

Chlamys opercularis (Linnaeus) **Queen Scallop** Shell up to 90mm long and about the same depth; outline typical of scallops; valves unequal, lower right less convex than upper left; anterior 'ear' slightly longer than posterior one; outside of shell sculptured by about 20 ribs. **Colour** outside variable, sometimes spotted and striped. **Habitat** when young attached to substrate by byssus threads; as adults lying free on gravel and sand from the shore down to about 200m. This species can swim by flapping the valves. **Similar species** several.

Chlamys varia (Linnaeus) **Variegated Scallop** Shell up to 60mm long and slightly deeper; valves not quite equal; hinge lacks teeth in adults; posterior ear between one-third and half the length of anterior ear; outside of shell sculptured by about 28 ribs which bear scale-like teeth although these are often rubbed way, especially near the umbones; edge of shell indented by teeth; one adductor scar towards the posterior. **Colour** outside variable: purple-red-white-yellow-brown, sometimes patterned. **Habitat** living free or attached by byssus threads to substrate, from the shore down to 80m. **Similar species** several.

Chlamys flexuosa (Poli) Shell up to 30mm long and about the same depth; outline rounded; valves unequal, anterior and posterior ears almost equal; hinge straight; shell somewhat folded into 5–6 ribs; one adductor muscle scar. **Colour** outside white to pale brown with brownish-red flecks. **Habitat** on soft substrates down to deep water. **Similar species** several.

Chlamys distorta (da Costa) **Hunchback Scallop** Shell up to 40mm long, slightly deeper; outline oval; ears not quite equal; outside of shell sculptured by about 70 fine ribs. **Habitat** attached by byssus when young, then living cemented to hard substrates by the right valve from the shore down to 100m. **Similar species** few.

Pecten jacobaeus Linnaeus **Fan Shell Scallop** or **St James's Shell.** Shell up to 150mm long, not quite so deep; valves unequal: upper (left) flat and plate-like, lower (right) concave or dished; ears almost equal; hinge straight; valves sculptured by radiating ribs which are square in section and not rounded (as in *P. maximus* of the Atlantic); one adductor scar. **Colour** upper (left) valve red-brown outside, lower (right) pinkish outside. **Habitat** on sand and gravel in quite deep water; this species can swim by flapping its valves. **Similar species** none.

Lima lima (Linnaeus) **Spiny Lima** or **File Shell** Shell up to 50mm long but much deeper; asymmetrical, rounded in outline; valves similar; anterior ear larger than posterior; ligament internal, sunk in a pit; hinge lacks teeth in adults; edge of shell indented by about 20 ribs which bear scales; scales better developed towards shell edge; single adductor muscle scar. **Colour** outside and inside of shell white. **Habitat** found on rocks and under stones from shallow water down, often attached by byssus and sometimes surrounded by grains of sand. **Similar species** *L. hians.*

Lima hians (Gmelin) **Gaping File Shell** Shell up to 25mm long; like *L. lima,* but when viewed from the front there is a conspicuous gap in between the valves; there are also many very conspicuous non-retractable orange-coloured tentacles around the shell edge. **Habitat** from lower shore down to 100m, sometimes living in a 'nest' of stones or among weeds and attached by byssus threads.

Spondylus gaederopus Linnaeus **Thorny Oyster** Shell up to 100mm long, rather deeper; valves unequal: right (upper) flattish, bearing many delicate ribs armed with sharp thorn-like projections; lower (left) concave and covered with wide spines of variable width. **Colour** outside of shell brown to violet, often encrusted by various invertebrates; inside white. **Habitat** attached to rocks and other solid seabed objects by lower valve. **Similar species** none, but do not confuse with *Echinochama lazarus.*

Chlamys opercularis

Pecten jacobaeus

Chlamys varia

Chlamys flexuosa

Chlamys distorta

Lima lima

ima hians

Spondylus gaederopus

Anomia ephippium Linnaeus **Common Saddle Oyster** Shell up to 60mm long, not quite so deep; valves unequal: upper (left) fairly thick and convex, bearing 1 adductor and 2 byssus muscle scars; lower (right) flat and with a conspicuous opening through which the chalky byssus attachment passes from the upper valve to the substrate, and bearing 1 adductor scar. **Colour** outside of upper valve rough and scaly and coloured brown to white. **Habitat** attached to rocks and other shells, its shape often being affected by the form of the substrate; from the lower shore down to quite deep water. **Similar species** none.

Ostrea edulis Linnaeus **Common European Oyster** Shell up to 100mm long and about as deep; generally rounded in outline but often variable; valves unequal, quite heavy; internal ligament; hinge lacking teeth; upper (right) valve flattened and sculptured, lower (left) saucer-like and sculptured; edges sometimes crenulate; single adductor muscle scar; periostracum thin and inconspicuous. **Colour** outside grey-brown, inside pearly-white. **Habitat** from shallow water down to about 50m. Commercially fished. **Similar species** none, but 3 forms are found: var. *adria*, var. *lamellosa* and var. *tarentina* (see line diagrams).

Cardita trapezia (Linnaeus) Shell up to 6mm long, less deep; outline trapezoid; valves equal; hinge with 2 cardinal teeth on the left valve and one on the right valve; umbones anterior to the midline; outer sculpturing of about 12 ribs; edge not indented by ribs. **Colour** outside brownish-white, inside white. **Habitat** on rocky substrates attached by byssus threads. **Similar species** *C. sulcata* and *C. squamigera*.

Cardita sulcata Bruguière **(= Venericardia sulcata)** Shell up to 30mm long, outline more triangular than *C. trapezia*; valves equal; outside of shell with well developed diagonal knobby ribs; edge of valves rounded; inside with fine grooves. **Colour** outside pale to dark brown, sometimes spotted white to red, rarely all white; inside pale. **Habitat** on sandy bottoms. N.B. not attached by byssus. **Similar species** *C. trapezia* and *C. squamigera*.

Cardita squamigera Lamarck **(= C. aculeata)** Shell up to 7mm long, outline triangular; valves similar; hinge at apex; outside of shell with radial ribs ornamented by conspicuous thorns. **Colour** outside white to brown, inside pale. **Habitat** on hard substrates, under pebbles or attached to rocks. **Similar species** *C. trapezia* and *C. sulcata*.

Astarte fusca Poli (Not illustrated) Shell up to 25mm long, less deep; outline triangular, compressed sideways; valves equal; ligament external; hinge with 3 teeth under each umbo; beaks anterior of midline; umbones facing forward; outside of shell sculptured with concentric ridges and grooves; anterior and posterior adductor scars similar. **Colour** outside of shell brown to yellow, inside yellow. **Habitat** on sandy substrates. **Similar species** none.

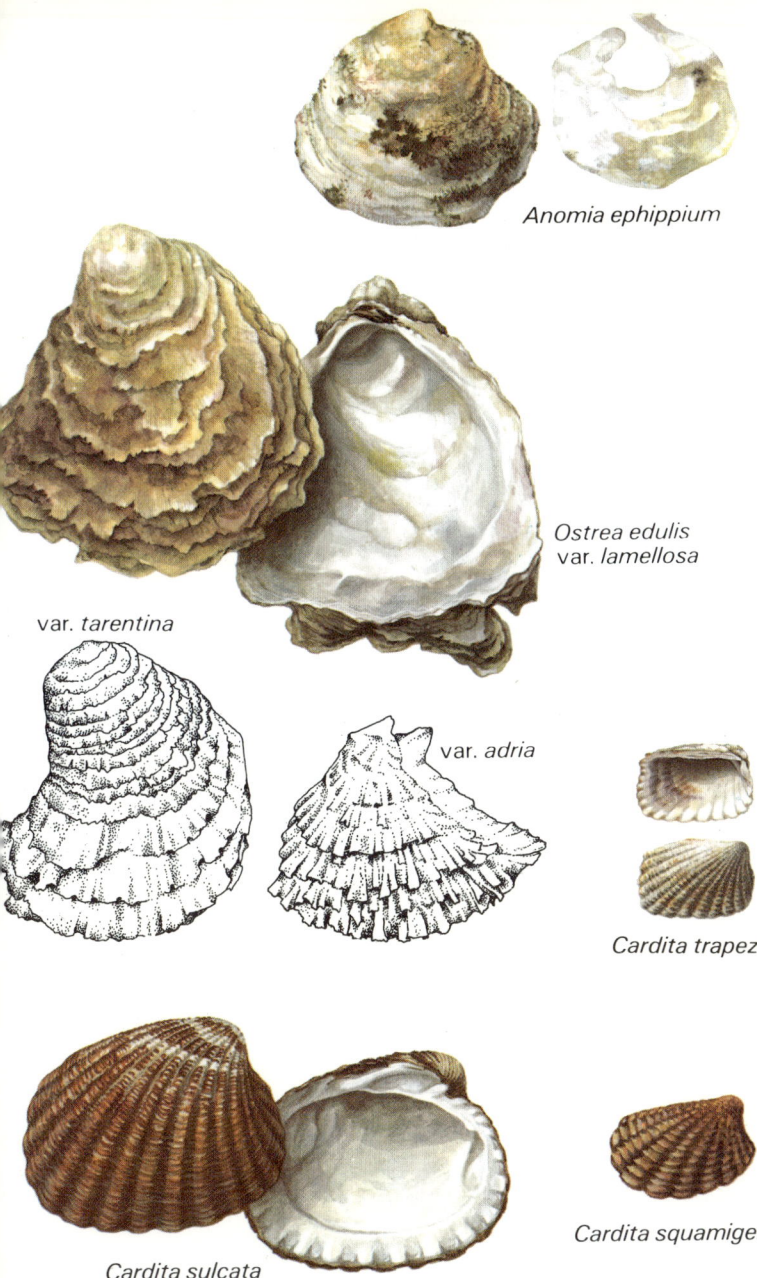

Anomia ephippium

Ostrea edulis
var. *lamellosa*

var. *tarentina*

var. *adria*

Cardita trapezia

Cardita sulcata

Cardita squamigera

Glossus humanus (Linnaeus) *(=Isocardia cor)* Shell up to 100mm long and about the same depth, wide and plump; outline circular; valves equal, heavy; strongly convex, spirally coiled and pointing away from the hinge. **Colour** outside dark brown, inside pale. **Habitat** on soft substrates below 10m. **Similar species** none.

Loripes lacteus (Lamarck) (Not illustrated) Shell up to 25mm long and about as deep; outline rounded; valves equal; ligament external; hinge with 2 small cardinal teeth and 2 minute lateral teeth in each valve; umbones about in midline and forward-pointing; posterior adductor muscle scar shorter and rounder than anterior. **Colour** outside of shell white and translucent; inside white. **Habitat** burrowing in soft substrates, down to deep water. **Similar species** *L. lucinalis, Lucina squamosa* and *Myrtea spinifera*.

Loripes lucinalis (Lamarck) Shell about 20mm long and the same depth; valves equal, thin and light, rounded; ligament external; hinge with 2 small cardinal teeth and 2 very small lateral teeth in each valve; umbones about in midline and pointing forward; outside of shell sculptured with concentric lines. **Colour** outside yellowish-white, inside white. **Habitat** on soft substrates down to 150m. **Similar species** *L. lacteus, Lucina squamosa* and *Myrtea spinifera*.

Lucina squamosa Linnaeus Shell up to 120mm long and about the same depth; outline rounded; valves equal, solid; hinge below umbo; umbo about in midline. **Colour** outside white to yellow background with fine concentric brown marks and radiating patterns; inside paler. **Habitat** from shallow water down. **Similar species** *Loripes lacteus, Loripes lucinalis* and *Myrtea spinifera*.

Myrtea spinifera (Montagu) Shell up to 25mm long and less deep, outline oval; valves equal; hinge with 2 cardinal and 2 lateral teeth in left valve, 1 cardinal and 2 lateral teeth in right; beaks pointing forward; umbones in midline; edges smooth; posterior adductor scar smaller than anterior; periostracum slight. **Colour** outside cream to white, inside white. **Habitat** burrowing in silt or gravel, down to about 100m. **Similar species** *Loripes lacteus, L. lucinalis* and *Lucina squamosa*.

Montacuta ferruginosa (Montagu) Shell up to 8mm long, less deep; valves equal; beaks and umbones well forward; edges smooth; periostracum thin and reddish. **Colour** outside reddish, inside white-purple. **Habitat** commonly associated with the spines of *Echinocardium* species (see page 240); burrowing in sand on the shore and in shallow water. **Similar species** several.

Chama gryphoides Linnaeus **Mediterranean Jewel Box** Shell about 40mm long and the same depth; valves unequal and thick; lower (left) strongly dished and attached to substrate, upper (right) flatter and acting like a lid; outside of valves sculptured by growth lines giving a layered appearance. **Colour** outside whitish, inside violet to brown and finely grooved. **Habitat** attached to rocks in shallow water. **Similar species** *Pseudochama ferruginosa* Reeve which has the right valve attached, and *Echinochama lazarus* which is spiny. N.B. do not confuse with *Spondylus gaederopus*.

Laevicardium crassum Gmelin Shell up to 60mm long and about the same depth; valves equal; outside sculptured by about 40 very weak ribs so that it seems almost smooth. **Colour** outside white to brown, interior pale. **Habitat** burrowing in soft substrates, often shell gravel. **Similar species** *L. oblongum*.

Laevicardium oblongum Gmelin **Oblong Cockle** Shell about 50mm long but considerably deeper; valves equal; somewhat asymmetrical in outline; outside of shell ornamented by fine ribs; edge grooved. **Colour** outside white to yellow. **Habitat** burrowing in soft substrates in quite shallow water. **Similar species** *L. crassum*.

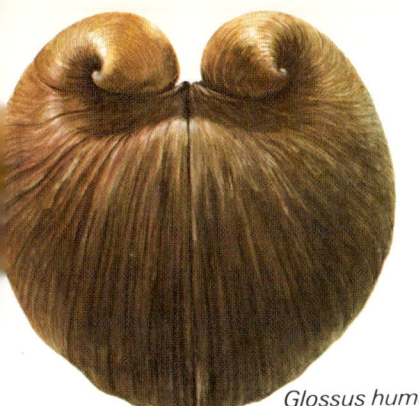

Loripes lucinalis

Glossus humanus

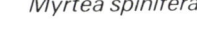

Myrtea spinifera

Montacuta ferruginosa

Lucina squamosa

Chama gryphoides

Laevicardium crassum

Laevicardium oblongum

Acanthocardium aculeata (Linnaeus) *(=Cardium aculeatus)* **Spiny Cockle** Shell up to 100mm long but not quite so deep; valves equal and convex, giving a wide plump appearance; outside and inside of shell furrowed by about 22 ribs which on the outside each carry a row of spines; the largest spines are nearest the edge of the valves and the spine bases do not meet those of their neighbours. **Colour** outside pale brown, inside white. **Habitat** burrowing in soft substrates, from about 10m down. **Similar species** *A. echinata.*

Acanthocardium echinata (Linnaeus) *(=Cardium echinata)* **Prickly Cockle** Shell about 75mm long, like *A. aculeata*; shell ornamented by 18–19 ribs which are better defined near the edges; rib spines on the outside are blunter and have broad bases which often join with those of their neighbours. **Habitat** burrowing in soft substrates, from about 10m.

Cerastoderma edule (Linnaeus) *(=Cardium edule)* **Common Cockle** Shell up to 50mm long, oval in outline; valves equal; umbones slightly anterior of midline; outside of shell lacks spines and is sculptured by about 24 ribs which make crenulations where they reach the edge; internally the grooves from the ribbing run only a short way; periostracum reduced. **Colour** outside brownish, marked with concentric growth lines; inside white with brown marks. **Similar species** *C. lamarcki* and others.

Cerastoderma lamarcki (Reeve) *(=Cardium glaucum)* Shell up to 40mm long, like *C. edule* but more triangular in outline; edge crenulate anteriorly and smooth posteriorly; grooves on the inside run nearly to the umbones. **Habitat** burrowing in soft substrates, usually in brackish water such as lagoons; permanently submerged.

Parvicardium exiguum (Gmelin) **Little Cockle** Shell up to 12.5mm in length, somewhat triangular in outline; valves equal; beaks well forward; outside of shell sculptured with about 20 ribs on each valve; anterior ribs themselves bearing bumps or tubercles; edges crenulate. **Colour** outside of shell brownish. **Habitat** usually in shallow water burrowing in sand. **Similar species** *P. papillosum* (below).

Parvicardium papillosum (Poli) Shell up to 12.5mm long; outline rounded, valves equal; beaks just anterior of midline; outside of shell ornamented with 25 ribs all of which bear tubercles; edge deeply crenulated. **Colour** outside white, grey to yellow with red-brown marks; inside smooth, pink to white. **Similar species** *P. exiguum* (above).

Cardium tuberculatum Linnaeus *(=C. nodosum)* **Rough Cockle** Shell up to 60mm long; valves equal, heavy; sculptured with coarse ribs carrying tubercles which are poorly developed towards the posterior; especially in silty places; empty shells may be washed ashore. **Colour** outside white to brown with darker bands. **Habitat** burrowing in soft substrates below 10m especially in silty places: empty shells may get washed ashore. **Similar species** *Parvicardium exiguum P. papillosum* and sometimes *Acanthocardium echinata.*

Cardium pauciostracum Sowerby Shell up to 30mm long; valves similar, sculptured by 16–17 coarse ribs each carrying many small blunt tubercles. **Colour** outside brownish with brighter bands. **Habitat** offshore in soft substrates. **Similar species** few.

Acanthocardium aculeata

Acanthocardium echinata

Cerastoderma edule

Cerastoderma lamarcki

Parvicardium papillosum

Cardium tuberculatum

Cardium pauciostracum

Parvicardium exiguum

Dosinia lupinus (Linnaeus) Shell up to 37.5mm long and almost as deep, outline rounded; valves equal; ligament external; hinge with 3 cardinal teeth on each valve and an extra small anterior tooth on the left valve; beaks pointing forward and lying adjacent to the characteristic heart-shaped depression called the 'lunule' straddling the valves; umbònes forward of the midline, outside sculptured with concentric lines; edges smooth; siphons fused with a horny covering and separate at the tips only. **Colour** outside of shell whitish; inside white. **Habitat** burrowing in sand and shell gravel down to 125m. **Similar species** several.

Venus verrucosa Linnaeus **Warty Venus** Shell up to 62.5mm long and of similar depth. Like *Dosinia lupinus* in many respects including the lunule; hinge has 3 cardinal teeth on each valve; concentric lines on outside of shell coarse; edges crenulate; siphons lack horny covering. **Colour** outside of shell grey, white or yellowish with brown markings; inside white. **Habitat** burrowing just below the surface of sand or gravel, down to 100m. **Similar species** several.

Venus striatula (da Costa) **Striped Venus** Shell up to 45mm long, outline somewhat triangular, rounded; valves similar; ligament external; hinge with 3 cardinal teeth on each valve; beaks, umbones and lunule well forward; outside sculptured with conspicuous concentric ridges which have fine ridges between; edges smooth; siphons fused. **Colour** outside of shell yellow, cream or greyish-white with brown marks; inside greyish-white. **Habitat** burrowing in sand, from the shore down to 55m. **Similar species** several.

Venus ovata Pennant **Oval Venus** Shell up to 20mm long, outline somewhat triangular, rounded; valves equal; ligament somewhat hidden but external; hinge with 3 cardinal teeth on each valve; beaks, umbones and lunule all forward of midline; outside of shell sculptured with about 50 ribs intersected with fine concentric lines; edges crenulate except opposite ligament; siphons fused. **Colour** outside of shell pale brown to white, interior white, orange or purple. **Habitat** burrowing in sand or gravel down to 180m. **Similar species** several.

Venerupis rhomboides (Pennant) **Banded Carpet Shell** Shell up to 60mm long, outline oval; valves equal, smooth; ligament external; beaks and umbones forward of midline, no lunule; outer surface sculptured with concentric lines; no ribs; edges smooth. **Colour** outside pink to brown with red-brown zig-zag marks arranged in radiating bands; inside white and slimy. **Habitat** burrowing in sand and gravel down to 180m. **Similar species** several, especially *V. pullastra*.

Venerupis pullastra (Montagu) **Pullet Carpet Shell** Shell up to 50mm long, very like *V. rhomboides*; outline oval, rear edge flattened; valves equal; outside of shell mildly sculptured with fine radiating and concentric lines. **Colour** outside brownish, grey or cream with patterns; inside white and shiny, sometimes with purple marks. **Habitat** burrowing in sand and mud; may be attached by byssus. **Similar species** several, especially *V. rhomboides* and *Tapes decussatus*.

Tapes decussatus (Linnaeus) *(= **Venerupis decussata**)* **Cross-cut Carpet Shell** Shell up to 60mm long, like *V. rhomboides* and *V. pullastra* but slightly deeper and more rounded; valves equal and heavy; outside of shell sculptured by conspicuous concentric lines and radial ridges. **Colour** outside variable: green, yellow, red or blackish with radiating purple colour bands; inside pale. **Habitat** burrowing in sand and mud. **Similar species** several (see above).

Cytherea chione (Linnaeus) *(= **Pitaria chione**)* **Brown Venus** Shell up to 80mm long; outline somewhat triangular; outside of shell sculptured with concentric lines. **Colour** outside of shell has concentric yellow to red-brown lines alternating with paler ones, and radiating bands; inside pale. **Habitat** burrowing in sand and mud down to quite deep water.

Dosinia lupinus

Venus verrucosa

Venus striatula

Venerupis rhomboides

Venus ovata

Tapes decussatus

Venerupis pullastra

Cytherea chione

Notirus irus (Linnaeus) *(=Irus irus)* Shell up to 25mm long, not so deep; valves equal; ligament external but somewhat recessed; umbones forward of midline; lunule lacking; outside of shell sculptured by about 15 concentric ridges; periostracum thin and brownish. Colour outside of shell brown to white with radiating pigment marks; inside yellow-white. Habitat in holes and crevices in rocks. Similar species *Petricola lithophaga*.

Petricola lithophaga Retzius Shell up to 30mm long; valves equal; hinge has 2 cardinal teeth on each valve; umbones forward of midline; outer surface sculptured by radiating ribs intersecting irregular concentric growth lines. Colour outside of shell white to grey; inside similar. Habitat boring into limestone or mud, usually in shallow water. Similar species *Notirus irus*.

Mactra corallina (Linnaeus) **Rayed Trough Shell** Shell up to 70mm long; outline triangular-ovoid; valves equal, light; ligament thin, external, behind umbones; hinge with complex arrangement of teeth; umbones almost in midline; outside of shell with fine concentric lines; edges smooth; siphons short; periostracum fine. Colour outside pinkish to brown or greenish with delicate rays running from umbones to edge; inside white to purple. Habitat burrowing in sand and gravel, down to 100m. Similar species several.

Mactra glauca Born **Grey Trough Shell** Shell up to 90mm long, less deep; outline ovoid; valves equal; ligament thin, external; umbones almost in midline; edges smooth; siphons short; periostracum yellowish. Colour outside grey with grey-blue concentric markings; inside violet to blue-grey. Habitat burrowing in clay or sand. Similar species several.

Spisula subtruncata (da Costa) *(=Mactra subtruncata)* Shell up to 30mm long, outline somewhat triangular; valves equal; umbones anterior to midline; edges smooth; periostracum brownish. Colour outside white, grey or bluish-brown with concentric brown or brown-red bands; inside pale. Habitat burrowing in sand or mud, down to quite deep water. Similar species several.

Lutraria lutraria (Linnaeus) **Common Otter Shell** Shell large, up to 125mm long, half as deep; outline ovoid; valves equal; ligaments external and internal; hinge complex; edges smooth; valves not meeting at either end when closed; periostracum brown-green. Colour outside brownish with concentric lines; inside white. Habitat burrowing in mud and sand, from the shore down to about 100m. Similar species none.

Donax vittatus (da Costa) **Banded Wedge Shell** Shell up to 37.5mm long and about half as deep, outline triangular; valves equal; ligament external; hinge with 2 main teeth below each umbo and 2 lateral teeth in each valve; umbones posterior of midline; outside of shell has very fine lines radiating from the umbones; edges conspicuously toothed; siphons separate and short; periostracum polished. Colour outside white, yellow, brown or purple with some pigment concentrated into radial bands; inside purplish. Habitat burrowing in sand, down to 20m. Similar species several.

Donax trunculus Linnaeus Shell up to 40mm long and about half as deep, outline triangular; valves similar; ligament external; hinge has 2 main teeth below each umbo and 2 lateral teeth in each valve; umbones well posterior of midline; edges toothed; siphons separate and short. Colour outside white, yellowish or brown, occasionally with a violet tinge, usually with contrasting rays; inside violet with a white edge. Habitat burrowing in sand, down to quite deep water. Similar species several.

Notirus irus

Petricola lithophaga

Mactra corallina

Mactra glauca

Spisula subtruncata

Lutraria lutraria

Donax vittatus

Donax trunculus

Tellina distorta (Poli) Shell up to 20mm long, slim; sideways flattened; valves not quite equal, delicate; ligament external; beaks and umbones posterior to midline; 2 siphons. Colour outside of shell white to grey with pink stripes; inside similar. Habitat burrowing in sand and mud, from shallow water down. Similar species several.

Tellina solidula Lamarck (Not illustrated) Shell up to 25mm long, slim, sideways flattened; valves not quite equal; ligament external; umbones about in midline; 2 siphons. Colour outside background white to pink with concentric brown and yellowish bands. Habitat burrowing in sand, from shallow water down. Similar species several.

Tellina planata Linnaeus Shell up to 40mm long, slim, sideways flattened; outline oval; slight point on posterior; valves not quite equal; ligament external; beaks and umbones about in midline; edges smooth; 2 siphons. Colour outside of shell yellow, white or pink, sometimes banded, with a brown lip; inside pale pink. Habitat burrowing in sand. Similar species several.

Tellina tenuis da Costa **Thin Tellin** Shell up to 30mm long, flattened sideways, slim, outline oval; valves almost equal, light and delicate; ligament external; hinge with 2 cardinal teeth on each valve, right valve bearing conspicuous lateral teeth; outside of shell sculptured with fine concentric lines; edges smooth; 2 long separate siphons; periostracum shiny. Colour outside yellow, orange, pink or white, often with contrasting pigment bands; inside similar. Habitat burrowing in sand, from the shore down to shallow water. Similar species several.

Tellina incarnata Linnaeus Shell up to 30mm long, thin, sideways flattened; outline ovoid but tapering towards rear; valves not quite equal; ligament external; beaks and umbones about in midline. Colour outside of shell reddish with stripes on right valve. Habitat burrowing in sand and mud, from shallow water down. Similar species several.

Tellina crassa Pennant Shell up to 50mm long, almost as deep, outline ovoid; valves not quite equal; ligament external; umbones in midline; outside of shell slightly sculptured. Colour outside white-grey background with pink stripes; umbo reddish; inside yellow-red. Habitat burrowing in sand. Similar species several.

Arcopagia balaustina (Linnaeus) (Not illustrated) Shell up to 15mm long and almost as deep, outline somewhat triangular; valves not quite equal; ligament external; beaks and umbones in midline; outside of shell sculptured with concentric lines. Colour outside of shell whitish with pink radial marks. Habitat burrowing in sand and mud, from shallow water down. Similar species several.

Gastrana fragilis (Linnaeus) Shell up to 45mm long, less deep; valves equal; hinge has 2 cardinal teeth on each valve but lacks lateral teeth; umbones anterior of midline; shell sculptured with concentric lines; edges smooth; periostracum brownish. Colour outside of shell greyish-yellow; inside paler. Habitat burrowing in mud, clay or sand in shallow water. Similar species several.

Scrobicularia plana (da Costa) **Peppery Furrow Shell** Shell up to 62.5mm long, ovoid in outline, sideways flattened; valves equal, light; hinge with 2 cardinal teeth on left valve and 1 on right; umbones about in midline; outside of shell sculptured with concentric lines; long siphons; periostracum brown. Colour outside greyish-yellow; inside whitish. Habitat burrowing in sand and mud on the shore and in shallow water. Similar species several.

Abra alba Wood Shell up to 20mm long, a little less deep, thin, outline oval; valves equal; anterior edge rounded, posterior edge slightly pointed; hinge with small teeth; umbones just posterior of midline. Colour outside and inside white. Habitat burrowing in mud, from shallow water down. Similar species several.

Tellina distorta

Tellina tenuis

Tellina planata

Gastrana fragilis

Tellina incarnata

Abra alba

Scrobicularia plana

Tellina crassa

Abra tenuis (Montagu) (Not illustrated) **Shell** up to 15mm long, not quite so deep, thin, brittle, oval outline, anterior edge rounded, posterior slightly pointed; hinge with small teeth; umbones in mid-line. **Colour** outside dirty white with iridescent, pale grey or pale brown periostracum. **Habitat** mud flats and estuaries, sometimes in brackish water. **Similar species** several.

Gari fervensis (Gmelin) *(=Psammobia fervensis)* **Faro Sunset Shell** **Shell** up to 45mm long, less deep; outline rounded at posterior, angular at anterior with blunt points; valves almost equal, light, not closing fully at the ends; ligament external; umbones posterior to midline; periostracum brownish. **Colour** outside pink to white with slightly darker rays; inside light bluish-red. **Habitat** burrowing in sand, from about 10m down. **Similar species** several.

Gari depressa (Pennant) *(=Psammobia vespertina)* **Large Sunset Shell** **Shell** up to 62.5mm long, less deep, outline rounded at posterior and anterior but with straight upper edge; valves not quite equal, light; ligament external; umbones slightly posterior of midline; periostracum brown-green. **Colour** outside of shell pinkish; inside purple-white. **Habitat** from the shore down to about 50m. **Similar species** several.

Solenocurtus strigillatus (Linnaeus) *(=Solecurtus strigillatus)* Shell up to 80mm long and about half as deep; outline oblong with rounded ends; valves equal; ligament external; hinge with 2 cardinal teeth on right valve, 1 on left; beaks and umbones forward of midline; periostracum green-yellow. **Colour** outside of shell pinkish with 2 conspicuous paler rays; inside paler. **Habitat** burrowing in sand and gravel in shallow water. **Similar species** none.

Pharus legumen (Linnaeus) Shell long and narrow, reaching 125mm in length; very shallow; top and bottom edges not quite parallel; valves equal; ligament external and black; beaks and umbones slightly forward of midline, inconspicuous; periostracum brown-yellow. **Colour** outside of shell whitish with concentric lines. **Habitat** burrowing on the shore and in shallow water. **Similar species** of *Solen* and *Ensis* have beaks reduced and umbones well towards anterior.

Solen marginatus Montagu **Grooved Razor Shell** **Shell** long and narrow, up to 125mm, very shallow; top and bottom edges virtually parallel; valves equal; ligament external, dark; beaks and umbones well towards anterior end; a conspicuous groove just behind anterior edge; short united siphons; perostracum pale brown. **Colour** outside of shell pale yellow. **Habitat** burrowing in sand, from the shore down to about 35m. **Similar species** *Pharus* and *Ensis* spp.

Ensis ensis (Linnaeus) **Shell** long and narrow, up to 125mm long, very shallow; top edge concave, bottom edge convex; valves equal; ligament external, dark; beaks and umbones well towards anterior and no conspicuous groove just behind anterior edge; short united siphons; periostracum green-yellow. **Colour** outside of shell whitish with brown to red patterns. **Habitat** burrowing in sand and mud on the shore and in shallow water. **Similar species** *Pharus, Solen* and *Ensis* spp.

Ensis siliqua (Linnaeus) **Pod Razor Shell** **Shell** long and narrow, up to 200mm long, very shallow; top and bottom edges almost parallel; valves similar; ligament external, dark; beaks and umbones well towards anterior edge and inconspicuous; no conspicuous groove behind anterior edge; short united siphons; periostracum shiny green. **Colour** outside of shell white with fine horizontal and vertical lines and reddish patterns. **Habitat** burrowing in sand on the shore and down to about 35m. **Similar species** *Pharus, Solen* and *Ensis* spp.

Gari depressa

Solenocurtus strigillatus

Gari fervensis

harus legumen

Solen marginatus

Ensis ensis

Ensis siliqua

Hiatella arctica (Linnaeus) Shell up to 37.5mm, about half as deep; irregular outline; valves similar; ligament external; hinge teeth usually worn away in adults; beaks and umbones well forward; outside marked by concentric lines; edge smooth; anterior adductor scar rounded and slightly smaller than posterior one; periostracum brownish. **Colour** outside white-yellow; inside white. **Habitat** boring into soft rock or living in holes already there, anchored by byssus threads, often where there is a lot of muddy silt; on the shore and down to quite deep water. **Similar species** few.

Corbula gibba (Olivi) **Common Basket Shell** Shell up to 15mm long, a little more than half as deep; outline asymmetrical; valves thick and unequal: right valve domed and larger than the left; ligament accommodated in groove in left valve; short hinge; periostracum brownish. **Colour** outside white with red marks. **Habitat** usually burrowing in clay, often near weeds. **Similar species** none.

Gastrochaenea dubia (Pennant) **Flask Shell** Shell up to 25mm long and about half as deep, plump and rounded in section; outline asymmetrical; valves similar; external ligament; hinge without teeth; beaks and umbones well forward; outside smooth; gap at anterior on lower side conspicuous; posterior adductor scar longer than anterior; siphons united. **Colour** outside white to pale brown, inside white. **Habitat** boring into soft rocks or even firm sand on the shore and in shallow water. **Similar species** *Barnea candida* and *Pholas dactylus*.

Barnea candida (Linnaeus) **White Piddock** Shell up to 70mm long, about half as deep at deepest point; valves equal, light; ligament lacking, valves held together by muscles; umbones away from midline; outside of shell sculptured by toothed ribs; one ascending shell plate near the umbones; periostracum grey. **Colour** outside of shell white. **Habitat** boring into limestone, clay or wood, usually in shallow water. **Similar species** *Gastrochaenea dubia* and *Pholas dactylus*.

Pholas dactylus Linnaeus **Common Piddock** Shell up to 150mm long, less than half as deep; plump and rounded in section, outline asymmetrical; valves equal, light; ligament lacking, valves held together by muscles; umbones well forward and turned over; 4 accessory plates situated dorsally; shell sculptured and bearing concentric and radiating ribs which are roughest at the front where they assist with drilling; interior of shell has 2 free teeth or 'apophyses' (1 per valve); siphons long, covered by horny sheath; periostracum pale yellow. **Colour** outside white-grey, inside white. **Habitat** boring into soft rock, shale, peat or sand in shallow water. **Similar species** *Gastrochaenea dubia* and *Barnea candida*.

Teredo navalis Linnaeus **Common Ship Worm** Remarkably modified worm-like animal forming calcareous tubes up to 200mm long and 8mm in diameter. The shell of 2 valves is relatively small and encloses only part of the animal at the innermost end of the tube; the shell functions as a drill and the chalky tube, which is secreted by the mantle, follows the course of the animal as it bores through the wood; a special pair of hard chalky pallets close off the end of the tube when the siphons have been withdrawn; pallets up to 5mm long. **Colour** shell white, animal brown. **Habitat** drilling into submerged timbers, e.g. ships and piles. **Similar species** several.

Pandora albida (Röding) **Pandora Shell** Shell up to 37.5mm long, about half as deep at maximum point, outline assymetrical; valves unequal: right (upper) flat, left (lower) trough-like; ligament internal; beaks and umbones well forward; shell sculptured with concentric lines; adductor scars about equal; periostracum pale brown. **Colour** outside and inside white. **Habitat** living on soft substrates, often in shallow water. **Similar species** none.

Thracia papyracea (Poli) **Paper Thracia** Shell up to 35mm long, about half as deep; valves unequal, right larger than the left, delicate, ligament internal and external; hinge lacks teeth; beaks and umbones behind midline; outside sculptured with concentric fine lines, sometimes spiny. **Colour** outside white. **Habitat** living on soft substrates, often near weeds. **Similar species** none.

Hiatella arctica

Corbula gibba

Pholas dactylus

Gastrochaenea dubia

dorsal view

Barnea candida

Pandora albida

Teredo navalis

Thracia papyracea

Class Cephalopoda
Cuttlefishes, squids and octopuses

Uncoiled cylinder-shaped or sac-shaped molluscs. The foot is divided into a number of suckered tentacles surrounding the mouth and joining the head. The head bears conspicuous eyes and is immediately connected with the abdomen (visceral region). A shell is sometimes present internally, but rarely externally. The mouth leads to a pair of horny jaws resembling a parrot's beak; the radula is like a tongue; the sexes are separate. Further information is given in Naef, I. 1921–28.

Order Decapoda Cuttlefishes and squids

Cephalopods with cylinder-shaped bodies bearing lateral fins. There is an internal shell, or 'cuttlebone'. The mouth is surrounded by 10 tentacles of which 2 are usually much longer than the remaining 8 and which may be retracted. The disposition and form of the suckers on the tentacles is important for identification. Mediterranean recorded species = 5, but other oceanic species may appear.

Sepia elegans d'Orbigny (Not illustrated) **Length** up to 120mm. **Body** fairly slender and somewhat flattened; dorsal surface of abdomen produced into pointed shield over head; conspicuous funnel on underside of head; paired fins run from behind head to tip of body; 8 short tentacles surround mouth, bearing suckers arranged in 2 straight or zig-zag rows, 2 longer retractable tentacles bear suckers on flattened tips. **Colour** variable. Habitat over sand, in **Zostera** beds.

Sepia officinalis Linnaeus **Length** up to 300mm; similar to *S. elegans* but broader; shield over head rounded; oral tentacles bear suckers arranged in 4 rows; 2 long retractable tentacles have up to 6 suckers conspicuously larger than the rest. **Colour** mottled grey-brown above, but varies with mood and background.

Sepiola rondeleti Gesner **Little Cuttle** **Length** up to 50mm; miniature cuttlefish with two wing-like rounded fins towards tip of body. **Colour** brownish above.

Loligo vulgaris Lamarck **Long-finned Squid** **Length** up to 500mm. **Body** torpedo-shaped with paired fins running about half-way along and joined together at the tip; small shield-like projection over the head; internal shell horny and quill-like. **Colour** very variable; often pink-white with purple-brown mottling above. **Habitat** often found in open water. **Similar species** *Ptodarodes sagittatus*.

Ptodarodes sagittatus Lamarck **Sagittal Squid** **Length** up to 600mm. **Body** torpedo-shaped with paired fins running about one-third of the way along and joining at the tip; no shield projecting over head; internal shell horny and quill-like; the two long tentacles are non retractile and can always be seen unlike in *Loligo vulgaris*. **Colour** upper side violet to dark brown with dark red spots. **Habitat** found in surface waters, often by night; **Similar species** *Loligo vulgaris*.

Order Octopoda Octopuses and their allies

Cephalopods with bag-like bodies and no internal shell and rarely with external shells; fins absent but the 8 suckered arms may be linked at their bases by a membrane or web of skin. Mediterranean recorded species=about 4.

Argonauta argo Linnaeus **Paper Nautilus** **Length** adult femals up to 200mm; adult males up to 10mm. **Body** female encased in a white paper-thin shell; 2 of the 8 arms appressed to the shell and folded back over the body; male dwarf and lacking shell. **Colour** highly variable according to mood and situation; with and without spots. **Habitat** creeping over the bottom and swimming.

Octopus vulgaris Lamarck **Common Octopus** **Length** up to 1m but often less. **Body** strong with conspicuous siphon; 2 rows of suckers on strong arms. **Colour** highly variable according to mood and background: grey-yellow-brown-green. **Habitat** among rocks and stones. **Similar species** *Eledone moschata*.

Eledone moschata Lamarck **Length** up to 400mm. **Body** with conspicuous siphon; one row of suckers on slender arms. **Colour** shades of brown with patterns. **Habitat** as for *Octopus vulgaris*.

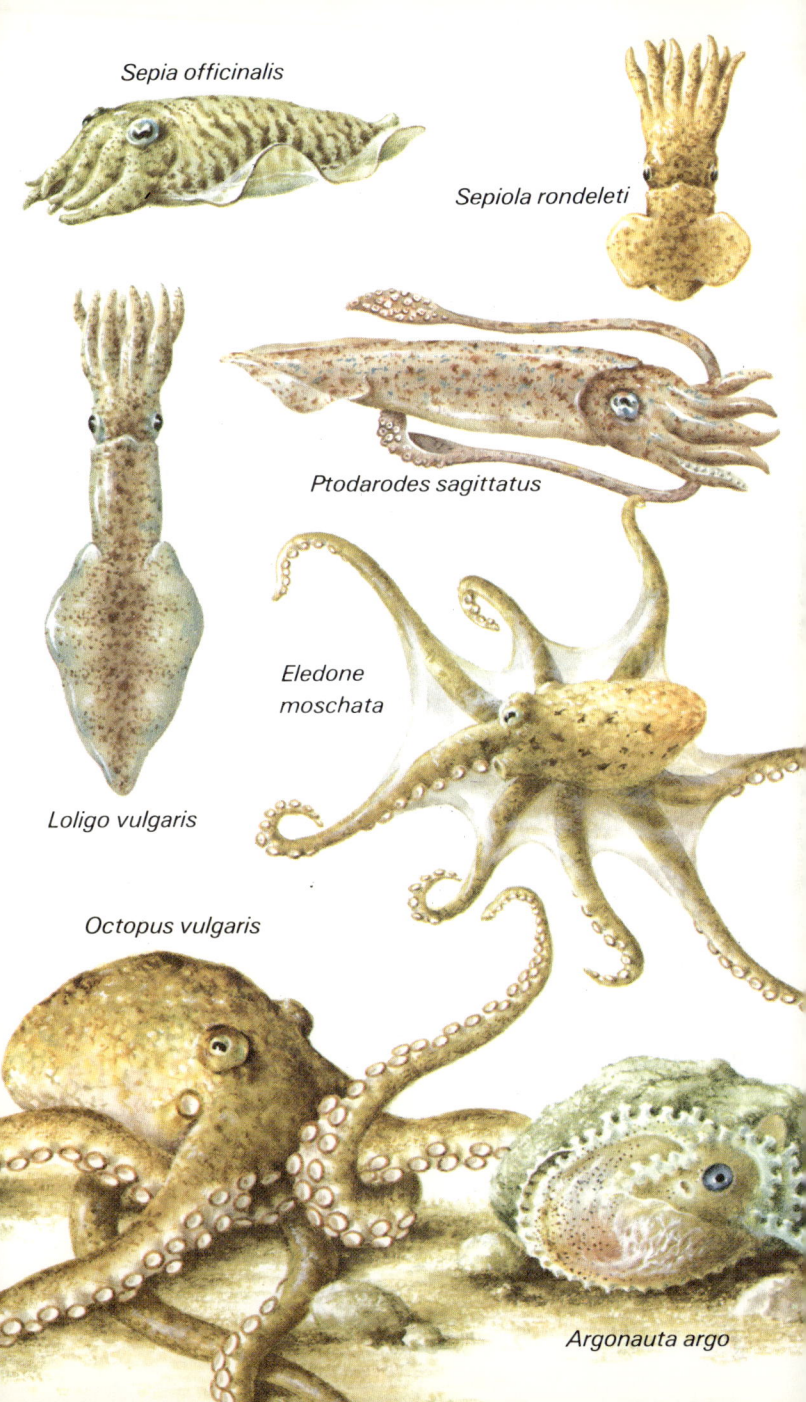

Sepia officinalis

Sepiola rondeleti

Ptodarodes sagittatus

Eledone moschata

Loligo vulgaris

Octopus vulgaris

Argonauta argo

Phylum Arthropoda (here treated as a single phylum)

These are bilaterally symmetrical, segmented animals whose bodies are composed of three cell layers; the reduced body cavity forms a true coelom, and a large blood space may be present. Paired appendages are borne on at least some segments, and one pair functions as jaws. The exoskeleton is composed of chitin and other materials, being hard and jointed. A well-developed nervous system is present, but true nephridia are lacking. Sexes are usually separate.

This very important phylum is the largest in the animal kingdom. Its members owe their success to the combined effects of their exoskeleton, and to their particular style of metabolic processes. These factors permit ease of locomotion (including flight in insects) and maintain the correct balance of water and salts in the body even in quite hostile environments. The phylum is divided into a number of classes which together have invaded almost every type of environment on earth. Some of these classes (for instance the insects and the pseudoscorpions), are almost wholly terrestrial and so lie largely outside the scope of this book.

The major class of marine arthropods is the Crustacea, and it is very well represented in most marine habitats. A much smaller, but nevertheless interesting group, is the class Pycnogonida (sea-spiders). Pycnogonids may be closely related to the true land spiders. Members of the class Crustacea have a body divided into three parts – head, thorax and abdomen. The exoskeleton is generally supported by calcium salts and is often heavy. The head bears two pairs of antennae (one pair on each of the second and third segments), and mandibles (jaws) on the fourth segment. Although the group is almost entirely aquatic, there are a few terrestrial exceptions, for instance, the familiar woodlouse. The Crustacea shows a diverse

Fig. 32 Planktonic crustacean larvae

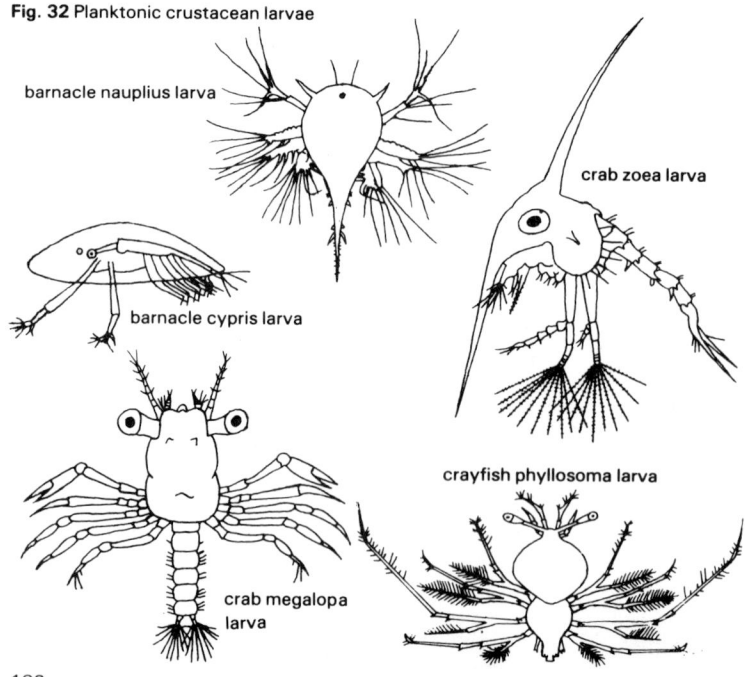

barnacle nauplius larva

crab zoea larva

barnacle cypris larva

crab megalopa larva

crayfish phyllosoma larva

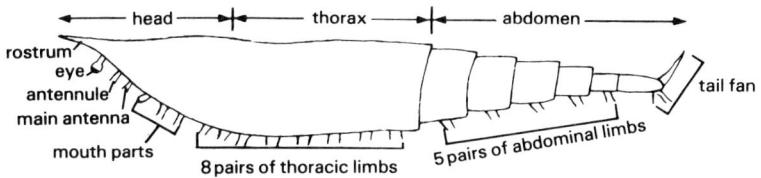

Fig. 33 Basic plan of the malacostracan body

range of body forms and life styles, and is divided up into several subclasses and a number of orders. Some of these comprise microscopic planktonic organisms, others are found solely in fresh water. Representatives of two subclasses (Cirripedia and Malacostraca) are commonly met with on the shore and in the shallow sea.

The basic crustacean body plan has been modified in a number of ways. In the Cirripedia (barnacles and their allies) the adults do not closely resemble other crustaceans, and for a long time they were classified elsewhere. Their larval development follows the typical crustacean form, however, and a small, free-swimming organism passes through two typically crustacean stages (nauplius and cypris) before leaving the plankton to settle and metamorphose on suitable rock or shell where it grows into an adult barnacle. Fig. 32 shows a number of different types of crustacean larvae which may be taken from the plankton at particular times of the year. Whilst they are clearly crustaceans, these larvae do not closely resemble the adult form into which they will grow. Crustaceans are found in almost all types of marine environment. The barnacles have evolved to lead a sessile life as adults, with their bodies protected by a number of massive plates of calcium carbonate which form the 'shell'. The type of shell base which attaches them to the rocks may help with identification; it may be calcareous or membranous, and can be seen if a specimen is removed from the rock. Effectively they are attached to the rocks by their head ends, and their thoracic appendages have become modified as filter-feeding organs. These can be projected outside the 'shell' and swept backwards and forwards through the water to collect any appropriate food particles.

The Malacostraca is a large and important subclass. It includes a great variety of organisms whose basic body plan is indicated in fig. 33. Many malacostracan appendages are biramous, i.e. they branch in two. Where this is the case one branch of the limb may have developed to carry out one function – for instance walking – and the other a different function – for instance respiration. This particular situation occurs in the thoracic legs of many malacostracans, where the shorter branch of the limb is modified to form a gill.

Members of the Malacostraca are generally free living and actively seek out their food using their well-developed eyes and chemoreceptors. This is especially true of many members of the order Decapoda such as crabs and lobsters, which often lead a scavenging or predatory life on the seabed. (They are arranged as a suborder Reptantia = creepers). The prawns and shrimps also belong to this order, but their way of life is different in that they can swim as well as walk. Indeed, some members of this group (arranged as a suborder Natantia = swimmers) live entirely by swimming in the waters of the ocean, and never touch the seabed.

A swimming life is generally the rule for the malacostracan larvae of all groups. After fertilization, the female may carry the developing embryos attached to the outside of the body in a manner that allows her to keep them well aerated. When the larvae emerge from the egg cases they find their way to the surface waters and become members of the plankton where they feed and grow. During this period they may change their form once or more before they are finally ready for metamorphosis, and the juveniles develop to begin life as adults on the seabed or elsewhere.

Many references deal with specific groups of crustaceans, and these are given as appropriate. See also Demetropoulos, A. and Neocleous, D. 1969.

Class Crustacea
Subclass Cirripedia Barnacles

Crustaceans which as adults have a calcified exoskeleton composed of several chalky plates, forming a shell. The thoracic appendages are used for straining suspended particles of food from the sea-water and are usually grouped together to form a basket-like feeding structure. Darwin, C. 1851–1854 gives more details. Mediterranean recorded species=about 18.

Lepas anatifera Linnaeus **Goose Barnacle** Shell about 50mm long with smooth translucent, nearly white plates which may have a bluish-grey tinge. **Stalk** 100–200mm long, somewhat retractile and with a brown-grey skin. **Habitat** pelagic; normally attached to boats and driftwood. **Similar species** *L. anserifera*, in which the larger side plates have a number of lines radiating from the junction with the stalk; also *L. hillii* and *L. pectinata*.

Scalpellum scalpellum (Linnaeus) Shell about 20mm long with 14 small grey-white plates. **Stalk** covered with small hairs. **Habitat** attached to hydroids, bryozoans and worm tubes from 30–100m. **Similar species** none.

Verruca stroemia (O.F. Müller) Sessile. Shell asymmetrical, flattened, up to 5mm in diameter with 4 unequal ribbed plates which may be grey, white or brown; base membranous if detached. **Habitat** under stones and attached to shells, from the shore down to 70m. **Similar species** none.

Chthamalus montagui Southward Sessile. Shell slightly asymmetrical, up to 12mm in diameter with 6 grooved plates arranged as a cone; lateral plates overlap the terminal plates so that these appear narrow (in order specimens the fusion of plates may make their delineation difficult); kite-shaped opening encloses 4 opercular plates, the horizontal sutures of which are less than one-third of the way down the vertical suture as shown in the fig; base membranous if detached. **Habitat** on rocks on the shore. **Similar species** *C. stellatus* and *Euraphia depressa*.

Chthamalus stellatus (Poli) Sessile; very similar to *C. montagui* (above) and occurring together with it but a little lower on rocky shores, and perhaps above in extremely wave-beaten places. The two species overlap and are not always easy to distinguish. See Crisp D.J., Southward A.J. and Southward E.C. 1981. Fig. 34 gives the salient differences in the appearances of the opening, which in *C. stellatus* is oval and the horizontal sutures are more than one-third of the way down the vertical one. The base is membranous if detached.

Euraphia depressa (Poli) *(=Chthamalus depressum)* Sessile. Shell up to 15mm in diameter but rather flat; less conical than *C. montagui* or *C. stellatus*. The opening is large, ovoid or roughly hexagonal; the 'horizontal' sutures form an angle of 50–75° with the vertical suture; base membranous if detached; rim of inside moveable flaps brownish. **Habitat** on the shore in crevices and on rocks; kept wet by wave action.

Balanus perforatus Bruguière Sessile. Shell fairly symmetrical, up to 30mm in diameter, tall and conical with 6 grey-purple-brown smooth or lined plates, often separated at the apex leaving a jagged lip; opening of shell off-centre; base calcareous if detached. **Habitat** on the shore.

Fig. 34
Form of shell plates

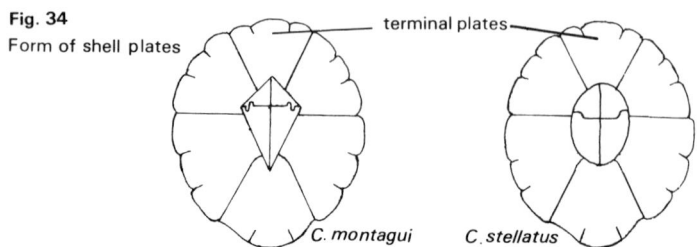

terminal plates

C. montagui   *C. stellatus*

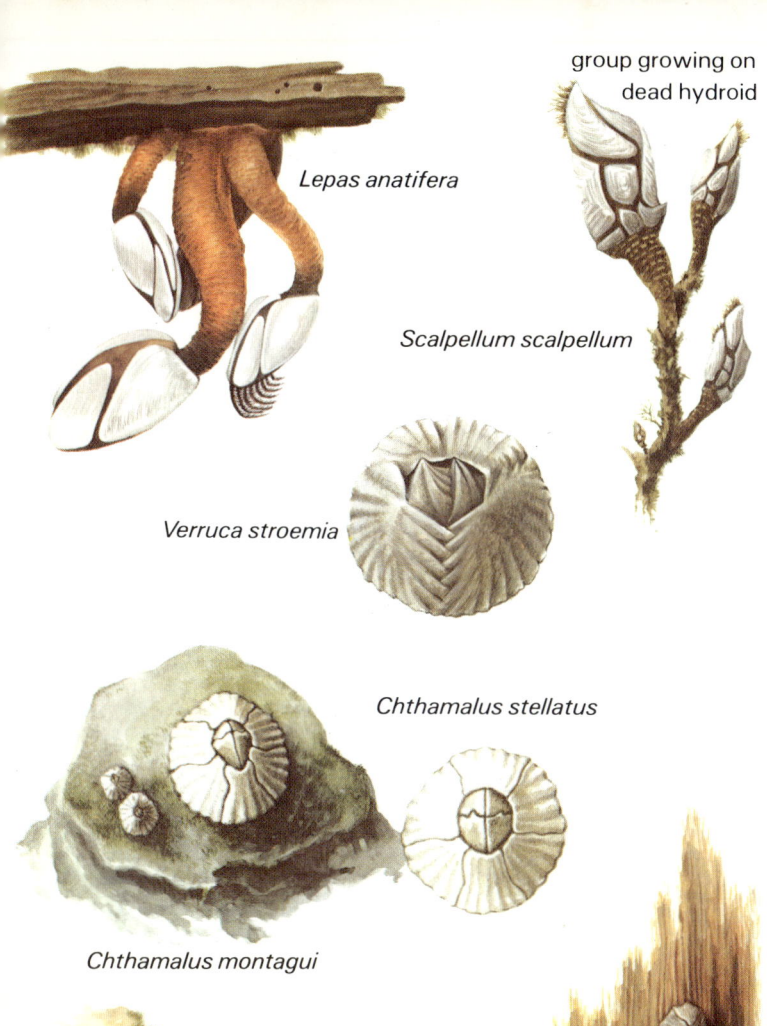

group growing on
dead hydroid

Lepas anatifera

Scalpellum scalpellum

Verruca stroemia

Chthamalus stellatus

Chthamalus montagui

Euraphia depressa

Balanus perforatus

Balanus improvisus Darwin Sessile. **Shell** up to 15mm in diameter, symmetrical; diamond-shaped opening; rim of inside moveable flaps white with purple bands; base calcareous if detached. **Habitat** in brackish water such as estuaries and lagoons.

Balanus eburneus Gould Sessile. **Shell** strongly conical and up to 30mm in diameter; plates smooth and cream or ivory-coloured with finely toothed summits; rim of inside moveable flaps banded and speckled brown on a cream or white background; base calcareous if detached. **Habitat** on ships and on the shore on rocks and among mussels and calcareous algae, in brackish water such as estuaries and harbours. **Similar species** Balanus amphitrite Darwin (not illustrated) in which the shell plates are thin, smooth and whitish, carrying patterns of grey, brown, pink or purple stripes; in harbours and estuaries especially on ships, buoys, piers and pilings as well as rocks.

Acasta spongites (Poli) Sessile. **Shell** up to 12mm in diameter; easy to identify as it is always partly or totally embedded in a sponge, and because the base is not flat but convex. **Habitat** associated with a sponge (usually Dysidia fragilis); from the shore down.

Boscia anglicum (Sowerby) (=**Pyrgoma anglicum**) (Illustrated on page 80) Sessile. **Shell** up to 3mm in diameter; easy to identify because always attached to a coral; colour grey-green-brown. **Habitat** as for the coral (usually Caryophyllia but occasionally Cladocora, Dendrophyllia or Balanophyllia – see page 80).

Conchoderma virgatum (Spengler) **Shell** 5 small skeletal plates on thick skin; overall height up to 50mm; purplish-red to dark red-brown with stripes, leathery. **Stalk** widening from attachment point to main body. **Habitat** on various mobile objects including the bottoms of boats and ships. **Similar species** C. auritum (Linnaeus) (not illustrated) which has ear-like tubes on the extremity of the body.

Chelonibia testudinaria (Linnaeus) Sessile. **Shell** flat with oval outline, up to 25mm in diameter with 6 stout smooth whitish parts; only the tips of the plates reach the opening and there are pieces of exposed skin between. **Habitat** attached to turtle shells. **Similar species** C. patula (Ranzani) (not illustrated) which occurs on the shells of crustaceans.

Sacculina carcini Thompson Parasitic and quite unlike the other barnacles: appears as a conspicuous lump covered with pale yellow-brown skin attached under the abdomen of the crab Carcinus, and sometimes other crab species, in such a way as to prevent the abdomen from folding closely under the carapace; readily distinguished from the crab's own eggs which may be brooded in this position by the female crab: the egg-mass is granular in texture while the barnacle is smooth. **Habitat** as for the host.

Peltogaster paguri Rathke A parasite, like Sacculina carcini, but appearing as a smooth elongated growth on the side of the abdomen of hermit crabs; not visible unless the hermit crab is removed from its gastropod shell; found on Pagurus bernhardus and P. cuanensis (see page 210). **Habitat** as for the host. **Similar species** few inhabiting other species of hermit crabs.

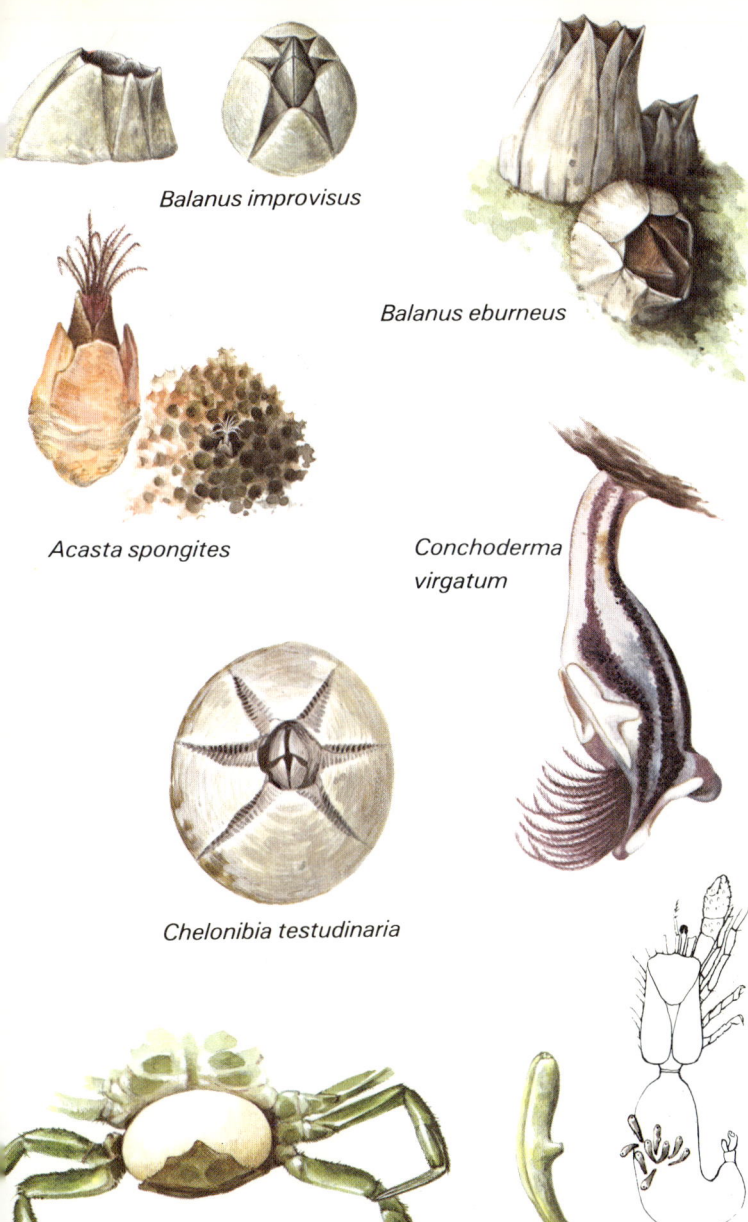

Balanus improvisus

Balanus eburneus

Acasta spongites

Conchoderma virgatum

Chelonibia testudinaria

Sacculina carcini

Peltogaster paguri

Subclass Malacostraca

Crustaceans with compound eyes, 2 pairs of antennae and with the head and thorax fused and usually covered by a carapace. The thorax has 8 segments, each bearing appendages (the first 3 pairs assist with feeding and the remaining 5 pairs usually develop as legs). The abdomen usually comprises 6 segments.

Order Nebaliacea

Small malacostracans with stalked eyes separated by a rostral spine. The carapace enfolds all the thoracic and 4 pairs of the abdominal appendages. Mediterranean recorded species = 1.

Nebalia bipes (Fabricius) Length up to 10mm; uppermost pair of antennae short, lowest longer. Habitat on shore and water to 30m, often in detritus.

Order Stomatopoda

Moderately-sized malacostracans possessing a short shield-shaped carapace. The last joint of the 2nd pair of thoracic legs forms a spiny crusher, and the number of spines on the inside of this is a key character to identification. Lewinsohn, C. and Manning, R. B. 1980 give more details of Mediterranean forms.

Meiosquilla desmaresti (Risso) Length up to 120mm; last join to 2nd thoracis legs bears 4 consecutive spines and terminates in a sharp point. Habitat burrowing in sand and mud, from low water down to 50m or more. Similar species several.

Squilla mantis (Linnaeus) Length up to 250mm; as for *Meiosquilla desmaresti* but 2nd thoracic legs bear 5 consecutive spines and terminate in a sharp point. Similar species several. N.B. tail has 2 dark spots.

Platysquilla eusebia Risso (Not illustrated) Length up to 60mm; as for *Meiosquilla desmaresti* but 2nd thoracic legs bear about 9 fine consecutive spines and terminate in a sharp point. Similar species several.

Order Cumacea

Small malacostracans with bodies compressed sideways. The distinctive carapace covers only part of the thorax. Mediterranean recorded species=about 10.

Pseudocuma longicornis (Bate) Length up to 5mm; anterior part of the carapace appears ridged. Habitat burrowing in sand or mud on the shore and in shallow water; sometimes swarming in surface waters. Similar species many.

Order Tanaidacea

Small malacostracans with bodies compressed dorso-ventrally. The reduced carapace covers only the first two segments of the thorax; 2nd pair of thoracic appendages bears small pincers and these are followed by 6 pairs of similar legs. Abdominal appendages bear swimmerets, but a tail fan is lacking. Mediterranean recorded species = about 5.

Apseudes latreillei (Milne-Edwards) Length up to 6mm; antennae and terminal appendages branched. Habitat under stones or in mud among seaweeds at low water and down to about 40m. Similar species several.

Order Euphausiacea

Small pelagic malacostracans with conspicuous eyes; the carapace bears a rostrum covering the thorax but not shielding the thoracic appendages; the last 7 pairs of thoracic appendages are clearly branched near their points of origin; frequently they bear no pincers. Allen, J. A. 1967 and Riedl, R. give a fuller account of the euphausiaceans. Mediterranean recorded species=about 5.

Meganyctiphanes norvegica (M. Sars) Krill Length up to 15mm; carapace bears rostrum and is about one-third of the body length; lower margins of carapace not toothed or horned as in other species; characteristic outgrowths on either side of head behind eyes. Habitat pelagic. Similar species several.

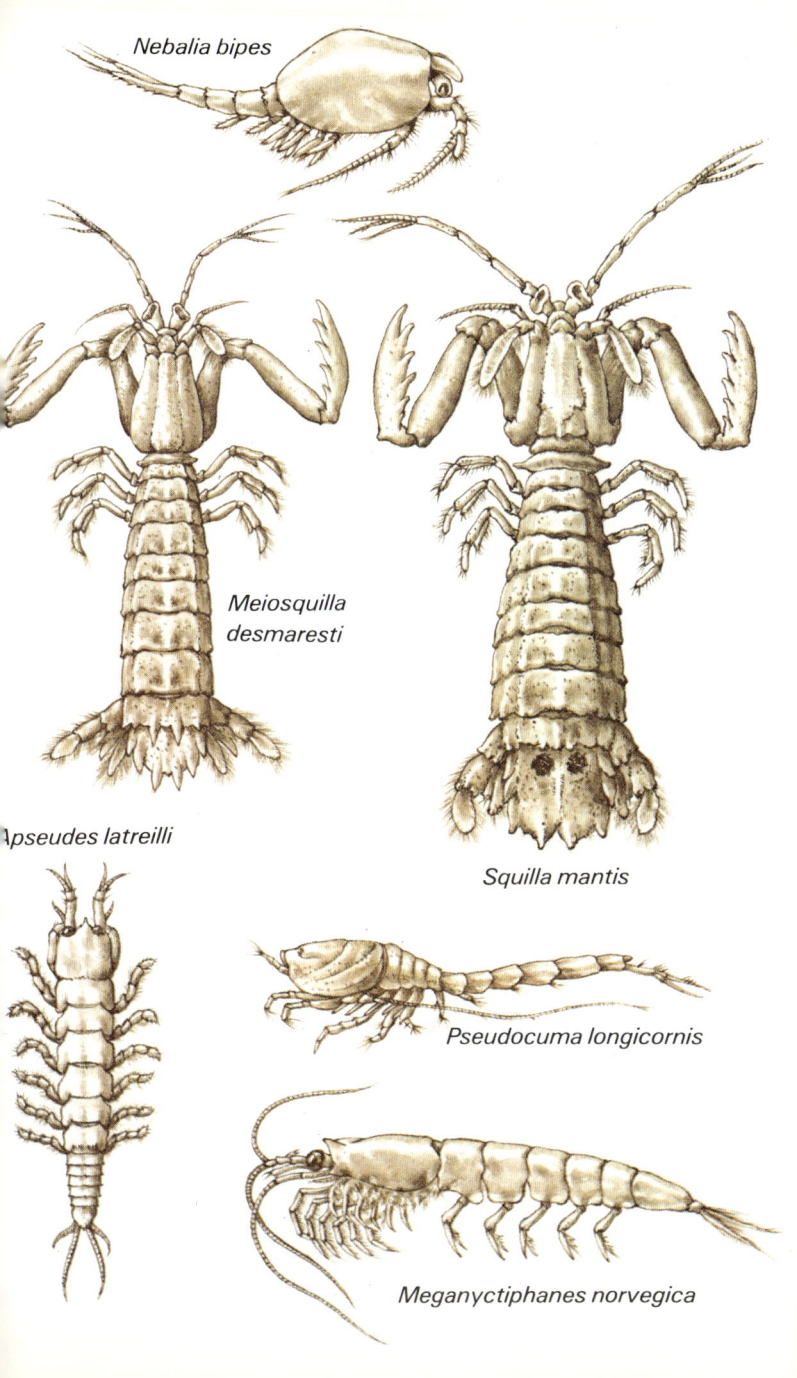

Nebalia bipes

Meiosquilla desmaresti

Apseudes latreilli

Squilla mantis

Pseudocuma longicornis

Meganyctiphanes norvegica

Order Mysidacea Opossum shrimps

Small, swimming malacostracans which seldom reach more than 30mm in length. They possess a thin carapace and a long abdomen of 6 segments. The branched thoracic appendages lack pincers and the abdominal appendages are short. A tail fan is borne on the terminal segment, the precise form of which is important for identification: this is shown from a dorsal aspect in each case. The eyes are conspicuous and the animals are often transparent. They may remain suspended in the water by their swimming movements. The females often carry a brood pouch on their under surfaces.

Further details of many mysids occurring in the Mediterranean are given by Tattersall, W.M. and Tattersall, O.F. 1951. Mediterranean recorded species not less than 32.

Lophogaster typicus M. Sars Length up to 22mm; carapace relatively large, covering the thorax at the sides but with a cleft in the dorsal aspect so that the last two thoracic segments are exposed to view from above; body surface covered with minute tubercles (visible with a lens), especially in the midline of the ventral region. Habitat from shallow water down to 200m or more, often on muddy bottoms. Similar species probably few.

Siriella clausii G. O. Sars Length up to 11mm; carapace very short, not completely covering the last three thoracic segments and equal to slightly more than half the length of the abdomen. Habitat in shallow water, in rock pools and among weeds, usually over gravel and sand, down to about 35m. Similar species few.

Anchialina agilis (G. O. Sars) Length up to 9mm; carapace relatively large, covering the whole thoracic region and equal to more than half the length of the abdomen; outer body surface covered with minute bristles or spines, especially noticeable on the terminal segment. Habitat on the seabed by day, from shallow water down to about 60m or deeper, migrating to the surface for a short period at night. Similar species none.

Leptomysis gracilis (G. O. Sars) Length up to 13mm; carapace short, not covering the whole thoracic region and very little wider than the abdomen; entire body surface covered by minute scales (visible with a strong hand lens). Colour transparent and nearly colourless; abdominal region may have a slight yellow-red tinge. Habitat in pools on the shore and on the seabed in shallow water; may be taken in plankton at certain times of the year. Similar species L. mediterranea G.O. Sars (not illustrated): length up to 10mm but the fine scales are absent from the body surface; sometimes occurs in great numbers from the shore down to 100m; there are also a few other similar species.

Mysidopsis gibbosa G. O. Sars Length up to 7mm; carapace short, not completely covering the 5th and 6th thoracic segments which are especially visible from above; when viewed sideways-on there are 2 conspicuous tubercles on the mid-dorsal line of the carapace. Colour dark grey to opaque white, sometimes variable; additional pigment as bands of minute spots. Habitat in shallow water on the bottom, usually from 1–20m but sometimes deeper. Similar species none.

Paramysis helleri (G. O. Sars) Length up to 11mm; carapace not entirely covering the last thoracic segment and not quite equal to half the length of the abdomen. Colour transparent with branching yellow, brown and pink markings. Habitat found in shallow water, usually close to the coast and sometimes in estuaries. Similar species P. arenosa (G.O. Sars) (not illustrated) which is up to 7mm long and transparent with branching pigments; it appears relatively thicker than P. helleri.

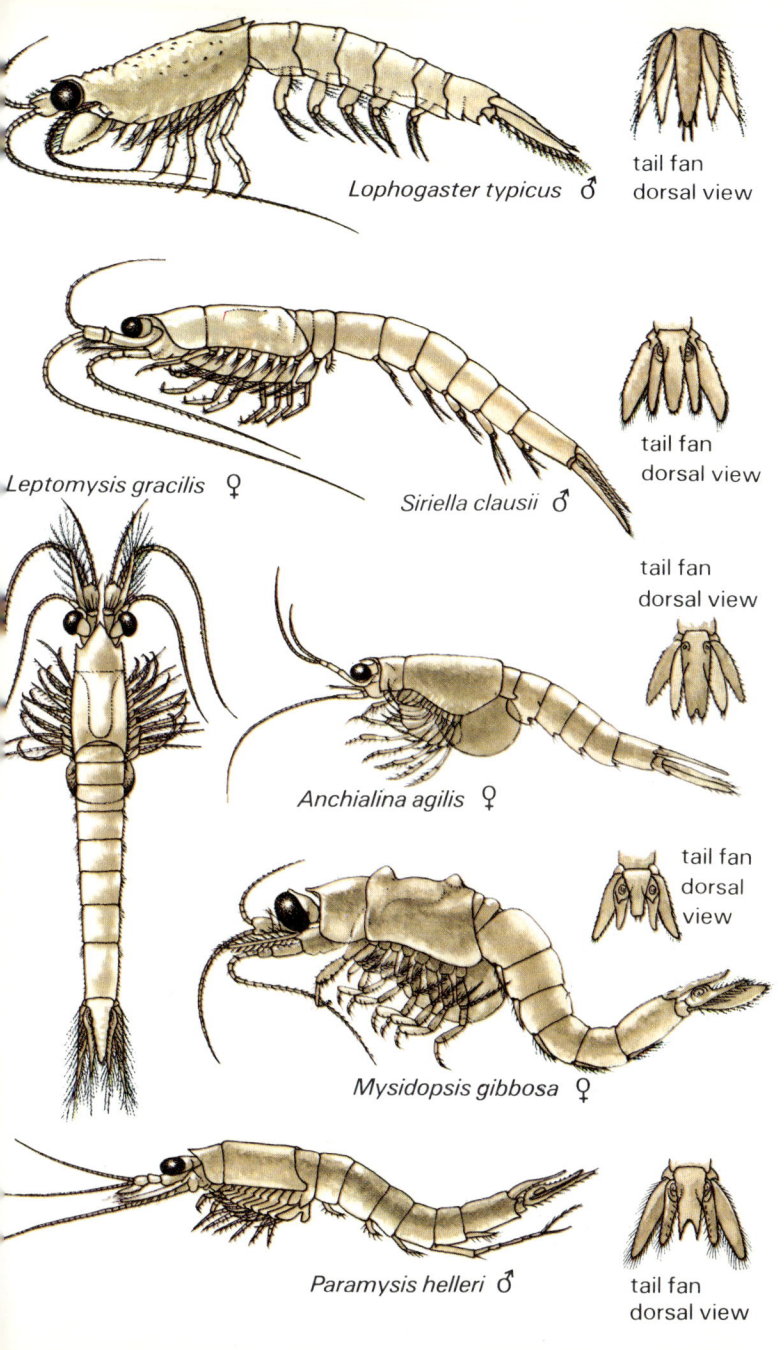

Lophogaster typicus ♂

tail fan
dorsal view

Leptomysis gracilis ♀

Siriella clausii ♂

tail fan
dorsal view

tail fan
dorsal view

Anchialina agilis ♀

tail fan
dorsal view

Mysidopsis gibbosa ♀

Paramysis helleri ♂

tail fan
dorsal view

Order Isopoda

Small flattened malacostracans resembling woodlice. Head usually bears small inner antennae and longer outer antennae. There are 7 pairs of similar unbranched thoracic limbs. The 1st pair may assist with feeding and the rest are used for crawling. The abdomen bears 5 pairs of feathery branched respiratory appendages and 1 terminal branched pair which usually lies either side of the tail-piece or telson. These assist with swimming. Many species are free-living but some live as parasites of other crustaceans or fish. Not less than 28 species have been recorded from the Mediterranean. See Chardy, P. 1970 and Naylor, E. 1972.

Gnathia maxillaris (Montagu) Length up to 6mm; inner antennae longer than outer; sexes very different. Male head has large hooked jaws between antennae; viewed from above head has a shallow recess between jaws with a slight forward process in midline; body widest between eyes; head and thorax wide, abdomen minute and tapering. Female head small, lacks jaws; thorax bulbous, widest in mid-thorax; abdomen small. Habitat in crevices and among weeds, on shore and in shallow water. Similar species *G. vorax* (Lucas) (not illustrated): up to 8mm; and *G. oxyuraea* (Lilljeborg) up to 6mm.

Anthura gracilis (Montagu) Length male up to 4mm, female up to 11mm. Head outer antennae long and feathery in male, short and inconspicuous in female; eyes large and conspicuous. Body long thin thorax, 1st leg pincer-like, abdomen short and fan-like. Habitat in crevices and among weeds, on the shore and in shallow water. Similar species *Cyanthura carinata* (Kröyer) (not illustrated): up to 14mm with small eyes.

Limnoria tripunctata (Menzies) Length up to 4mm. Head both pairs of antennae short. Body thorax slightly longer and wider than abdomen, which has almost parallel edges; 6th pair of abdominal appendages extending just beyond edge of telson; rear part of telson carries 3 small tubercles (visible with a lens). Habitat boring in wood. Similar species *L. quadripunctata* Holthuis (not illustrated) which has 4 small tubercles on the rear part of the telson.

Eurydice affinis Hansen Length up to 6mm. Head inner antennae short, outer nearly as long as head and thorax together. Body thorax oval, abdomen not as wide and about two-thirds as long; 6th pair of abdominal appendages short and not extending much beyond edge of telson; rear of telson convex (visible with a lens); small black spots on dorsal surface only. Habitat in sand on shore and in shallow water. Similar species *E. spinigera* Hansen (not illustrated): up to 9mm.

Dynamene bidentata (Adams) Sexes different. Male: up to 7mm. Head rounded, convex, with conspicuous eyes. Body 6th thoracic segment has one pair of small lateral processes which project back over 7th segment and 1st two abdominal segments; telson rough (visible with a lens); 6th pair of abdominal appendages with conspicuous outward-pointing terminal branch. Female up to 6mm. Body 6th thoracic segment lacks backward projections. Habitat rock crevices and empty barnacle shells; juveniles among algae on shore. Similar species none, but females may be confused with *Sphaeroma* species.

Anilocra physodes Linnaeus Length up to 24mm. Head 'segment' extended backward in midline; inner antennae fractionally longer than outer. Body thorax ovoid, with legs modified for gripping on to hosts; abdomen tapering; telson squarish; posterior edge rounded; 6th abdominal appendage conspicuous. Habitat parasites of fish, especially wrasses. Similar species *Nerocilla bivittata* Risso: up to 22mm; squarish telson with a blunt point on its rear edge; also a fish parasite.

Sphaeroma serratum (Fabricius) Length up to 12mm. Head inner antennae about half as long as outer antennae which are one-third of the body length. Body oval; 6th pair of abdominal appendages terminate in conspicuous oval flat joints with serrated outer edges; upper surface of telson smooth (visible with a lens). Habitat in crevices and under stones on the shore. Similar species *S. hookeri* Leach (not illustrated): up to 10.5mm long; and *Cymodoce truncata* (Leach) (not illustrated): up to 14mm long.

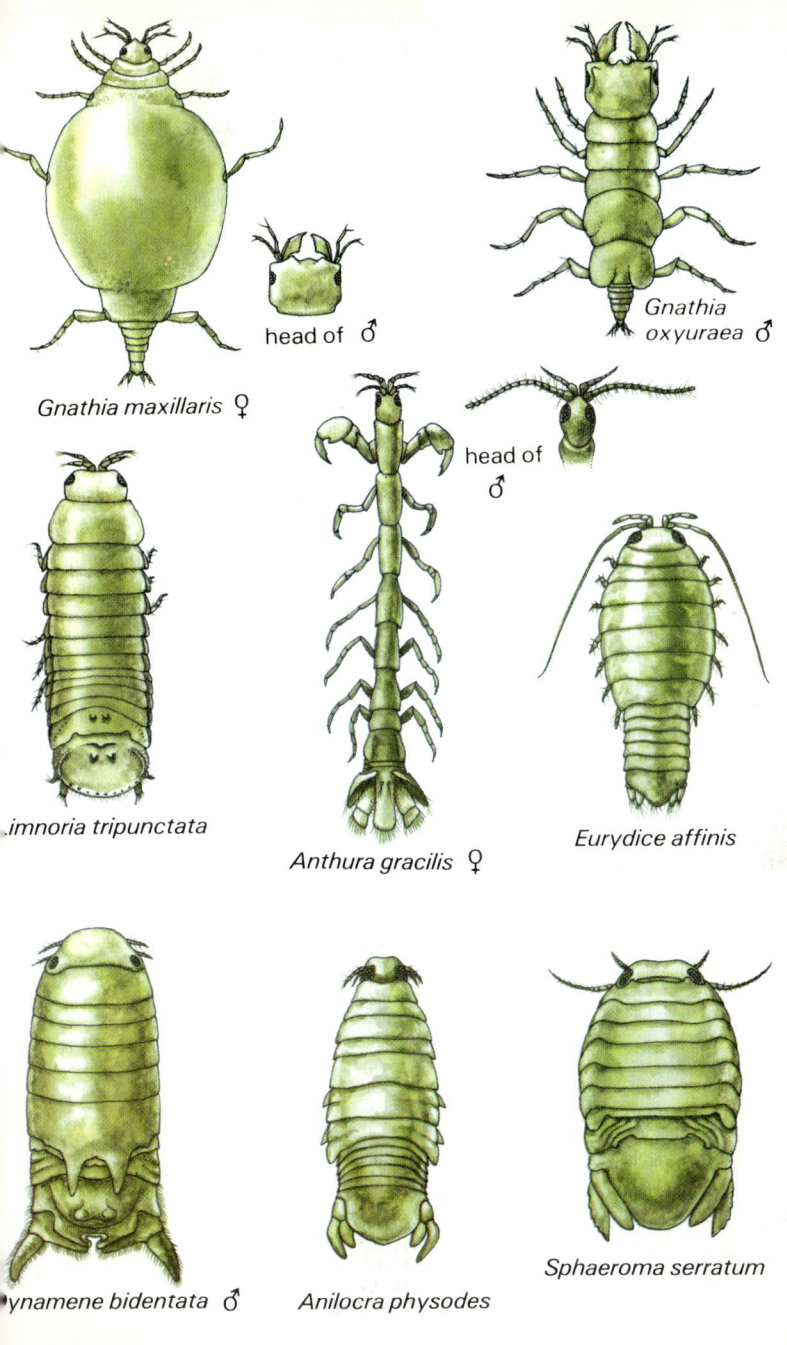

head of ♂

Gnathia maxillaris ♀

Gnathia
oxyuraea ♂

head of ♂

imnoria tripunctata

Anthura gracilis ♀

Eurydice affinis

ynamene bidentata ♂

Anilocra physodes

Sphaeroma serratum

Idotea baltica (Pallas) Length male up to 30mm, female up to 18mm. **Head** inner antennae reach about as far as start of 3rd joint of outer antennae; outer antennae quarter to one-third body length. **Body** oblong thorax with slightly convex sides; abdomen narrower than thorax and terminating in a long telson with nearly straight edges; upper surface keeled, rear edge of telson drawn out into 3 teeth. **Habitat** on weeds on the shore and in shallow water. **Similar species** *I. metallica* Bosc (not illustrated): length up to 13mm; also (rarely) 2 other species.

Idotea linearis (Linnaeus) Length male 40mm, female less. **Head** inner antennae reach just beyond start of 2nd joint of outer antennae, which themselves reach about half-way along the slender body; rear edge of telson concave. **Habitat** in shallow water near shore. **Similar species** none.

Jaera nordmanni (Rathke) Length male up to 4.5mm, female up to 3.5mm. **Head** with small eyes; inner antennae very short, outer antennae about half body length. **Body** ovoid and flat with edges very hairy; deep notches apparent between the rear thoracic segments; abdomen rounded, with terminal appendages set in a shallow notch. **Habitat** usually under stones on shore, especially in freshwater streams. **Similar species** *J. hopeana* da Costa in which the terminal appendages of the abdomen are not set in a notch; may live as an ectocommensal on *Sphaeroma serratum* (page 192).

Ligia italica Fabricius Length up to 12mm. **Head** outer antennae reach about two-thirds of body length, inner antennae minute. **Body** flat and oval; 6th abdominal appendages widely spread, long and trailing behind the telson; outer terminal branches not as long as inner ones; base-piece fairly slender. **Habitat** on rocks above the water level, often in crevices by day. **Similar species** *L. oceanica* (Linneaus) which has 6th abdominal appendages not widely spread.

Tylos sardous (Arcangeli) Length up to 15mm. **Head** outer antennae a little more than a quarter of body length; head itself small and almost surrounded by 1st thoracic segment. **Body** flat and ovoid; abdomen small, its appendages scarcely visible when viewed from above. **Habitat** usually above the waterline near shells, stones and coarse sand. **Similar species** *T. europaeus* (Arcangeli) (not illustrated).

Order Amphipoda

Small sideways flattened malacostracans lacking a carapace. Head bears variously developed upper and lower antennae; thorax bears 7 pairs of unbranched limbs; the 1st two pairs terminate in gripping claws, the 3rd and 4th pair lack gripping claws and are long and slender; the 5th, 6th and 7th are similar but stouter. Thoracic limbs 3–7 serve various functions including climbing in weeds, burrowing etc. The abdomen carries 6 pairs of branching limbs of which the first 3 pairs have a stout base-piece supporting 2 delicate hairy filaments. These appendages fan the gills and assist in swimming. The remaining 3 pairs of abdominal limbs each have a stout base-piece supporting 2 stubby end pieces which grip the substrate during hopping. The tailpiece or telson is sometimes reduced. Mediterranean recorded species = over 57. See Chevreux, E. and Fage, L. 1925, Chardy, P. 1970 and Lincoln, R. J. 1979.

Talitrus saltator (Montagu) **Sand Hopper** Length up to 16mm. **Head** upper antennae not branched and usually reaching a little beyond the joint of the 1st and 2nd segments of the lower antennae, which themselves terminate in many small rough, toothed, hairy joints (visible with a lens). **Body** 2nd thoracic limb ends in a claw-like joint. **Habitat** on the shore, often associated with rotting algae. **Similar species** no close allies but *Orchestia* and *Gammarus* are somewhat alike.

Orchestia gammarella (Pallas) **Sand Hopper** Length up to 20mm. **Head** upper antennae usually reach as far as the 2nd joint of the lower antennae; lower antennae terminate in a number of small joints which appear smooth under a lens. **Body** 1st thoracic leg terminates in a small claw; 2nd thoracic leg terminates in a large pincer, the subterminal joint being greatly expanded; 1st 3 pairs of abdominal appendages bear delicate hairy terminal filaments. **Habitat** among stones and algae on the shore. **Similar species** 2 others.

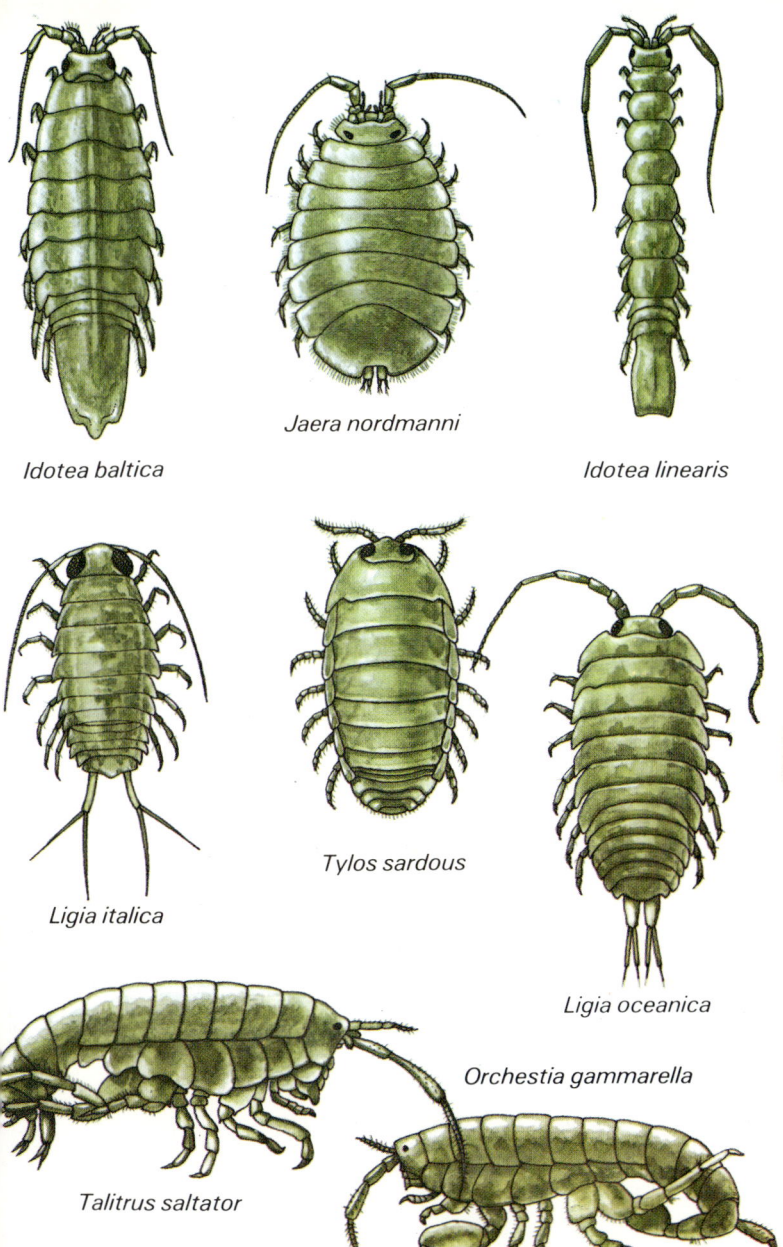

Jaera nordmanni

Idotea baltica

Idotea linearis

Ligia italica

Tylos sardous

Ligia oceanica

Orchestia gammarella

Talitrus saltator

Hyale schmidti (Heller)Sexes different. **Length** up to 6mm male, 5mm female. **Head** upper antennae unbranched and with 15 joints to the terminal part, and reaching more than half the length of the antennae: lower antennae with 31 joints in the terminal part. **Body** in the male has 1st and 2nd thoracic limbs terminating in a small claw, 2nd limb 2x length of 1st: in the female both limbs similar size and the female claws are bent back against the sub-terminal joint: 6th abdominal limb of both sexes bears 6 spiny hairs on its terminal joint and 3 on its subterminal joint. **Colour** white-brownish-orange, eyes pink. **Habitat** on shore and in shallow water. Similar species about 6.

Lysianassa ceratina (Walker) **Length** up to 10mm. **Head** upper antennae mounted on a thick base; lower antennae not much longer than upper in the female but nearly twice as long in the male. **Body** 1st thoracic limb with very slender short pointed tip, 2nd slightly broader, blunt, flat. **Habitat** among algae in shallow water. Similar species *L. plumosa* Boeck (not illustrated) which has a slightly concave end to the subterminal joint.

Leucothoë spinicarpa (Abildgaard) **Length** up to 18mm. **Head** lower antennae slightly shorter than upper antennae. **Body** first thoracic limb formed into pincer using the terminal (6th) joint closing against a fixed projection of the 4th joint; terminal joint itself about one-third to half length of 5th joint; 2nd thoracic limb much larger with terminal (6th) joint closing to make pincer against side of the expanded subterminal (5th) joint; telson 3 times as long as it is wide. **Habitat** from shore down to 600m among algae, sponges and sea-squirts. Similar species *L. incisa* Robertson and *L. richiardii* Lessona: see Lincoln, R.J. 1979.

Dexamine spiniventris (da Costa)Sexes different. **Length** up to 7mm male, 5mm female. **Head** bears 2 pairs of antennae in male, upper antennae slightly shorter than the lower, but *vice versa* in females; in male lower antenna is same length as the body; in female upper antenna is two-thirds body length; large pink eyes. **Body** is robust, 1st and 2nd thoracic limbs terminate in minute claws: 1st thoracic limb smaller than the 2nd; dorsal surface of 3rd abdominal segment has 3 teeth. **Colour** translucent white, sometimes with brown and red marks. **Habitat** among algae on the sea bed. Similar species few.

Tritaeta gibbosa (Bate) **Length** up to 6mm. **Head** with roughly equal upper and lower antennae in the female, upper a little shorter than lower in the male. **Body** 1st and 2nd thoracic limbs with terminal joints acting as fine pincers against subterminal joints. **Habitat** on the shore and down to 150m, often associated with sponges and sea-squirts.

Gammarus locusta (Linnaeus) **Length** male up to 30mm, female up to 20mm. **Head** upper antennae quite long and bearing a small branch which may only be visible under water; lower antennae shorter. **Body** strongly compressed sideways and curved; posterior edges of last 3 abdominal segments have small spinelets, as does the telson; abdomen carries 3 longer swimming appendages and 3 shorter jumping appendages; inner branch of the last pair of appendages more than half the length of the outer branch. **Habitat** under stones on the shore. Similar species *G. insensibilis* Stock and *G. crinicornis* Stock (not illustrated): see Lincoln, R.J. 1979.

Elasmopus rapax da Costa **Length** up to 10mm. **Head** with upper antennae nearly half as long as body; lower antennae shorter. **Body** 1st thoracic limb terminates in a fine pincer-like claw, 2nd similar but much larger; both quite hairy on posterior edges; 5th, 6th and 7th thoracic appendages with stout bases; remaining parts of limbs relatively slender. **Habitat** from the shore down to 100m, often among algae. Similar species *Melita palmata* (Montagu): see Lincoln, R.J. 1979.

Maera inaequipes (da Costa) **Length** up to 8mm. **Head** with upper antennae bearing a side branch, lower antennae not so long. **Body** 2nd thoracic appendage bears extensively flattened paddle-like subterminal joint against which a fine claw-like terminal appendage closes to form a pincer; inside of this is finely serrated on the subterminal side in the female. **Habitat** in shallow water among sand and algae. Similar species few.

196

Lysianassa ceratina

Hyale schmidti

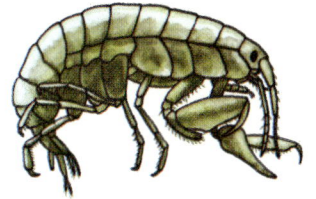

Leucothoë spinicarpa

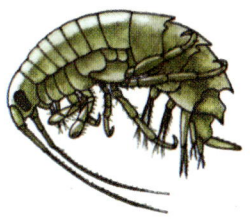

Dexamine spiniventris

Tritaeta gibbosa

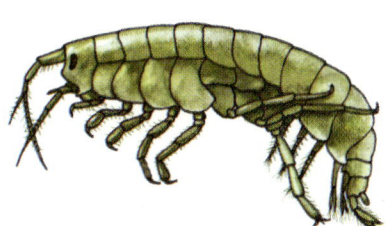

Gammarus locusta

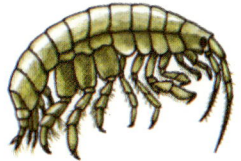

Elasmopus rapax

Maera inaequipes

Aora typica Kröyer Length up to 9mm. **Head** upper antennae bear a short side branch visible with a lens when animal is immersed; lower antennae shorter and more robust. **Body** (male) in 1st thoracic limb the 3rd joint is extended into a large pointed process forming a spine which reaches the subterminal joint; terminal joint claw-like; female lacks these features. Body of both sexes relatively shallow. **Habitat** shore down to 50m among algae and hydroids. **Similar species** few.

Lembos websteri Bate (Not illustrated) Length up to 6mm. **Head** upper antennae longer than lower antennae with a side branch visible with a lens when in water. **Body** 1st thoracic limb of male has flattish, broad, hairy 4th and 5th joints; the 5th is spined while the 6th is small and claw-like. The 2nd thoracic limb is similar but the 4th and 5th joints are not so expanded. In the female the 1st thoracic limb has less expanded and less hairy 4th and 5th joints, the 5th joint with a small spine; in the 2nd limb there is a large spine and recess on the 5th joint just before the articulation with the 6th claw-like joint; the 7th thoracic limb of both is very elongated. **Habitat** on shore and in shallow water. **Similar species** few.

Gammaropsis maculata (Johnston) (Not illustrated) Length up to 10mm. **Head** bears 2 pairs of roughly equal antennae. **Body** 1st and 2nd thoracic limbs have hairy, flattened and expanded 4th and 5th joints supporting claw-like 6th terminal joints. In the female the 5th joint of the 2nd limb has 2 teeth where it forms a pincer with the 6th joint; the male has 3 teeth here. **Habitat** from shore down to 250m.

Jassa falcata (Montagu) Length up to 8mm. **Head** upper antennae have a minute side branch (visible under water with a lens) and are about three-quarters the length of the thicker lower antennae. **Body** 1st thoracic leg terminates in a joint which is bent back as a pincer; 2nd thoracic leg similar but terminal joint pincer is larger. In the male the subterminal joint of this limb has a conspicuous side branch; the female lacks this. **Habitat** living in tubes among weeds and stones on shore or with floating weeds, or in deeper water on wrecks, pilings etc. **Similar species** J. ocia Bate (not illustrated).

Ericthonius brasiliensis (Dana) Length up to 10mm. **Head** shallow with upper and lower antennae of almost equal length. **Body** shallow; 1st thoracic limbs in both sexes terminate in a small claw working as a pincer against the subterminal joint; 2nd thoracic limb in female has 4th joint with side pincers, larger 5th joint and terminal claw; in male 4th joint is very large and has 2 teeth working against smaller 5th joint; terminal claw. **Habitat** in small tubes amongst algae and hydroids. **Similar species** few.

Chelura terebrans Philippi Length up to 6mm. **Head** upper antennae with a minute branch half-way along (may be seen with a hand lens under water); lower antennae flat and about twice as long with hairy joints. **Body** not flattened sideways; strange tail segment ending in a sharp point; of the last 3 pairs of abdominal appendages the middle one is very long, terminating in an oval plate (female) and a larger spine (male). **Habitat** in bore holes in wood. **Similar species** none.

Hyperia hydrocephala Vosseller Length up to 4mm. **Head** large with enormous eyes and minute antennae in female. **Body** quite deep and short, tapering strongly to tail; thoracic appendages all fairly simple. **Habitat** living free on seabed or associated with planktonic organisms. **Similar species** several.

Phtisica marina Slabber Length up to 15mm. **Head** upper antennae about twice as long as lower antennae. **Body** long and thin; abdomen reduced and vestige of tail present. **Habitat** associated with algae and other organisms in shallow water. **Similar species** Pseudopotella phasma (Montagu) (not illustrated), 2 conspicuous dorsal spines on the head and 2 on the 1st thoracic segment.

Caprella acanthifera Leach Length up to 9mm. **Head** upper antennae considerably longer than lower; both antennae of male about twice as long as respective parts of female. **Body** long and slender; 1st thoracic appendage pincered, long in male, short in female; female with broad pouch; thoracic segments bear dorsal spines (head not spiny); abdomen vestigial. **Habitat** on hydroids, bryozoa etc. **Similar species** C. aequilibra Say (not illustrated).

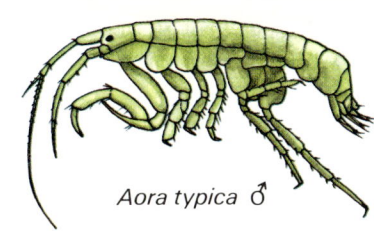

Aora typica ♂

Jassa falcata ♂

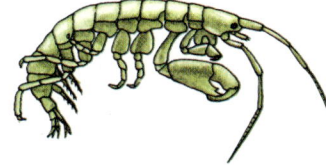

Ericthonius brasiliensis ♂

Chelura terebrans ♂

Hyperia hydrocephala ♀

Phtisica marina ♂

Caprella acanthifera ♀

Order Decapoda

Malacostracans with the head and thorax fused to form a shield-like carapace drawn out between the eyes as a 'rostrum'; the abdomen is clearly defined. There are 8 pairs of thoracic appendages: the 1st and 3rd pairs are developed as mouth parts; the 4th, 5th, 6th, 7th, and 8th pairs are used for walking and may terminate in a claw-like joint. Five pairs of abdominal appendages (swimmerets) are used for swimming (and for brooding eggs in the female), the last pair forming the tail fan or terminal segment. Fig. 33 shows the general malacostracan characters.

Suborder Natantia Prawns and shrimps

These decapods can swim and have light exoskeletons. Sometimes their bodies are compressed sideways. The 1st pair of antennae are obviously branched into 2–3 filaments, while the 2nd pair are much longer and have an unbranched filament and a flat basal process. In most prawns the rostrum is conspicuous and the precise form may be very important in identification; in the shrimps it is generally greatly reduced. Mediterranean recorded species = not less than 31. N.B. Smaldon, G. 1979 describes a number of shrimps and prawns to be found in the Mediterranean Sea. In these illustrations the appendages of one side of the body only are drawn.

Lucifer acestra Dana Length up to 10mm; carapace small and sideways flattened; 2nd antennae longer than 1st but not as long as body; eyestalks very long; thoracic legs about twice length of abdominal swimmerets; tail fan reduced. **Colour** transparent. **Habitat** pelagic. **Similar species** none.

Pennaeus trisulcatus (Leach) *(=P. kerathurus)* Length up to 20mm, often smaller; carapace relatively small with conspicuous rostrum; 1st antennae very short; 2nd antennae longer than body and with broad flat inner basal process; dorsal side of carapace furrowed either side of rostral crest. **Colour** yellow-brown with red tints. **Habitat** in estuaries and on muddy bottoms, usually not deeper than 40m. **Similar species** several, mainly from very deep water.

Pandalina brevirostris (Rathke) Length up to 33mm, may be less; carapace about quarter of body length; rostrum straight and up to half the length of the carapace, and bearing 7–8 teeth on dorsal side of which 4–5 are posterior to eye, and 2–3 on ventral side; 1st antennae short, 2nd antennae not as long as body. **Colour** translucent white with orange and red pigment spots. **Habitat** on sand and shell gravel from 5–100m. **Similar species** none.

Hippolyte varians Leach **Chameleon Prawn** Length up to 25mm; carapace about quarter of body length; rostrum straight, almost as long as carapace and terminating in a single point; 2 widely spaced teeth above and 2 below, which are closer together; 2nd antennae half as long as body; 3rd pair of walking legs is the longest. **Colour** carapace green, red or brown by day, transparent at night. **Habitat** on the shore among rocks and weed and down to 50m. **Similar species** *H. inermis* Leach (=*H. prideauxana*) (rostrum only illustrated) in which the rostrum is longer than the carapace; body length up to 42mm; green-brown to crimson in colour, sometimes with a white dorsal stripe. **Habitat** on the shore down to 54m.

Alpheus glaber (Olivi) *(=A. ruber)* **Snapping Prawn** Length up to 65mm, often much less; carapace about one-third of body length completely shielding the eyes when viewed from above; rostrum short, untoothed; 2nd antennae longer than body; 1st pair of walking legs bears large unequal pincers (either right or left may be the larger). **Colour** pink-red above, paler on sides. **Habitat** usually in mud and silt from 30–100m; often located by the loud clicks generated by the pincers: these sounds are used to stun prey and ward off predators. **Similar species** *A. dentipes* Guerin. **Habitat** on stony bottoms from 2–40m (below); length up to 22mm. See also *Synalpheus laevimanus* and *Athanas nitescens* (page 202).

Synalpheus laevimanus (Heller) **Snapping Prawn** Length up to 20mm; carapace about one-third body length; rostrum short with a large tooth on either side; 1st pair of walking legs unequal; left greatly enlarged and bearing an enor-

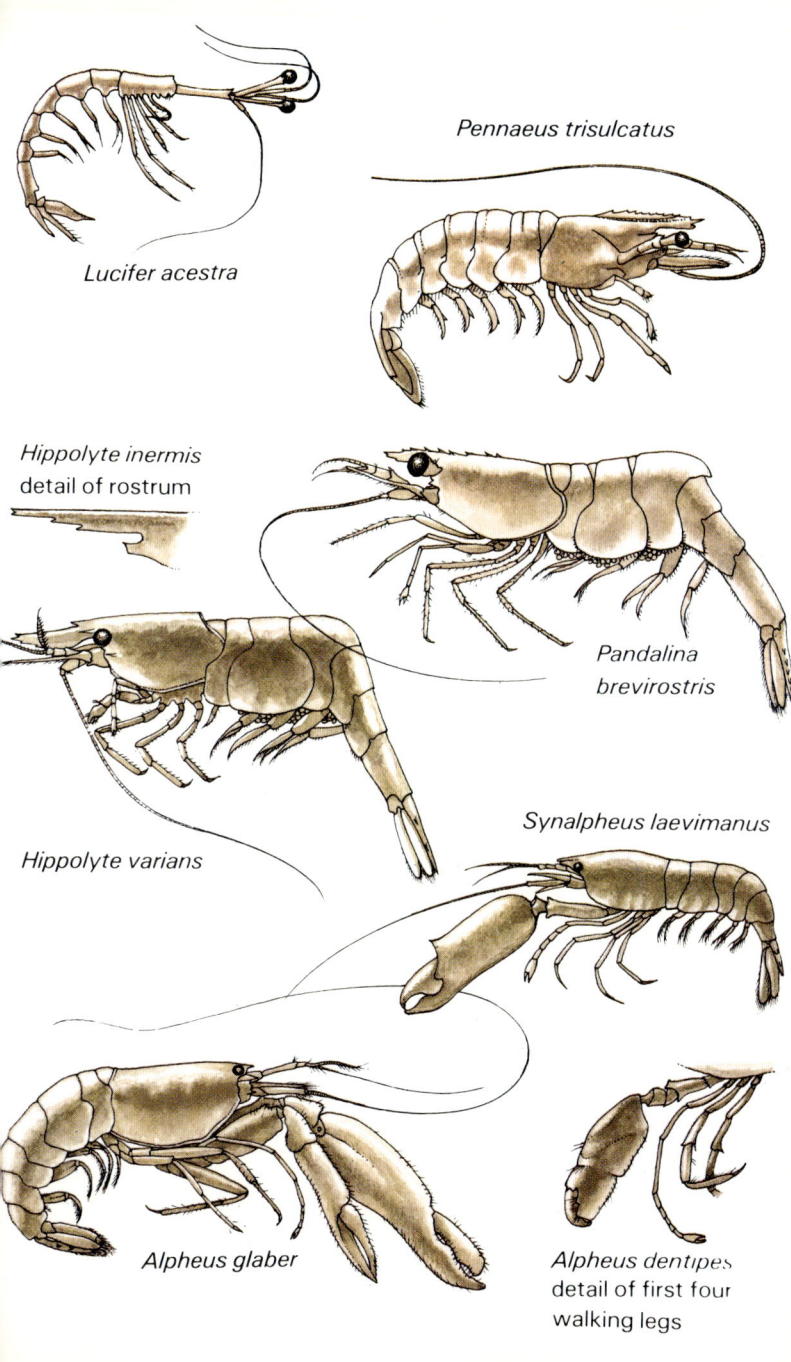

Lucifer acestra

Pennaeus trisulcatus

Hippolyte inermis
detail of rostrum

*Pandalina
brevirostris*

Hippolyte varians

Synalpheus laevimanus

Alpheus glaber

Alpheus dentipes
detail of first four
walking legs

mous pincer; 2nd pair bears minute pincers. **Habitat** 15–30m, often with plants and other animals. **Similar species** *Alpheus* spp. and *Athanas nitescens*.

Athanas nitescens (Leach) Length up to 20mm; carapace about one-third of body length; small spines project forward beside the eyes which are partly covered by the overhanging leading edge of the carapace; rostrum straight, short and pointed with no teeth; 2nd antennae about as long as body; 1st walking legs large and bearing pincers, often unequally developed in the male; 2nd walking legs quite long and bearing minute pincers. **Colour** variable: red, brown, green or blue, often with a white dorsal stripe. **Habitat** among rocks and weeds on the shore and in shallow water. **Similar species** *Alpheus* spp (page 200).

Palaemon elegans Rathke (=**Leander squilla**) **Prawn** Length up to 63mm; carapace one-third to quarter of body length, with 2 spines on the front on each side; rostrum straight or curved slightly upwards, terminating in a sharp tooth or 2 small teeth; 7–9 teeth on dorsal side of which 2–3 are posterior to eye; generally 3 teeth on ventral side, rarely 2 or 4. First antennae with 3 branches; 1st walking leg smaller than 2nd, which bears pincers. **Colour** carapace and tail may be dark with yellow bands; rostrum colourless or with red pigment spots. **Habitat** on the shore and in shallow water. **Similar species** *P. serratus* (Pennant) (rostrum only illustrated): length up to 110mm; rostrum distinctly upcurved; terminating in 2 small teeth with 6–7 dorsal teeth not reaching distal third, 2 behind eye, 4–5 ventral teeth; habitat: in pools on the shore and down to 40m. *P. adspersus* Rathke: up to 70mm; rostrum straight, 5–6 dorsal teeth extending to distal third, 3 (rarely 2 or 4) ventral teeth; habitat: shallow water, often in estuaries. *Leander xiphas* (Risso): up to 60mm; rostrum slightly upcurved with 6–8 teeth on dorsal side on the proximal two-thirds and 4–5 teeth ventrally; habitat: in shallow water.

Pontonia pinnophylax (Otto) Length up to 40mm; carapace about one-third of body length; rostrum short, like a curved spine without dorsal or ventral teeth; 1st pair of walking legs bears small pincers; 2nd pair has larger, often unequal pincers. **Colour** pink to transparent. **Habitat** among sponges and bivalves below 5m. **Similar species** few.

Processa canaliculata Leach Length up to 74mm; carapace about one-third of body length; rostrum short, straight or slightly curved down, with 2 terminal teeth, the dorsal one shorter than the ventral; large eyes may appear higher than the rostrum; 2nd antennae longer than body; unequal development of both 1st and 2nd pairs of walking legs. **Colour** pinkish with red-orange patches; rostrum orange. **Similar species** one.

Pontophilus fasciatus (Risso) Length up to 19mm; carapace less than one-third of body length; short broad rostrum about two-thirds of the length of stalked eye, with a median spine on the carapace behind it; 1st antennae short, 2nd antennae about half body length. **Colour** carapace and tail brown, remainder pale; 2 dorsal blue spots behind carapace. **Habitat** in shallow water and down to 60m. **Similar species** *Crangon crangon*.

Pontocaris cataphracta (Olivi) Length up to 35mm; carapace spiny and quarter to one-third of body length; 2nd antennae about half body length; abdomen spiny; 1st pair of walking legs bears pincers, 2nd pair with very small pincers; 4th and 5th pairs longest. **Colour** pale pink with red spots. **Habitat** on sandy substrates from 10–50m. **Similar species** few.

Crangon crangon (Linnaeus) **Common Shrimp** Length up to 50mm, sometimes more; carapace about quarter of total length with 1 central spine and 1 on either side; rostrum reduced to a small tooth; 2nd antennae almost as long as body; 1st pair of walking legs bear the largest pincers, 2nd carry mintue pincers, 3rd and 4th pairs are the longest. **Colour** sandy brownish to transparent. **Habitat** on the shore and in shallow water as well as in estuaries.

Note on deep water species seen in fish markets Visitors to the western Mediterranean region will almost certainly find large red prawns in fish markets. These are generally caught at great depths and are beyond the scope of this book.

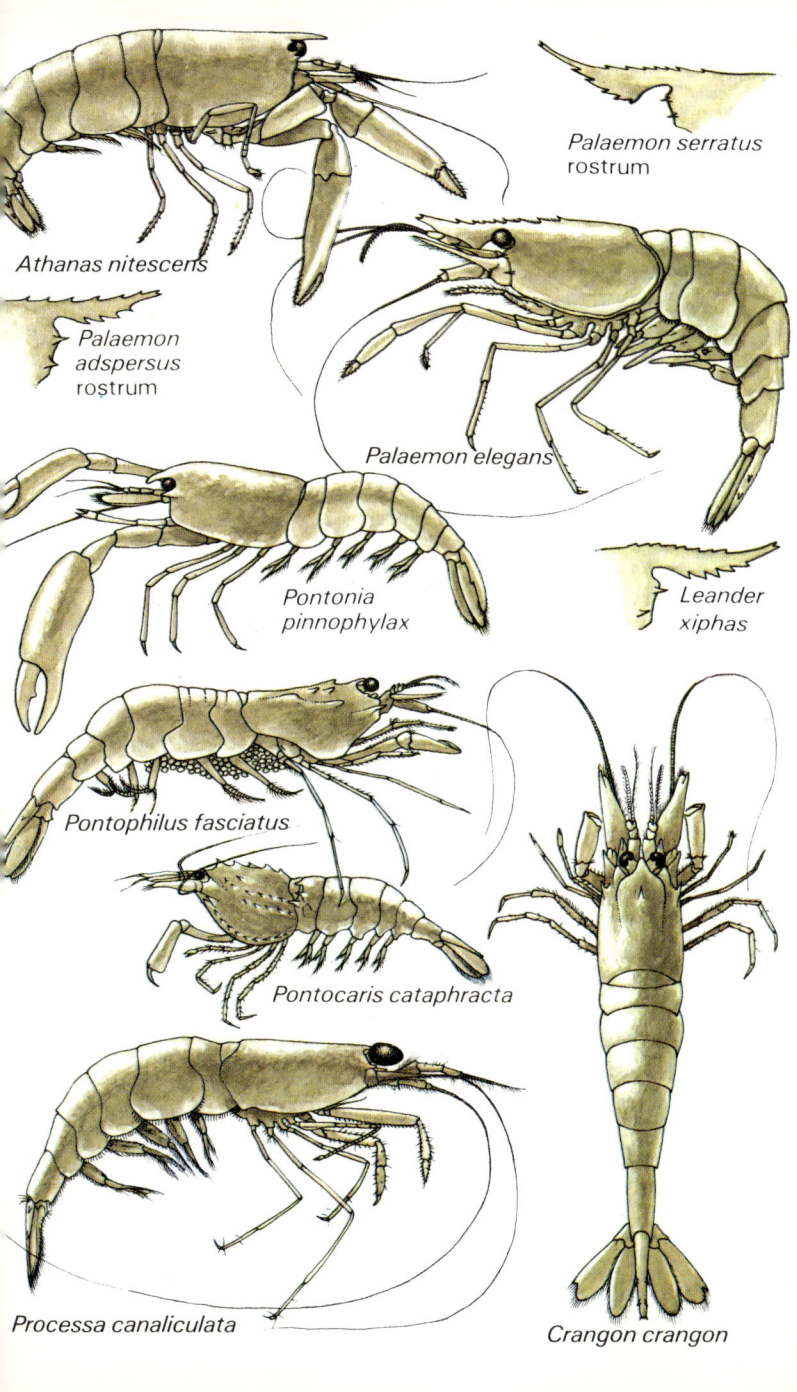

Athanas nitescens

Palaemon serratus rostrum

Palaemon adspersus rostrum

Palaemon elegans

Pontonia pinnophylax

Leander xiphas

Pontophilus fasciatus

Pontocaris cataphracta

Processa canaliculata

Crangon crangon

Suborder Reptantia

Division Macrura True lobsters and crayfish

Walking decapods which live on the seabed. They are strong and powerfully built, often with heavy exoskeletons which are not flattened laterally. The rostrum is reduced. The first pair of walking legs sometimes carries large and conspicuous pincers or terminates in a claw-like joint; the remaining four pairs of legs are strong with pincers or claws. The last pair of abdominal appendages is modified to form a tail fan known as the telson. Bouvier, E. L. 1940 and Zariquiez Alvarez, R. 1968 give more detail. Mediterranean recorded species = 5.

Palinurus elephas (Fabricius) *(=P. vulgaris)* **Crawfish** or **Spiny Lobster** Length 300–500mm; large crayfish lacking pincers on all walking legs except 5th pair in females; 2nd antennae considerably longer than body; carapace and abdomen spiny (can cause bad wounds if mis-handled). **Colour** red-brown background with striped antennae. **Habitat** among rocks and in crevices, occasionally on stony substrates, down to 70m. **Similar species** attention to colour and the form of the antennae and first walking legs should prevent confusion with the other species described on this page.

Scyllarides latus (Latreille) **Length** up to 350mm; large crayfish with 2nd antennae reduced to plate-like processes terminating in a point; lacking pincers on all walking legs except the 5th pair of the female; carapace and abdomen spiny. **Colour** red-brown to pink above, pale yellowish below. **Habitat** on rocks, stones and sand from 3m down. **Similar species** *Scylarus arctus* (see below); see also note for *Palinurus elephas* (above).

Scyllarus arctus (Linnaeus) **Length** up to 150mm; small crayfish with 2nd antennae reduced to plate-like processes terminating in about 5 lobe-like processes; lacking pincers on all walking legs except the 5th pair of the female; carapace and abdomen less spiny than preceding species; abdominal segments more rounded at the edges. **Colour** pink-red. **Habitat** among rocks with mud and on stony bottoms from 3m down. **Similar species** *Scyllarides latus* (above); see also note for *Palinurus elephas* above.

Nephrops norvegicus (Linnaeus) **Norway Lobster, Dublin Bay Prawn** or **Scampi** Length up to 150mm or more; slender animal with large, spiny, slightly unequal pincers carried on the first pair of walking legs; 2nd and 3rd walking legs also bear small pincers. **Colour** bright orange-red. **Habitat** on soft substrates from about 50m downward. **Similar species** none strictly similar (see note for *Palinurus elephas* above).

Homarus gammarus (Linnaeus) **Common Lobster** Length up to 450mm but often less; large lobster with massive pincers carried on the slightly unequal 1st pair of walking legs; 2nd and 3rd walking legs with small pincers; 2nd antennae about as long as body; carapace and abdomen not spiny. **Colour** blue-black on an orange background, paler and mottled underneath; turns red when boiled. **Habitat** among rocks in holes and in caves. **Similar species** none. N.B. the large pincers can be formidable defence weapons.

Division Anomura Squat lobsters, hermit crabs etc

Walking decapods which live on the shore or seabed. The abdomen is not greatly reduced but is often twisted, as in the hermit crabs, or folded under the caraspace, as in the squat lobsters. The last thoracic segment is free from the carapace. Bouvier, E. L. 1940 and Selbie, C.M. 1914 and 1921 gives further information on many anomurans found in the Mediterranean Sea. Mediterranean recorded species = not less than 23.

Scyllarides latus

Scyllarus arctus

Palinurus elephas

Nephrops norvegicus

Homarus gammarus

Fig. 35

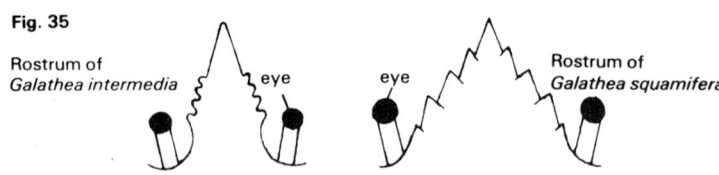

Rostrum of
Galathea intermedia   eye   eye   Rostrum of
Galathea squamifera

Galathea intermedia Lilljeborg Length up to 10mm; carapace bears rostrum with a slightly blunt tip; tip itself slightly longer than the 4 small spines on either side of the rostrum (see fig. 34); first pair of walking legs bears pincers and is about twice as long as body. Colour bright red with blue spots. Habitat among stones and rocks from the shore down to 80m. Similar species confusion can occur between immature specimens of the various *Galathea* species.

Galathea squamifera Leach Length up to 45mm; carapace bears rostrum with 4 pairs of side teeth, the rearmost being the smallest and the rest being about the same size (see fig. 35) first pair of walking legs bears pincers and is about one and a half times as long as the body; the pincer joints have scales and spines on their outer edges but the joints nearer the body have scales and spines on their inner or opposite edges. Colour usually green-brown, sometimes with reddish marks. Habitat under stones on the shore and down to 80m.

Galathea nexa Embleton Length up to 20mm; bears rostrum with 4 pairs of side teeth, the rearmost pair being very small; 1st walking leg bears pincers, is about the same length as the body and is hairy. Colour uniform red-brown. Habitat often on soft bottoms down to 30m. Similar species *G. squamifera*.

Galathea dispersa Bate (Not illustrated) Length up to 45mm, hairy rostrum has 4 points on each side. Similar to *G. squamifera*, Colour dull orange red and spotted with red and white. Habitat in shallow water among coralline algae.

Galathea strigosa (Linnaeus) Length up to 130mm but often smaller; carapace bears rostrum with pointed tip and 3 pairs of side spines; 1st pair of walking legs bears pincers and is about one and a half times as long as the body; pincers and 2nd, 3rd and 4th pairs of legs are spiny. Colour patterned bright red and blue. Habitat under stones and rocks from the shore down to 35m.

Munida bamffica (Pennant) *(=**M. rugosa)** Length up to 60mm; carapace bears 1 central and 2 lateral rostral spines; antennae not quite as long as first pair of walking legs, which bear delicate pincers and are about 4 times the body length. Colour reddish with yellow-brown marks. Habitat in deep water down to 150m. Similar species *M. tenuimana* Sars (not illustrated) which is coloured with yellowish-brown; habitat in very deep water down to about 1000m.

Note Several groups of anomurans occur apart from the squat lobsters. Some of them resemble other divisions such as the lobsters and true crabs, but their affinities with the squat lobsters and hermit crabs are indicated by their long and conspicuous antennae and by the miniature fifth pair of walking legs which is a key character of the porcelain 'crabs'.

Porcellana platycheles (Pennant) **Broad-clawed Porcelain Crab** superficially like a crab. Length up to 12mm; carapace round and squat with long antennae set wide apart; 1st pair of walking legs with thick flat conspicuous pincers, hairy on the outer edges; other legs hairy; abdomen small and folded tightly under the carapace. Colour yellow-brown-dirty grey. Habitat under stones and among mud and gravel on the shore and in shallow water. Similar species *Pisidia longicornis*.

Pisidia longicornis (Linnaeus) **Long-clawed Porcelain Crab** similar to *P. platycheles*: up to 6mm long; 1st pair of walking legs carries long slender hairless pincers; carapace not hairy but clean-looking; Colour red-brown.

Galathea squamifera

Galathea intermedia

Porcellana platycheles

Pisidia
longicornis

Munida bamffica

Galathea strigosa

Galathea nexa

Axius stirhynchus Leach Superficially prawn-like. **Length** overall up to 72mm; carapace bears a short flat triangular rostrum and is noticeably compressed sideways, being widest in the middle and tapering slightly towards the anterior and posterior; eyes small; long outer antennae more than twice the length of the inner ones; 1st pair of walking legs bears massive unequal pincers; 2nd pair has small flattened pincers; abdomen slender with wide tail fan. **Colour** pale red-brown. **Habitat** in shallow water or even on the shore, burrowing in mud and soft deposits. **Similar species** *Upogebia deltaura* and *Callianassa stebbingi* (below).

Calocaris macandreae Bell (Not illustrated) Superficially prawn-like. **Length** overall up to 50mm; carapace bears a slightly upturned rostrum which reaches nearly to the end of the base joint of the outer antenna; a ridge bearing spines passes back either side from the rostrum; carapace compressed sideways; abdomen tapering; animal may have an 'arched back' appearance; eyes large and not pigmented, with virtually no eye-stalk; outer antennae longer than body, inner antennae longer than carapace; 1st, 3rd, 4th and 5th pairs of walking legs lack pincers; pincers borne by 2nd pair. **Colour** delicate pink-rose. **Habitat** down to deep water on soft bottoms. **Similar species** few.

Jaxea nocturna (Chiereghin) Nardo Superficially prawn-like. **Length** up to 50mm; carapace bears pointed rostrum; eyes not apparent; long outer antennae; 1st pair of walking legs bears long slender hairy pointed pincers, 2nd pair has incomplete pincers, 5th pair reduced; abdomen may be folded under carapace and terminates in a tail fan. **Colour** white, pink or brown. **Habitat** in mud from 15m down. **Similar species** none.

Callianassa (Callichirus) tyrrhena Petagna Superficially prawn-like. **Length** up to 40mm; carapace bears reduced rostrum and reduced eyes; 1st pair of walking legs hairy and bearing unequal pincers; remaining walking legs slender and sometimes with spatulate terminal joints; abdomen delicate and often folded under carapace, terminating in a tail fan. **Colour** white-red-blue. **Habitat** burrowing in sand and mud from 2m down. **Similar species** *Axius stirhynchus* and *Upogebia deltaura*.

Upogebia deltaura (Leach) Superficially prawn-like. **Length** up to 100mm; carapace reduced at anterior; rostrum hairy; eyes reduced; 1st walking leg bears strange pincer with moving part longer than fixed part; remaining legs hairy but lacking pincers; abdomen slender and delicate, may be folded under carapace, terminating in a tail fan. **Colour** white-grey-yellow-green. **Habitat** burrowing in clay and mud down to deep water. **Similar species** *Axius stirhynchus* and *C. stebbingi*.

Hermit crabs

The anomura include a number of hermit crabs which are familiar seashore animals. They are easily recognized because they use empty gastropod shells to protect their delicate soft-skinned abdomens. Sponges, hydroids and sea-anemones often grow on the outside of shells occupied by hermit crabs. Mediterranean recorded species = not less than 12.

Diogenes pugilator (Roux) **Length** overall up to 25mm, carapace about 10mm; the animal is often smaller than this. Antennae are very hairy; eyes black; 1st pair of walking legs bears unequal pincers, the left nearly twice as large as the right; 2nd and 3rd pairs of walking legs bear claws, 4th and 5th pairs greatly reduced. **Colour** pincers have white tips; body pale. **Habitat** in gastropod shells, usually in shallow water, on sand and among sea-grasses. **Similar species** care must be taken to avoid confusion with other species.

Paguristes oculatus (Fabricius) *(=**Pagurus maculatus**)* **Length** overall about 40mm, carapace about 12mm: the animal is often smaller than this. Antennae are not hairy; 1st pair of walking legs bears slightly unequal pincers, left a little larger than right, and a conspicuous dark violet spot; other walking legs are as in *Diogenes pugilator* (above). **Colour** red-brown, antennae red, eyes clear blue, eye-stalks yellow, rostrum red. **Habitat** on various types of substrate down to 40m, often in shells of *Cerithium, Turbo* or *Murex* and associated with *Suberities domuncula* (sponge), *Podocoryne* spp. (hydroid) or *Calliactis parasitica* (anemone).

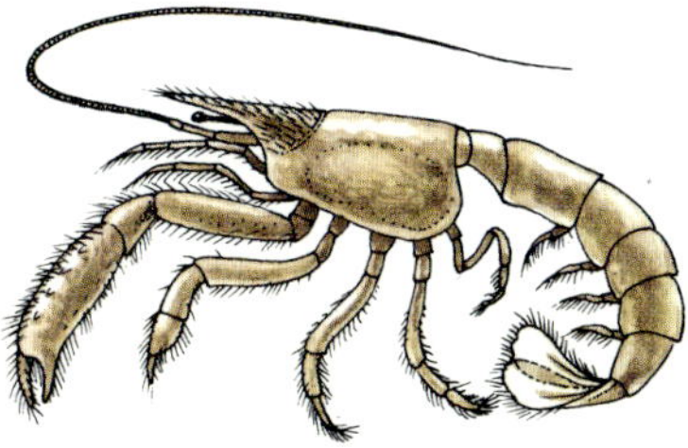

Axius stirhynchus

Upogebia deltaura

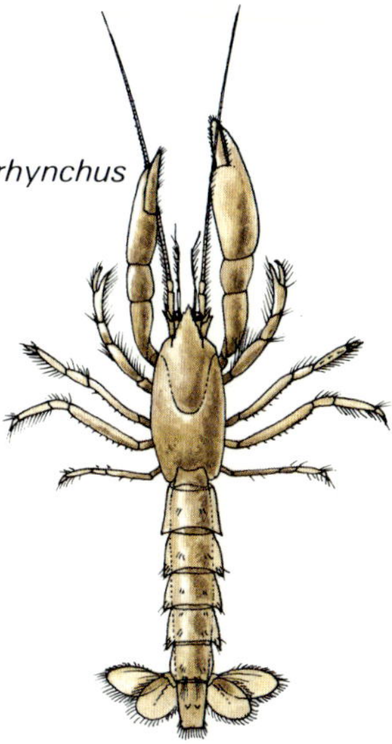

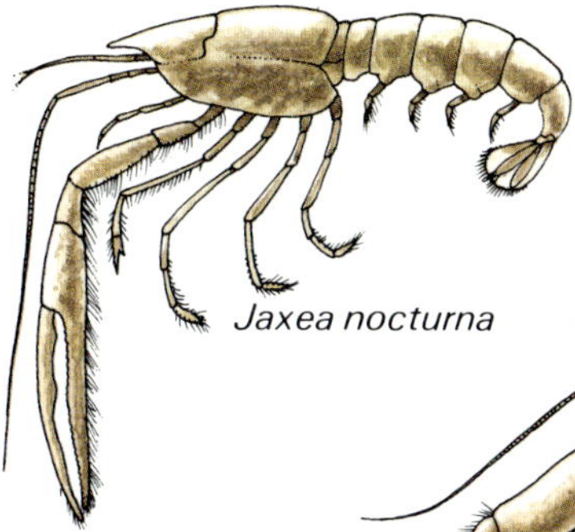

Jaxea nocturna

Calianassa (Callichirus) tyrrhena

Paguristes oculatus

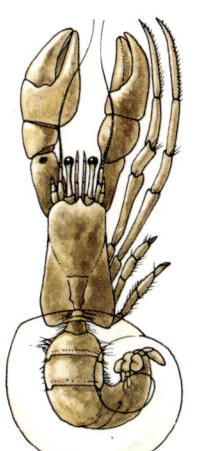

Diogenes pugilator

Dardanus arrosor (Herbst) *(=Pagurus arrosor)* Length overall 80mm, carapace about 15mm, may be larger; antennae not hairy; 1st pair of walking legs bears large pincers, the left bigger than the right, tips curved and pointed. Colour eyes brown-black; background of pincers and legs reddish, tips of pincers dark brown or black; body yellow-orange. Habitat on soft bottoms and amongst stones from about 30m down, usually in a large gastropod shell and associated with sponges or anemones. Similar species none.

Clibanarius erythropus Latreille Length up to 20mm, carapace about 5mm, often smaller; antennae not hairy; 1st pair of walking legs bears pincers of roughly equal size. Colour tips of walking legs and pincers black, remainder of these limbs spotted or striped pale blue or red; eye-stalks red; body red-brown-green. Habitat in shallow water under stones and on gravel, often very common. Similar species none.

Pagurus anachoretus (Risso) *(=Eupagurus anachoretus)* Length up to 25mm, carapace about 8mm; antennae not hairy; 1st pair of walking legs bears roughly equal pincers; upper surface of right pincer is hairy and smooth. Colour antennae and eye-stalks ringed in red; body yellow to pink-red. Habitat on rocks in shallow water, often by itself. Similar species *P. cuanensis*.

Pagurus cuanensis Thompson *(=Eupagurus cuanensis)* Length up to 25mm, carapace up to 6mm long; upper surface of right pincer is hairy and bears granules (see inset diagram). Colour yellow-brown. Habitat from 10–100m. Similar species *P. anachoretus*.

Pagurus alatus Fabricius *(=Eupagurus excavatus)* Length up to 40mm, carapace up to 8mm; 1st pair of walking legs bears unequal pincers: right may be nearly twice as large as left; pincers strong and with strong moving part or 'thumb'; pincer sculptured by 2 conspicuous furrows separated by a keel-like ridge and lacking hairs or spiny knobs. Colour yellow with red marks. Habitat usually between 30 and 100m. Similar species *P. anachoretus*.

Pagurus prideauxi Leach *(=Eupagurus prideauxi)* Length up to 60mm, carapace up to 15mm; 1st pair of walking legs bears unequal pincers, the right up to twice as large as the left; pincers almost hairless and with fine granules; pincer-bearing legs long in relation to body; abdomen small. Colour body brown-red, antennae yellowish, eyes dark grey. Habitat on muddy and sandy bottoms, associated with the sea-anemone *Adamsia carciniopados* below 10m. Similar species *P. alatus*, which may also carry *A. carciniopados*.

Pagurus bernhardus (Linnaeus) *(=Eupagurus bernhardus)* Length up to 100mm, carapace up to 25mm; 1st pair of walking legs bears large unequal, coarsely granulated pincers, the right being the larger. Colour carapace grey-red; pincers red-brown. Habitat among rocks on the shore and in shallow water; may be associated with *Suberites domuncula* (sponge), *Hydractinia echinata* (hydroid), *Calliactis parasitica* (anemone) and *Nereis fucata* (polychaete). Similar species *P. prideauxi*. May be infected with the parasitic barnacle *Peltogaster paguri*, a North Atlantic species which may be found in the Mediterranean.

Pagurus sculptimanus (Lucas) *(=Eupagurus sculptimanus)* (Not illustrated) Length up to 15mm, carapace up to 4mm. Similar to *P. prideauxi* but with pincers of 1st walking legs blunt-ended, slightly hairy and with a conspicuous keel on the upper side. Colour yellow to yellow-red; Habitat below 6m.

Anapagurus laevis (Thompson) Length up to 20mm, carapace up to 10mm; 1st pair of walking legs bears slightly hairy pincers, the right being much larger than the left. Colour carapace white, pincers banded orange. Habitat below 10m. Similar species none.

Cestopagurus timidus (Roux) (Not illustrated) Length up to 10mm; 1st pair of walking legs bears pincers, the right pincer slightly larger than the left; in the right leg the joint before the pincer is swollen and considerably larger than the pincer itself; abdomen relatively large. Colour grey-brown. Habitat from 6m down; not common. Similar species none.

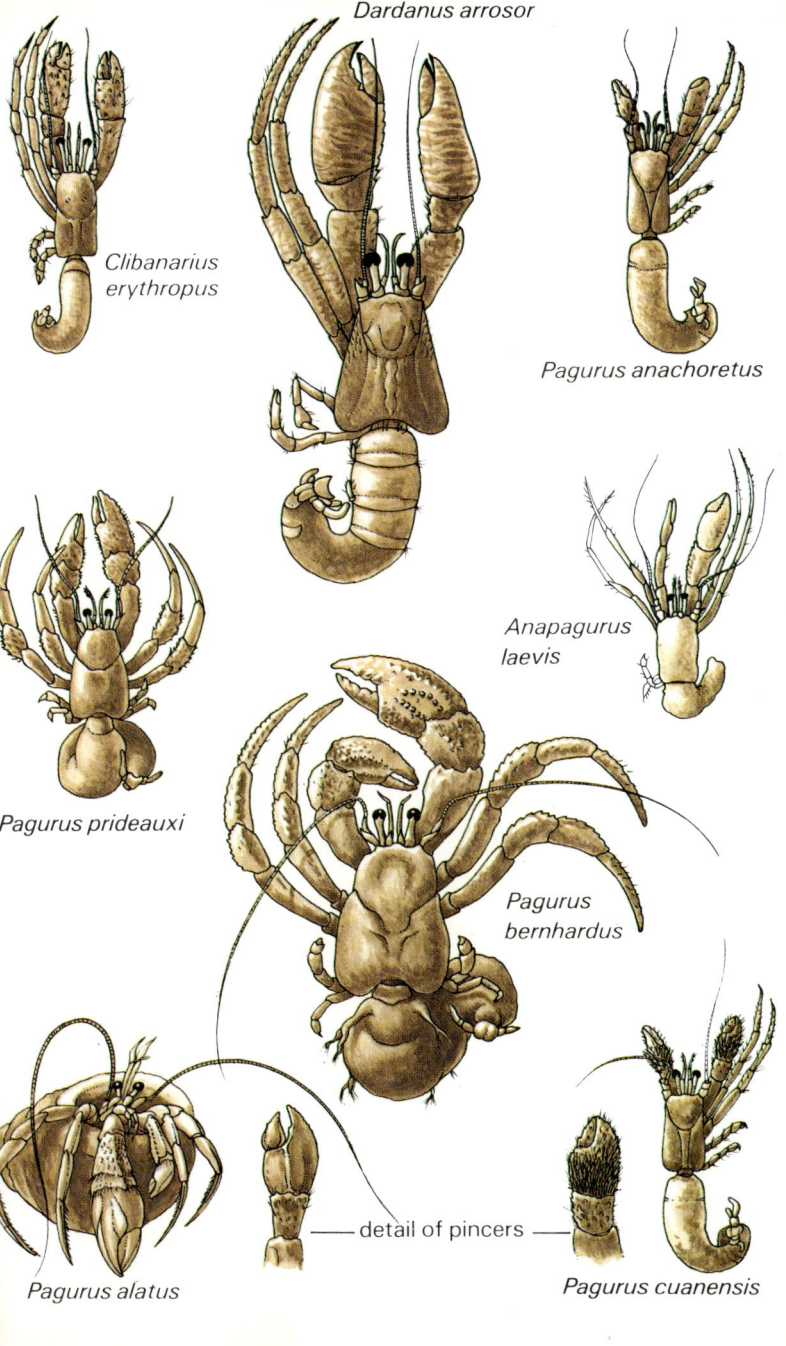

Dardanus arrosor

Clibanarius
erythropus

Pagurus anachoretus

Anapagurus
laevis

Pagurus prideauxi

Pagurus
bernhardus

Pagurus alatus

— detail of pincers —

Pagurus cuanensis

Division Brachyura True crabs

Decapods which as adults live on the shore or seabed but with a planktonic larval phase. The carapace (head/thorax) is generally flattened and rounded; the abdomen is smaller, being folded forward under the carapace. The antennae are short; the leading edge of the carapace bears the eyes and is often ornamented with teeth which may be important in identification; the first pair of walking legs is pincer-bearing, the other 4 pairs of walking legs normally terminate in a claw. Further information regarding crabs to be found in the Mediterranean is available in Bouvier, E. L. 1940, Christiansen, M.E. 1969; Ingle R.W. 1980 and Zariquiez Alvarez, R. 1968.

Dromia personata (Linnaeus) *(=D. vulgaris)* **Sponge Crab** Length up to 80mm. Carapace width to length ratio about 5:4, domed and rounded and largely covered by hairs giving it a furry appearance; 1st pair of legs bears big pincers, 4th and 5th pair displaced upwards; 5th pair bears small pincers. **Colour** hairs dark brown; pincers pink. **Habitat** on the shore and down to 30m. **Similar species** none. N.B. this species often carries pieces of sponge on its back.

Homola barbata (Herbst) Length up to 30mm. Carapace width to length ratio about 4:5; relatively straight-fronted with a number of spines, straight-sided and slightly curved at rear; upper surface spiny; 1st pair of walking legs bears small pincers. **Colour** brownish. **Habitat** on sandy and muddy seabeds from 50 to 100m.

Ethusa mascarone (Herbst) Length up to 16mm. Carapace width to length ratio roughly 1:1 at widest part; relatively straight-fronted with a conspicuous frontal notch and 3 teeth on either side between notch and eye; straight sides, slightly bulbous towards rear, rounded trailing edge; antennae quite long; legs not very hairy; 1st pair of walking legs short, bearing moderately developed pincers; 4th and 5th pairs of legs small and displaced upwards. **Colour** grey-brown. **Habitat** on sandy and muddy bottoms down to 30m, sometimes among weeds. **Similar species** *Dorippe lanata* (below).

Dorippe lanata (Linnaeus) Length up to 30mm. Carapace width to length ratio about 5:4 at its widest part; relatively straight-fronted with a small frontal notch, 2 teeth between notch and eye on each side and 2 more beyond eye; 1 conspicuous tooth halfway back on each side; rest of carapace pear-shaped and hairy; 1st pair of walking legs short, bearing long slender pincers; 4th and 5th pairs displaced upwards and conspicuously shorter. **Colour** carapace pink-brown, legs brownish. **Habitat** on muddy and sandy bottoms down to 50m. **Similar species** *Ethusa mascarone* (above).

Calappa granulata (Linnaeus) Length up to 110mm. Carapace width to length ratio about 4:3, curving back quite sharply from front; a pair of rounded spines between the eyes; left and right sides finely toothed; rear edge less strongly convex than front; 1st pair of walking legs long and massive, bearing pincers which are crested with stubby spines on the upper side; remaining pairs become progressively shorter towards the rear. **Colour** carapace grey-yellow, spotted red. **Habitat** on soft substrates where it burrows in search of prey. **Similar species** none.

Ebalia nux Norman Length up to 8mm. Carapace rounded, width to length ratio about 1:1; small central notch in leading edge and 2 small teeth either side between notch and eye; upper surface smooth and lacking spines or knobs; 1st pair of walking legs long and bearing slender pincers. **Colour** pinkish, reddish or brown. **Habitat** on muddy bottoms below 100m. **Similar species** four.

Ebalia cranchi Leach Length 7mm; like *E. nux* but with rhomboidal carapace with granulated surface and with 5 conspicuous knobs: 2 almost at the centre, 1 behind these and 1 each left and right. **Colour** yellow-red. **Habitat** on sand and gravel from 20 to 130m. N.B. a fuller account of similar Mediterranean species, e.g. *E. tumefacta* (Montagu), *E. granulosa* (Milne-Edwards) and *E. tuberosa* (Pennant) is given by Riedl, R. 1963.

Dromia personata

Homola barbata

Dorippe lanata

Ethusa mascarone

Calappa granulata

Ebalia cranchi

Ebalia nux

Corystes cassivelaunus (Pennant) **Masked Crab** Length up to 40mm. Carapace ovoid, width to length ratio about 4:6 at its widest; several teeth situated along sides; antennae very long and joined together by hairs to form a respiratory tube; 1st pair of walking legs very long in male and bearing relatively small pincers; in female they are about same length as carapace; remaining pairs become shorter towards rear. Colour yellow-brown. Habitat burrowing in sand on the lower shore and in shallow water. Similar species none.

Atelecyclus rotundatus (Olivi) Length up to 30mm. Carapace almost circular in outline; fringed with teeth; antennae about half as long as diameter of carapace; 1st pair of walking legs quite massive but bearing small pincers; remaining pairs roughly of equal length; carapace and appendages lightly dusted with hairs. Colour reddish-white with red spots, pincer tips black, hairs brown. Habitat on sandy and muddy bottoms from 15m down.

Liocarcinus puber (Linnaeus) *(=Macropipus puber)* Length up to 40mm (occasionally more). Carapace width to length ratio 5:4 at the widest point; anterior edge moderately convex and with 8–10 variously shaped, often fine teeth between the eyes; 1st pair of walking legs strong and bearing pointed pincers, others of roughly equal length, 5th bears flattened swimming paddle instead of terminal claw; surface of carapace covered with fine hair giving a felty appearance. Colour eyes red, body dirty brown; parts of limbs not covered by hairs are bluish. Habitat among rocks and stones from the shore down. Similar species four.

Liocarcinus corrugatus (Pennant) *(=Macropipus corrugatus)* Length up to 25mm (sometimes more). Carapace width to length ratio 5:4 at widest point; anterior edge moderately convex; central tooth between eyes and 2 shallow ones on either side; under a lens these will be seen to be very finely subdivided; upper surface of carapace horizontally corrugated. Colour reddish-brown, sometimes with red patches. Habitat on coarse sand and gravel down to 100m. Similar species four.

Liocarcinus depurator (Linnaeus) *(=Macropipus depurator)* Length up to 40mm. Carapace width to length ratio 5:4 at the widest point; anterior edge moderately convex and with 3 well developed teeth between the eyes; carapace moderately hairy. Colour red-brown. Habitat on various substrates from about 1m down. Similar species four.

Carcinus mediterraneus Czerniavsky **Common Shore Crab** Length 35mm, sometimes more. Carapace width to length ratio 5:4 or 4:3; leading edge convex, 3 blunt teeth between the eyes; 1st pair of walking legs with strong pincers, last pair bear a flattened claw rather than a paddle as in *Liocarcinus*. Colour variable: dark green-grey with white, yellow or orange marks. Habitat on sandy and rocky shores and in shallow water. The status of Mediterranean members of this genus is under debate. Some authorities believe that *Carcinus* is represented in the Mediterranean by both *C. mediterraneus* and *C. maenas* (Linnaeus), the Atlantic species.

Thia scutellata (Fabricius) *(=T. polita)* Length up to 22mm. Carapace slightly wider than it is long, convex at leading edge otherwise heart-shaped in outline; eyes very small and almost hidden in eye sockets; several poorly defined teeth along either side; limbs generally short, pincers stout. Colour pinkish. Habitat burrowing in sand around 20m deep.

Pirimela denticulata (Montagu) Length up to 25mm. Carapace width to length ratio 11:10; anterior edge convex and strongly toothed; 3 teeth between eyes, the outer 2 being triangular and flattened and the middle longer and round; 1st pair of walking legs bears small pincers, other legs terminate in pointed claws and not paddles. Colour variable: green, brown, purple-red; middle of carapace usually darker, paler periphery and a white cross at rear. Habitat among rocks, gravel and weed from the shore down to 60m.

incer ♂

Corystes cassivelaunus ♀

Atelecyclus rotundatus

Liocarcinus puber

Liocarcinus corrugatus

Liocarcinus depurator

Carcinus mediterraneus

Pirimela denticulata

Thia scutellata

Cancer pagurus Linnaeus **Edible Crab** Length up to 140mm, often less, occasionally more. Carapace width to length ratio 3:2; somewhat ovoid in outline, domed and lightly granulated above and fringed on either side from eyes back with about 9 edge lobes giving the characteristic 'piecrust' appearance; 1st pair of walking legs bear powerful black-tipped pincers; remaining pairs of legs hairy, 5th pair shortest. **Colour** pink-brown. **Habitat** from the shore down to 100m.

Goneplax rhomboides (Linnaeus) **Angular Crab** Length up to 27mm, often less. Carapace width to length ratio 7:4; carapace oblong, leading edge reasonably straight and drawn out into conspicuous teeth, with a second pair a little behind on either side; eyes carried on long eye-stalks; in the male the 1st pair of walking legs is very long, the pincer-bearing joint is about equal to carapace length; other legs about equal. **Colour** yellow-red. **Habitat** on soft substrates down to 100m. **Similar species** none.

Pilumnus hirtellus Linnaeus **Hairy Crab** Length up to 15mm, often less. Carapace width to length ratio almost 3:2; somewhat convex leading edge which bears a frontal notch in midline on either side of which is a shallow lobe bearing about 7 fine serrations and 1 tooth at base of antenna; outside eye are 5 teeth (including eye socket margin). Carapace tapers to posterior and is hairy; all legs hairy, 1st pair bear strong unequal pincers. **Colour** reddish-brown, pincers brown or purple. **Habitat** under stones and rocks on the shore and in shallow water. **Similar species** *P. spinifer* Milne-Edwards (not illustrated) which has coarser serrations on the frontal lobes and shiny pincer-bearing legs.

Xantho pilipes Milne-Edwards *(X. hydrophilus)* Length up to 24mm, often less. Carapace width to length ratio a little more than 4:3. Carapace smooth, flat in front, slightly undulating behind; few obtuse teeth on side of carapace outside of eye, the incisions between these quite sharply defined; 1st pair of walking legs carries dark brown-tipped pincers, right pincers slightly longer than left; other walking legs hairy and progressively shorter towards rear. **Colour** yellow-brown, patches of red. **Habitat** lower shore and down to 50m on hard and soft bottoms. **Similar species** *X. incisus* Leach *(= X. floridus)* (not illustrated): carapace furrowed; colour red-brown; carapace teeth separated by more rounded incisions.

Pinnotheres pisum (Linnaeus) **Pea-crab** Length of female carapace up to 13mm, width up to 14mm; length of male carapace up to 6mm, width up to 6mm. Female carapace soft, rounded and smooth, 1st pair of walking legs bears quite delicate pincers; last joint of 5th walking legs relatively short and hook-like. Male carapace hard, leading edge extended slightly between eyes, pincers of 1st pair of walking legs stronger; all legs stronger and more hairy than female; hook-like last joint to 5th pair. **Colour** female transparent, brown-yellow above bearing a yellow spot at front and yellowish patches at side; male yellow-grey. **Habitat** inside bivalve shells e.g. *Mytilus* and *Spisula*. **Similar species** *P. pinnotheres* (Linnaeus) with terminal joint of 5th walking leg long, straight and tapering, about same length as subterminal joint.

Pachygrapsus marmoratus (Fabricius) **Marbled Crab** Length up to 30mm, occasionally longer. Carapace slightly wider than it is long and squarish in outline; leading edge nearly straight with 3 slight hollows; antennae and eyes relatively far apart; 2–3 pointed teeth outside eyes; quite large pincers on 1st pair of walking legs; other legs hairy. **Colour** violet, brown, greenish or blackish with variable marbled pattern. **Habitat** in rocky cracks and crevices from the shore down. **Similar species** one.

Cancer pagurus

Goneplax rhomboides

Pilumnus hirtellus

Pinnotheres pisum ♀

♂

Xantho pilipes

Pinnotheres pinnotheres ♀

Pachygrapsus marmoratus

Inachus dorsettensis (Pennant) **Spider Crab** Length up to 35mm, often less. Carapace width to length ratio about 6:7, triangular; drawn out into short rostrum of 2 clearly separated horns; retractable eyes in eye-sockets; carrying 4 short spines in a horizontal row, behind which is a longer spine, and there are other spines behind these; 1st pair of walking legs short and quite plump compared with other pairs, carrying pincers; other pairs of legs very long; 2nd longest, 5th shortest. Colour grey to light brown, sometimes spotted and sometimes encrusted. Habitat on soft and hard bottoms from about 6m down. Similar species *I. phalangium* (Fabricius) (not illustrated) has a very narrow incision between rostral horns and only 2 short spines in horizontal row; and *I. leptochirus*.

Macropodia rostrata (Linnaeus) **Long-legged Spider Crab** Length up to 22mm. Carapace width to length ratio 5:7, triangular with anterior edge drawn out into a prominent rostrum composed of two rostral horns joined together and less than half the length of the antennae; eyes not retractable into sockets; body bears a number of prominent spines; 1st pair of walking legs short and relatively stout, bearing weak pincers; remaining legs very long and flimsy. Colour grey-brown, yellow-brown or red-brown, often encrusted. Habitat on hard and soft bottoms down to about 85m. Similar species *M. tenuirostris* (Leach) which has a rostrum more than half the length of the antennae.

Acanthonyx lunulatus (Risso) Length up to 25mm. Carapace width to length ratio 3:4, anterior edge drawn out into a number of coarse spines, 4 of which are between the eyes; eyes themselves carried on the sides of the outer two of these; outside the eyes there are about 3 spines on each side; rear of carapace curved; 1st pair of walking legs carries moderate pincers, other walking legs carry typically notched subterminal joints which assist with gripping weeds when the terminal claw flexes. Colour intense green or golden-brown. Habitat beautifully adapted for clinging to weeds e.g. *Cystoseira* from 1–20m, especially in turbulent water.

Pisa armata (Latreille) *(=P. gibbsi)* **Four-horned Spider Crab** Length up to 40mm, occasionally more. Carapace width to length ratio about 8:11, triangular, widening towards rear, domed above, slightly concave at sides, armed with several spines including 3 strong ones at rear: one pointing left, one behind and one right; 2 rostral horns, about same length as antennae, parallel in male and diverging in female; small spine in front of each eye; 1st pair of legs bears strong but not large pincers; other pairs shorter towards rear. Colour brownish, may be encrusted. Habitat usually below 20m on soft substrates with algae. Similar species *P. nodipes* Leach (not illustrated) has 3 blunt rear spines; *P. tetradon* (Pennant) (not illustrated) has no spines at rear but 4 on either side.

Lissa chiragra (Herbst) Length up to 40mm. Carapace width to length ratio about 4:5, pear-shaped, with a number of tubercles or nodules round the perimeter and 2 'humps' on top; 2 bent rostral spines united to produce a T-shaped rostrum with a small spine behind on either side; 1st pair of walking legs bears pincers and has knobbly joints, as do the other legs which get shorter towards the rear. Colour reddish, often encrusted. Habitat on soft bottoms and among other organisms from 30–80m. Similar species none.

Maja squinado (Herbst) **Spiny Spider Crab** Length up to 180mm. Carapace width to length ratio 6:7, like a rounded triangle in outline, fringed with large and small spines; 2 conspicuous spines between eyes; upper surface spiny and sometimes encrusted; 1st pair of walking legs bears strong but not large equal pincers; remaining legs hairy and long except the 5th pair. Colour carapace white, pink, red; sometimes spotted. Habitat rocks and sand from the shore to 50m. Similar species *M. verrucosa*.

Parthenope massena Roux *(=Lambrus massena)* Length up to 20mm. Carapace width to length ratio 7:8 and shaped like a compressed pear; upper surface not strongly tuberculous; rostrum not sharply pointed; 1st pair of walking legs relatively long, tuberculous and bearing moderate pincers; remaining walking legs shorter. Colour grey, green or yellowish, sometimes encrusted. Habitat among rocks and on sand from 5 to 200m. Similar species *P. angulifrons*: carapace width to length ratio almost 5:4; upper surface more strongly tuberculous.

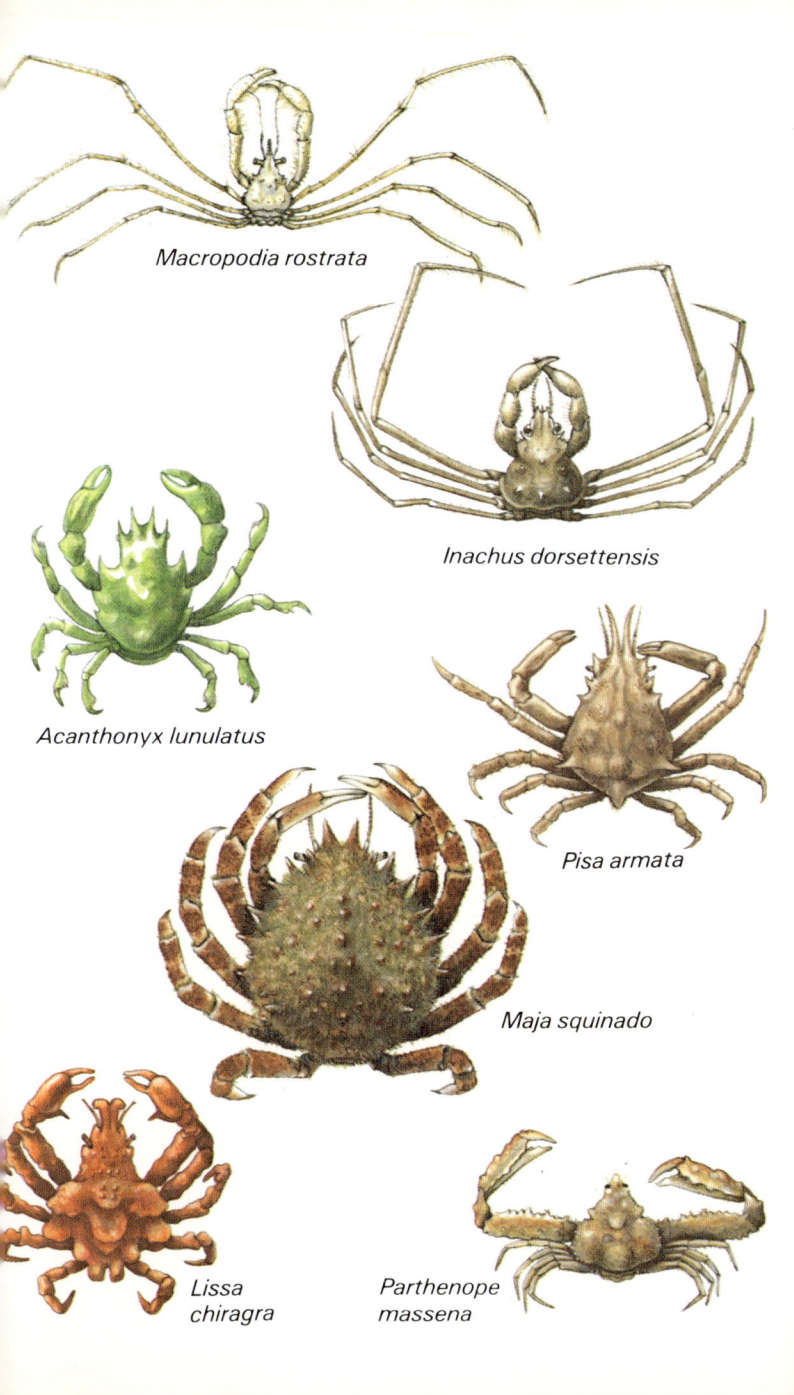

Macropodia rostrata

Inachus dorsettensis

Acanthonyx lunulatus

Pisa armata

Maja squinado

Lissa chiragra

Parthenope massena

Class Insecta Insects

Arthropods in which the adult body is divided into three distinct regions: a head with a single pair of antennae; a thorax with three pairs of legs and often one or two pairs of wings, and an abdomen without walking appendages. The insects are a major group of invertebrates, particularly successful on land. However very few have adapted to life in or beside the sea. Nevertheless because of the effects of winds and migratory instincts true terrestrial insects may visit the shore. The two species given here are considered the most likely to be encountered. They are true shore-dwellers but the visitor may find a variety of others.

Petrobius maritimus (Leach) **Bristle-tail** Length up to 12.5mm. Head has conspicuous antennae which are about the length of the body, and a pair of palps. Body thorax bears 3 pairs of legs but has no wings; abdomen terminates in a long bristle which is nearly as long as the body. Habitat towards the landward limits of the shore in rocky crevices and cracks. Similar species there are many truly terrestrial bristle-tails.

Lipura maritima (Laboulbène) Length about 3mm. Head bears short antennae. Body thorax and abdomen quite plump; thorax bears 3 pairs of short legs but no wings; abdomen broad towards the tail but tapering to a blunt point at the tip. Habitat generally floating on the surface film of water in rock pools towards the upper limit of the shore or crawling on rocks and seaweeds.

Class Pycnogonida (sometimes regarded as a subphylum) Sea-spiders

Exclusively marine arthropods with a cephalothorax (head and thorax) drawn out into an anterior proboscis which opens by a terminal mouth. The abdomen is reduced to a single segment. Four pairs of relatively long legs are borne by the thorax, into which the ovaries and digestive system extend. For a fuller account of the Mediterranean pycnogonids see Dohrn, A. 1881, King, P. E. 1974 or Reidl, R. 1963. These animals are often very difficult to identify precisely without the aid of a microscope. Mediterranean recorded species = 10.

Nymphon gracile Leach Length up to 10mm, sometimes longer; walking legs reach up to 25mm. Body slender cephalothorax; proboscis equipped on either side with a pair of pincer-like feeding appendages (chelicerae), behind which are a pair of 5-jointed palps; in addition to the 4 pairs of walking legs there is one pair of 'oviferous' legs held out below the body on which the male carries the eggs after they are laid. Habitat shore and shallow localities. Similar species several.

Callipallene brevirostris (Johnston) Similar in general shape to *Nymphon* but much smaller. Length up to 1.5mm; walking legs reach 6mm. Body cephalothorax has much shorter proboscis than *Nymphon*. Habitat on the shore and in shallow water, often associated with algae and hydroids. Similar species several.

Phylum Phoronida

The phoronids are a small phylum of exclusively marine sessile worm-like animals which bear a crown of hollow filter-feeding tentacles known as the lophophore. A true body cavity is present and the mouth is surrounded by the lophophores. They live in chitinous tubes secreted by the body. Individuals are small and cryptic but may be very abundant. They are thought to be related to the bryozoa (see page 222). For more details see Emig, C .C. 1979. Mediterranean recorded species = 3.

Phoronis mulleri Selys-Longchamps The crown of tentacles may be all that is visible and this may reach 5mm in diameter. Length body up to 12mm. Body swollen towards lower end. Colour translucent grey with pinkish tinge to bases of the tentacles; tube brownish-grey. Habitat growing on shells and pebbles or rocks, often embedded in sediment. Similar species two.

Petrobius maritimus

Lipura maritima

Nymphon gracile

Callipallene brevirostris

Phoronis mulleri

tube

Phylum Bryozoa (= Ectoprocta or Polyzoa)

Bryozoa are minute, sessile, colonial animals. Individuals grow inside a secreted case called the *zooecium*. A true body cavity (coelom) is present, and the mouth is surrounded by a ring of hollow, ciliated tentacles (collectively called a lophophore), which can be retracted inside the zooecium. The anus lies outside the lophophore.

Bryozoa are among the commonest animals inhabiting stony and rocky shores and seabeds, yet they are often neglected – probably on account of their small size and lack of commercial value. They are divided into three classes: Phylactolaemata, Stenolaemata and Gymnolaemata, of which the first class is exclusively freshwater and the second consists largely of fossil forms. The body plan is quite characteristic. Individuals dwell in a colony which has been developed by asexual budding from one ancestral animal. Colonies such as *Membranipora* often encrust rocks, shells or seaweeds, but other bryozoa like *Bugula*, grow largely unsupported, being attached only at their bases. The zooecium is sometimes hardened by chalky secretions from the body wall; and its shape may be flat and box-like, or tubular. The mouth leads to a U-shaped gut lying inside the zooecium, where the reproductive organs are also housed. Circulatory and excretory structures appear to be lacking. The general form of the bryozoan animal is shown below in fig. 36.

One striking feature of this phylum is the evolution of polymorphic individuals within a colony. Many members of the colony may be feeding individuals, but some are specialized to fulfil other roles. One type called *avicularia* resemble minute birds' beaks. They appear to defend the other members from small organisms which would otherwise creep over them and possibly clog them up. *Vibracula* are another variety of zooid; these bear miniature paddles, which probably aid the circulation of water around the colony and discourage the accumulation of silt and other particles. The principal type of feeding individuals filter small food particles from the surrounding seawater by means of the ciliated tentacles of the lophophore. Fig. 37 shows the form of avicularia and part of a colony.

Although each colony develops by budding from an ancestral individual, sexual reproduction leads to the development and dispersal of free larvae. After a period in the plankton these larvae metamorphose into new ancestral individuals assuming they can find a suitable place on which to settle, and then develop a new colony by asexual budding. In some species the developing embryos are brooded either inside the body or in special pouches called *ooecia*. Colonies usually contain both male and female zooids, but sometimes the zooids themselves are hermaphrodite. Bryozoa do not generally flourish in brackish water, but a few species, for

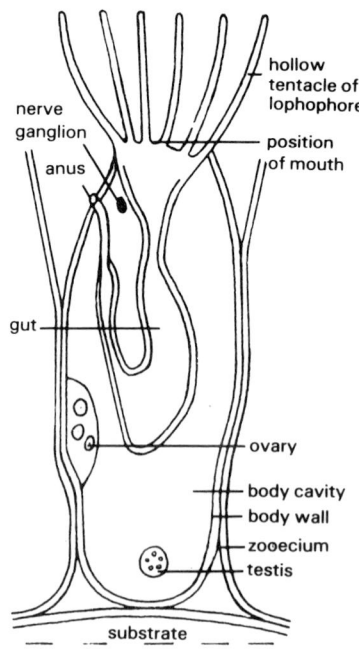

Fig. 36 Diagrammatic arrangement of an individual or zooid

instance *Electra crustulenta*, can tolerate low salinities.

Some clue to the identification of a bryozoan colony will immediately be given by its overall appearance – e.g. erect or branching; flat or encrusting, etc. – but a good hand lens will be essential in order to show the fine details of the zooids themselves which are necessary to establish final identification. The general shape of the zooecium, the position of the opening through which the lophophore is protracted, the presence of an operculum (effectively a lid) for occluding the opening, the presence of ooecia, aviculcariae, etc., are all very important in helping to make an identification. Some species also carry spines and bristles which also assist with their identification. Unfortunately, many more species of bryozoa occur in the Mediterranean area than can be described here. References which will help with further identification of many Mediterranean bryozoa are: Hayward, P. J. and Ryland, J. S. 1979; Hincks, T. 1880; Medioni, A. 1970; Prenant, M. and Bobin, G. 1956; Ryland, J. S. 1962, 1970 and 1974; and Ryland, J. S. and Hayward, P. J. 1977.

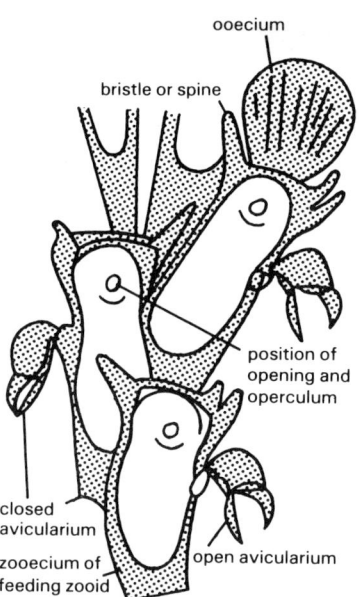

Fig. 37 Arrangement of a few individuals of *Bugula*

Class Stenolaemata

Bryozoa with tubular zooecia and with zooecial walls calcified. The lophophore is often circular.

Order Cyclostomata

Zooecial opening rounded. No operculum, avicularia or vibracula.

Crisia eburnea (Linnaeus) Branched jointed colony up to 20mm high, attached to substrate at base; a long spine grows behind each tubular zooid. **Colour** grey-white. **Habitat** on red seaweeds and on shells and rocks from the shore down to 50m. **Similar species** few.

Class Gymnolaemata

Zooecia tubular and box-like; the zooecial walls may be calcified. The lophophore is circular.

Order Ctenostomata

Zooecial walls not calcified; zooecial opening may be shut by a collar. No operculum, ooecia, avicularian or vibracula are present.

Alcyonidium polyoum (Hassall) Gelatinous smooth sponge-like colony of irregular shape; up to 200mm high; polyps set in mass of colony with no spines or ornaments on the surface. **Colour** yellow, green, grey or brown. **Habitat** from the shore down to 100m. **Similar species** 4, including *A. gelatinosum* (Linnaeus) (not illustrated) which has a more regularly shaped colony.

Zoobotryon verticillatum (Delle Chiaje) Branched tufted colony up to 500mm high; zooecia round or oval, arranged spirally on the branch tips. **Colour** opaque white, occasionally greenish. **Habitat** growing in clusters on submerged objects, especially in harbours. **Similar species** none. N.B. this is an important fouling organism.

Bowerbankia pustulosa (Ellis & Solander) Tufted branching colony up to 30mm high, attached at base; zooids living at intervals along the stems, often in groups. **Colour** brownish. **Habitat** in shallow and deeper water attached to weeds and rocks. **Similar species** *B. imbricata* (Adams), which reaches up to 70mm; *B. gracilis* Leidy, which has very long zooids.

Order Cheilostomata

Zooecia generally flattened; the zooecial walls are calcified. An anterior opening guarded by an operculum is present. Ooecia, avicularia and vibracula are present.

Aetea sica (Couch) Branching colony of tubular zooids, each with a part adhering to the substrate and a part standing erect and free, arising from a swelling. The adhering parts form a mass of root-like growths and the erect parts are marked with many close rings or annuli; height of individual zooids up to 1.8mm. **Colour** white. **Habitat** attached to various substrates down to 80m. **Similar species** two.

Membranipora membranacea (Linnaeus) **Sea Mat** Mat-like encrusting colony of varying size, outline often rounded, according to substrate; zooids rectangular and beset with a blunt bristle on each side; elevated growths called 'towers' may occur. **Colour** white-grey, often contrasting with substrate. **Habitat** on weeds and possibly shells on the shore and in shallow water. **Similar species** *Electra pilosa* (below).

Electra pilosa (Linnaeus) **Hairy Sea Mat** Similar to *Membranipora membranacea* (above) but colonies less rounded, more irregular and angular in outline; zooids bear 2 blunt bristles at one end and 1 conspicuous spine at the other, and various smaller bristles all round. **Colour** silver-grey. **Habitat** on weeds and stones on the shore and in shallow water. **Similar species** five, including *E. posidoniae* Gautier, which encrusts the leaves of *Posidonia* (see page 44).

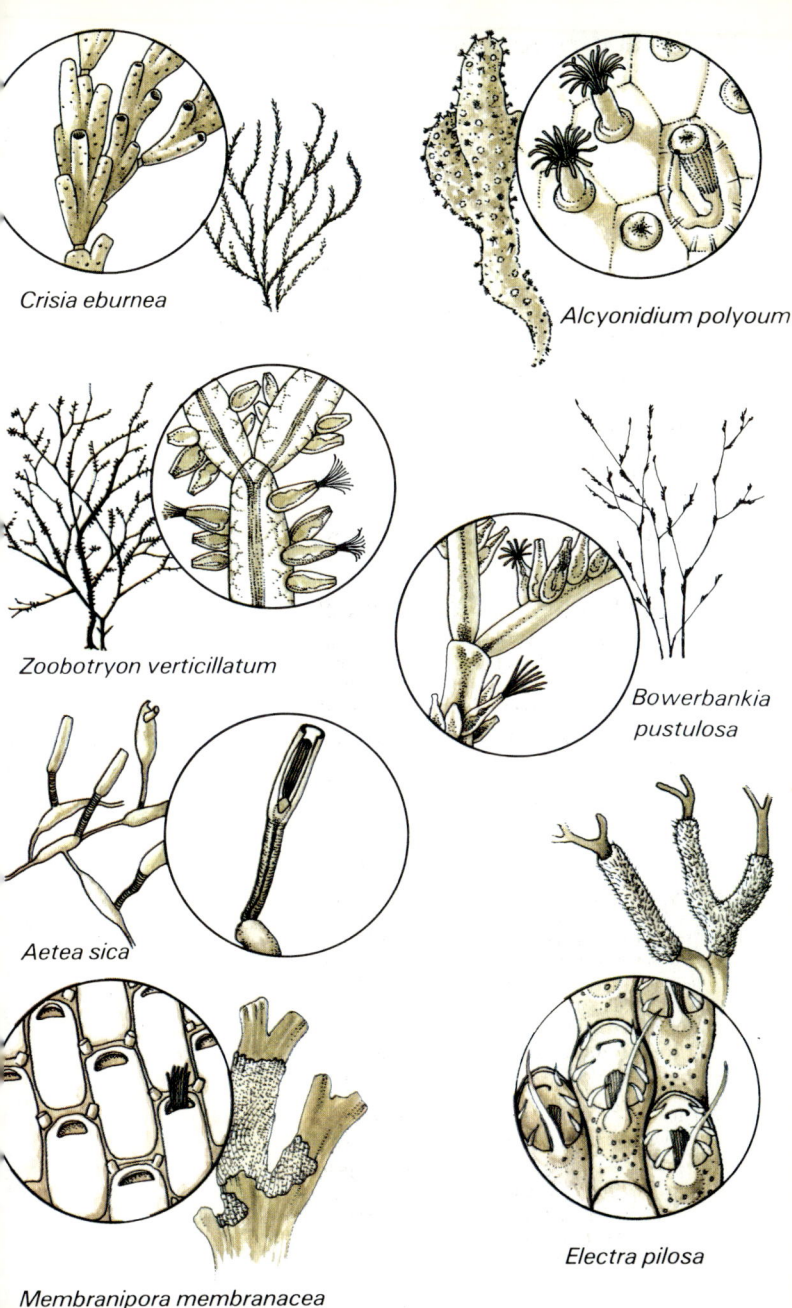

Crisia eburnea

Alcyonidium polyoum

Zoobotryon verticillatum

Bowerbankia pustulosa

Aetea sica

Electra pilosa

Membranipora membranacea

Carbasea papyrea (Pallas) Colony of flattened branching tufts, the branches expanding in width towards tips; height up to 50mm; zooids present on one side of colony only; zooids of irregular shape, lacking spines and ornaments. **Colour** brown-grey-white with pink developing embryos sometimes visible. **Habitat** on *Posidonia* (see page 44) and coralline algae, and on shells and sometimes the carapaces of crabs down to 100m. **Similar species** one. The genera *Flustra* and *Securiflustra* have zooids on both sides of the flat fronds.

Cellaria salicornioides Lamouroux Diffuse colony of delicate, dichotomously branching, tufted filaments with conspicuous joints, up to 50mm high. Zooids are oval or hexagonal in outline and arranged in alternating series. **Habitat** down to 100m on coarse gravel and sand. **Similar species** two.

Scrupocellaria scrupea Busk Erect branching colony up to 20mm high, with root-like rhizoids towards the base; zooids widen towards their distal ends and bear 3–4 spines on the distal outer point and 1–2 on the distal inner point; prominent avicularia. **Colour** orange-brown. **Habitat** on stones, rocks and shells down to quite deep water. **Similar species** eight.

Bugula neritina (Linnaeus) Thick tufted, dichotomously branching colony with branches showing a slight spiral twist near their tips, reaching up to 80mm high; zooids in the branches are arranged in 2 rows and are rounded; zooids lack spines and avicularia; frontal region is membranous; the outer distal corner projects slightly from the line of the branch. **Colour** brown-purple. **Habitat** a serious fouling organism growing on buoys, ships, piers and pilings as well as inside inlet pipes. **Similar species** ten, including *Bugula flabellata* (Thompson) which has low thick tufts up to 30mm in height, zooids with several spines and avicularia with hooked beaks; also *Bugula turbinata* Alder, which has stout bushy spiral fronds up to 50mm high with elongated zooids having 2 bristles at one end; avicularia present. **Colour** orange, but often covered in muddy sediment. **Habitat** under overhanging rocks, amongst weeds and sponges on the shore and in shallow water.

Myriopora truncata (Pallas) **False Coral** Colony resembling a coral: branches have flat tips but lack the septa of true corals; reaching up to 100mm high with rounded zooids embedded in the stems. **Colour** yellow-red. **Habitat** on rocks in crevices and caves, usually in shallow water in shaded places. **Similar species** few; for corals see pages 80–85.

Pentapora fascialis Pallas *(=Hippodiplosia fascialis)* Large conspicuous colonies reaching 200mm high; oval to rhomboidal zooids, tightly grouped. **Colour** orange-pink. **Habitat** on hard substrates among corals etc., down to 25m. **Similar species** none.

Margaretta cereoides (Ellis & Solander) *(=Tubucellaria opuntioides)* Branching colony up to 50mm high; oval zooecia pressed together forming the stem. **Colour** brown-yellow. **Habitat** in shallow water, often associated with roots of *Posidonia* (see page 44). **Similar species** few.

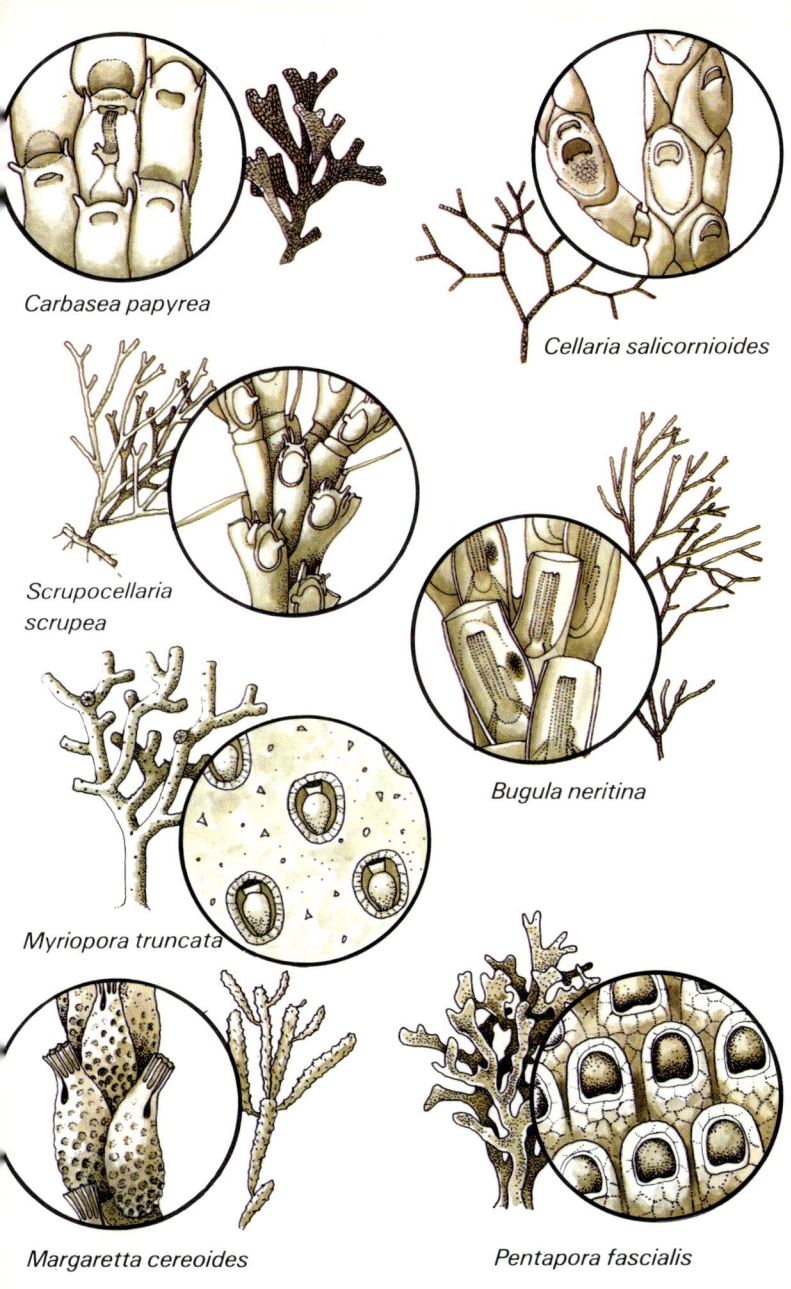

Carbasea papyrea

Cellaria salicornioides

Scrupocellaria
scrupea

Bugula neritina

Myriopora truncata

Margaretta cereoides

Pentapora fascialis

Phylum Echinodermata

Echinoderms (the word means 'spiny-skinned') are unique in the animal kingdom. They are exclusively marine and they display in their adult form a unique type of symmetry known as pentamerism. This symmetry is essentially radial, with a mouth in the centre on one side of the body and the anus normally in the centre on the opposite side. The body may be disc-shaped or globular as in the sea-urchins or it may be drawn out into five or more radii as in the starfishes and brittle-stars.

Present-day echinoderms are divided into five distinct classes, although fossil evidence shows that other groups previously existed. Each of these extant classes is represented in the Mediterranean Sea: they are the Crinoidea (sea-lillies and feather-stars); the Asteroidea (starfishes); the Ophiuroidea (brittle-stars); the Echinoidea (sea-urchins, sand-dollars and heart-urchins) and the Holothuroidea (sea-cucumbers). All of these animals have a number of features in common, which usually make it easy to recognize echinoderms as such, yet at the same time each class has features sufficiently different to allow one to decide which one it is without difficulty.

Echinoderms are triploblasts (i.e. their bodies are composed of three layers originating from ectoderm, mesoderm and endoderm). The skeleton is basically internal but it sometimes protrudes to the exterior as in the spines of a sea-urchin. This skeleton consists of many plates of calcium carbonate; some of these are linked together to form the test or so-called 'shell', while others are mounted on this test to form spines. In most species apart from the echinoids the test plates are loosely connected together so that the animals are relatively flexible. The outer surface of the test is covered by a thin layer of epidermal cells which are often highly pigmented. Most of the organ systems lie within the large body cavity carried inside the test. These include the digestive and reproductive systems as well as the greater part of the unique water-vascular system. There appears to be no distinct osmoregulatory system, and perhaps this is why the echinoderms cannot survive in water with reduced salinity. The sexes are usually separate and synchronous spawning often takes place at certain times of the year. External fertilization occurs in the sea and leads to the formation of a pelagic larva which passes through several stages in the plankton before settling to metamorphose into a juvenile.

The water-vascular system is manifest outside the test in the form of double rows of tube-feet. These tube-foot rows are known as ambulacra, and there are usually five (one for each radius of the body). Each tube-foot is elastic and can be extended by being filled with fluid under pressure from within the body. Muscles along the shaft can contract, and empty the fluid from the tube-foot. They can also cause the tube-feet to bend, thus allowing locomotory 'steps'. Most starfish as well as the sea-urchins and sea-cucumbers, have suckers on their tube-feet, and these can grip the substrate and serve as locomotory organs. This does not occur in the feather-stars or the brittle-stars which rely on flexion of their arms for movement. The internal anatomy of the water-vascular system is complex, and although it appears to open to the exterior via a special sieve-like test plate (the madreporite) there is little evidence for water movements in and out of it. In addition to locomotion the water-vascular system is involved with respiration and food collection.

In the crinoids the mouth and anus are borne on the same side of the disc and face away from the substrate. Cri-

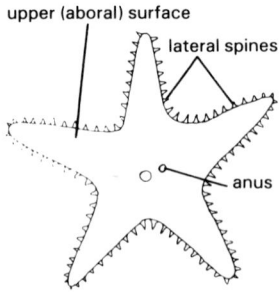

Fig. 38 External features and symmetry of the starfish *Astropecten irregularis*

noids anchor themselves to the sub-strate by means of special appendages (cirri) on their undersides, and use their tube-feet, which lack suckers and which occur in great numbers on the branching arms, to filter the seawater and collect from it small particles of suspended food matter. These particles are then passed down the arms to the mouth.

Asteroids hunt their prey by relying on a sense of smell to track it down. They frequently evert their stomachs over the prey or insert it into gaping shells, etc. Digestion then occurs and the products can be absorbed. The

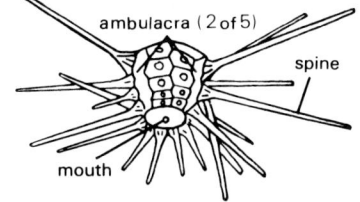

Fig. 39 External features of the sea-urchin *Cidaris cidaris* (some spines removed)

mouth is on the underside and is thus well positioned for such behaviour. The small spines of asteroids are not as freely moveable as those of the echinoids.

Ophiuroids are scavengers and use their suckerless tube-feet to pass food to their mouths. They move by means of their flexible arms. An anus is lacking and waste is passed out through the mouth.

Most round sea-urchins live by browsing on plant and animal growths which cover the rocks. A five-toothed chewing organ (the lantern of Aristotle) is carried inside the test; this also bears moveable spines which serve a variety of functions including locomotion, defence and sensory detection. In some of these functions the spines are assisted by minute pincer-like organs (pedicellariae) which help to keep the test surface clean, and which also assist with defence. The tube-feet have suckers and can usually be extended considerable distances; at such times they often resemble guy-ropes assisting in posture and locomotion. The sand-dollars and heart-urchins are modified for burrowing and derive their food from organic substances in the sand and gravel where they live.

The skeletal elements of the holothuroids are often considerably reduced, giving the animals a softer texture. Modified tube-feet are arranged around the mouth to gather food either by filtering sea water or by sweeping organic matter from the surface of sand. The sea-cucumbers usually progress with one end leading, in the same manner as the sand-dollars and heart-urchins. In some cases the remaining tube-feet are not developed and the animal moves in a worm-like fashion. Further information is contained in Demetropoulos, A. and Hadjichristophorou, M. 1976; Koehler, R. 1921; Mortensen, T. 1927; Nicholls, D. 1969 and Southward, E. 1972.

Class Crinoidea Sea-lilies and feather-stars

Echinoderms with a cup-shaped body bearing branching arms of which there are usually five pairs. These arms carry tube-feet which are used for filtering food from the seawater and for respiration, but not for locomotion. However the arms themselves, which are flexible, are used for creeping and swimming. The mouth and anus are situated on the upper side of the body and below it is either an attachment stalk (sea-lilies) or a number of short gripping appendages called cirri for holding on to the substrate (feather-stars). Mediterranean recorded species: sea-lilies = 0, feather-stars = 2.

Antedon mediterranea (Lamarck) **Feather-star** Diameter overall up to 200mm or more. **Body** small and inconspicuous; five pairs of branching arms which often wave slowly in the water. **Colour** yellow-red-brown. **Habitat** on rocks, stones and seaweed and sometimes on gorgonians (see page 84) holding on with its cirri; down to 40m. **Similar species** *Leptometra phalangium* O. F. Müller. (Not illustrated) Diameter overall up to 300mm or more. **Body** small and inconspicuous; five pairs of branching arms, cirri longer than *Antedon mediterranea*, reaching up to one-third of arm length. **Colour** green in life but fading after death. **Habitat** in deeper water, from 70m down; usually on hard substrates.

Class Asteroidea Starfishes

Echinoderms in which the body normally bears rays, usually 5 in number. The mouth opens on the underside and the anus on the upper side. Tube-feet for locomotion are carried on the underside of each ray and usually bear suckers. Mediterranean recorded species = 18.

Luidia ciliaris (Philippi) Diameter overall 400mm or more. **Body** small in relation to overall size; always has 7 flattened rays; tube-feet end in knobs rather than suckers. **Colour** orange-red above, white below. **Habitat** on sand and mud, sometimes buried, from the shore down to 150m. **Similar species** *Luidia sarsi* and several *Astropecten* species, but these all have 5 rays; *Coscinasterias tenuispina* may have 7 rays but these are rounded in section, spiny, and tube-feet have suckers (see below).

Luidia sarsi Düben and Koren (Not illustrated) Diameter overall 200mm. **Body** small in relation to overall size; always has 5 flattened rays; tube-feet end in knobs rather than suckers. **Colour** yellow-red-brown above, sides of rays may be darker; paler below. **Habitat** on soft substrates from 10m down; absent from the Adriatic. **Similar species** *Luidia ciliaris* (above) has 7 rays; *Astropecten* species (below) all have conspicuous rows of spines along the edges of their rays; *Hacelia attenuata* (page 232) has rounded rays with suckered tube-feet.

Astropecten species Note at least 5 species of *Astropecten* have been described from the Mediterranean. **Bodies** flattened, with 5 rays which are bordered by conspicuous rows of spines giving a comb-like appearance; bodies small in relation to overall size; tube-feet lack suckers. **Habitats** crawling over or burrowing in sand and mud from the shore down to 300m or more. Separation of the species depends on examining the dorsal and ventral marginal plates of the side of a ray and counting the spines (see inset diagrams). **Species data:** *A. aranciacus* (Linnaeus) Diameter overall 600mm, very conspicuous marginal plates; **colour** orange-red above; yellow below. *A. bispinosus* Otto Diameter overall 80mm; **colour** dark olive-brown above; paler below. *A. spinulosus* (Philippi) Diameter overall 70mm; **colour** dark brown-red-green; paler below. *A. irregularis* (Linck) Diameter 80mm; **colour** yellow-brown above; paler below. Only *A. aranciacus* illustrated. The polychaete worm *Acholoë astericola* often occurs in the ambulacral groove of *Astropecten*.

Ceramaster placenta (Müller & Troschel) Diameter 160mm. **Body** flat, pentagonal and very solid. **Colour** brown-yellow-red. **Habitat** soft substrates from 30m downward.

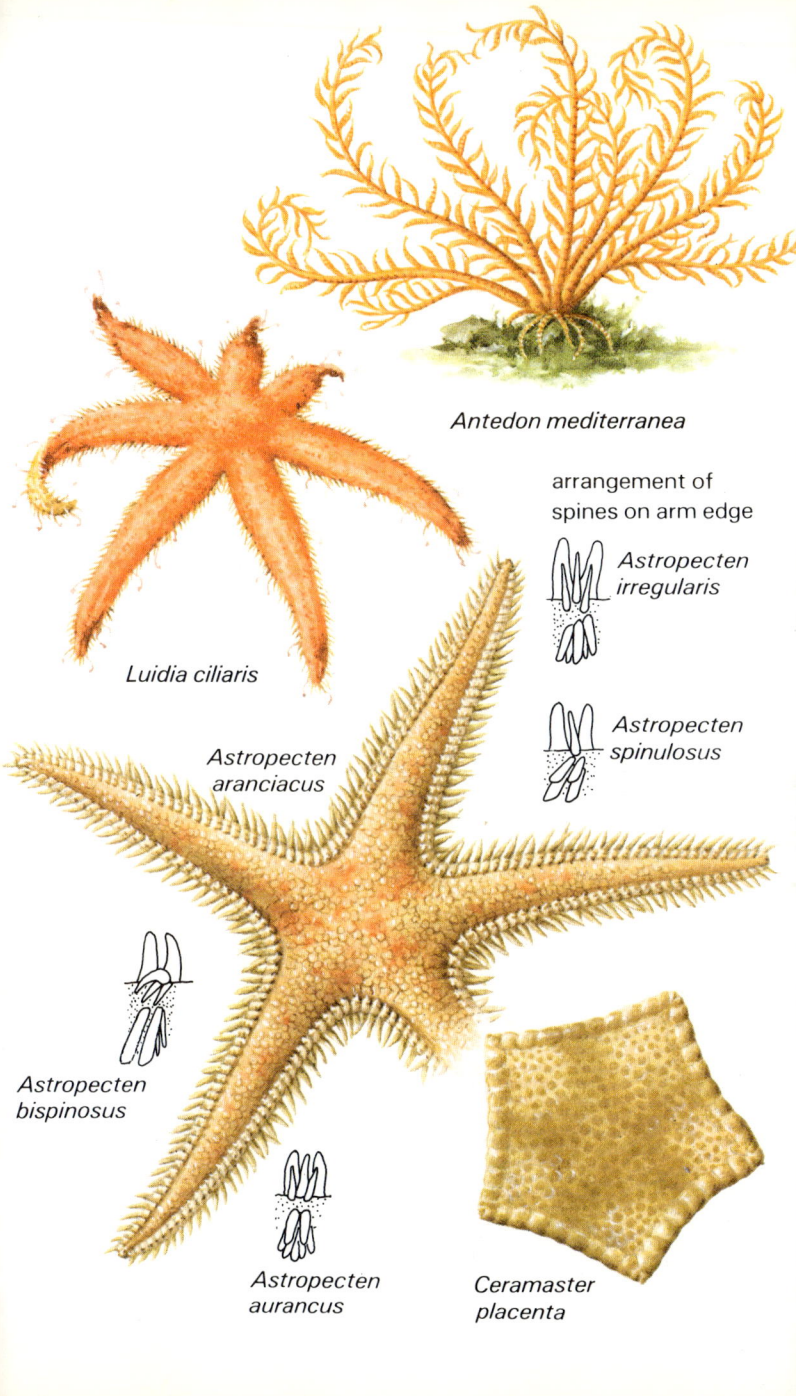

Antedon mediterranea

arrangement of
spines on arm edge

Astropecten
irregularis

Astropecten
spinulosus

Luidia ciliaris

Astropecten
aranciacus

Astropecten
bispinosus

Astropecten
aurancus

Ceramaster
placenta

Ophidiaster ophidianus Lamarck Overall diameter up to 200mm. Body very small in relation to rays; long rays rounded in section; not tapering until tips; tube-feet suckered, ambulacra guarded by rows of short spines. Colour bright violet-red above, may be paler below. Habitat on rocks from 1m downward in warmer regions of the Mediterranean, e.g. Southern Italy; absent from Adriatic. Similar species *Echinaster sepositus* (below).

Hacelia attenuata Gray Overall diameter up to 200mm. Body small in relation to overall size, with rows of 'pock-marks' on upper surface; rounded tapering rays; tube-feet suckered. Colour brown-red-scarlet. Habitat on rocks from 1 to 150m. Similar species *Echinaster sepositus* (below).

Echinaster sepositus Gray Overall diameter up to 200mm. Body not large, covered in soft skin; rounded, gradually tapering rays; tube-feet suckered; upper surface covered in irregularly arranged 'pock-marks'. Colour scarlet. Habitat on rocks and softer substrates from 1 to 250m. Similar species *Ophidiaster ophidianus*, which has rays almost parallel until the tip; *Hacelia attenuata* which has 'pock-marks' arranged in rows.

Anseropoda placenta (Pennant) *(=Palmipes membranaceus)* **Goose-foot Star** Overall diameter up to 150mm. Body flat, 5-sided, with rays not clearly distinguishable; often looks rather tattered with patterns of red and white; tube-feet suckered. Colour red-white above, yellow-white below; detailed distribution of pigment is variable. Habitat on sand and mud from 10 to 100m. Similar species only *Asterina* species which are brown-green.

Asterina gibbosa (Pennant) **Cushion-star** or **Starlet** Overall diameter up to 50mm, occasionally more. Body 5-sided, with rounded tips to the rays; not as flat as *Anseropoda placenta*; tube-feet suckered; cryptic colour and habits prevents identification at a distance. Colour brownish, patterned with red-brown and paler markings when viewed under a lens. Habitat under rocks and stones from the shore down to 100m. Similar species *A. phylactica*.

Asterina phylactica Emson and Crump (Not illustrated) Overall diameter up to 15mm. Body 5-sided with rounded tips to the rays, tube feet suckered, Body not as flat as that of *Anseropoda*. Colour greenish with a dark-brownish red over the radii. Habitat in rock pools and shallow water. Similar species *A. gibbosa* which usually has 2 small spines on each of the 5 angles surrounding the mouth. These are lacking in *A. phylactica*.

Marthasterias glacialis (Linnaeus) **Spiny Starfish** Overall diameter up to 800mm but frequently much smaller. Body upper surfaces spiny; small in relation to overall size; 5 rays, rounded in section, gradually tapering and bearing conspicuous spines surrounded by rings of small pincer-like organs (pedicellariae) discernible with a hand lens; tube feet suckered; cryptic coloration may hinder identification at a distance. Colour dark brown-yellow with greenish-grey markings above, paler below. Habitat on rocks and stony substrates from the shore down to 180m. Similar species *Coscinasterias tenuispina*, which always has more than 5 arms.

Coscinasterias tenuispina (Lamarck) Overall diameter up to 150mm. Body small in relation to overall diameter; between 6 and 10 rays, not all of same length, not very flattened like *Luidia* and bearing spines surrounded by pedicellariae (see *Marthasterias glacialis* above); tube-feet suckered. Colour variable: basic shade may be white, red-brown or purple with blue or brown spots. Habitat on rocks and stones from the shore down to 30m. Similar species *Luidia ciliaris* and *Marthasterias glacialis* (above).

Hacelia attenuata

Ophidiaster ophidianus

Asterina gibbosa

Anseropoda placenta

Echinaster sepositus

Marthasterias glacialis

Coscinasterias tenuispina

Class **Ophiuroidea** Brittle-stars

Echinoderms normally with 5 unbranched fragile jointed arms bearing spines, and suckerless tube-feet on the underside. Flattened rounded body disc bears mouth on underside but lacks anus and pedicellariae. The identification of brittle-stars can be very difficult and frequently depends on the recognition of small structures best seen with a hand lens. Koehler R. 1921, Mortensen, T. 1927 and Southward, E. 1972 provide detailed accounts of many Mediterranean species. The number and character of spines carried on one side of each arm joint (arm spine number, or ASN) is a useful identification guide. Mediterranean recorded species = about 20.

Ophiomyxa pentagona Müller & Troschel Disc diameter up to 25mm; somewhat 5-sided in appearance, devoid of small spines or conspicuous plates; arm length about 4 times disc diameter. **Arms** may be coiled and may show signs of regeneration; arm spines short and inconspicuous, arranged close together in groups; ASN 4–5. **Colour** brownish, sometimes with markings. **Habitat** on substrates, among algae, from the shore down to 100m. **Similar species** many.

Ophiothrix fragilis (Abildgaard) Disc diameter up to 20mm; often pentagonal; points visible between the arms or at base of arms; upper surface of disc has many minute spinelets and some longer ones arranged in 5 radiating V-shaped groups over each arm and on either side of 2 naked triangular plates. **Arms** not more than 5 times disc diameter, fragile and often regenerating; arm spines conspicuous and finely thorned; lowermost arm spine hooked; ASN 7. **Colour** very variable: red-brown-purple-green and patterned. **Habitat** under rocks, seaweeds and shells from the shore down to 350m. **Similar species** many.

Ophiothrix quinquemaculata (Delle Chiaje) Disc diameter up to 15mm; similar to *O. fragilis*. **Arms** about 8 times as long as disc diameter; arm spine hooked; ASN 6. **Colour** red-pink-grey with patterns. **Habitat** usually between 20 and 100m on soft substrates and among algae. **Similar species** many.

Amphiura chiajei Forbes Disc diameter up to 8mm; pentagonal but rounded over the origin of each arm; upper surface of disc covered with scale-like plates; 2 large plates arranged like a V above the origin of each arm; arms long and delicate, about 7 times the diameter of the disc; arm spines not long and conical; ASN 5. **Colour** orange-red. **Habitat** on soft bottoms and among algae from the shore down to 50m or more. **Similar species** many; see references above (about 18 species recorded for Europe as a whole).

Amphiura filiformis (O. F. Müller) (Not illustrated) **Disc** diameter up to 10mm; similar to *A. chiajei* but second arm spine has characteristic flattened head; ASN 5. **Colour** grey-brown-red. **Habitat** on sandy bottoms and among pebbles between 5 and 50m and deeper. **Similar species** many (about 18 in Europe).

Amphipholis squamata (Delle Chiaje) Disc diameter up to 5mm; rounded, with 2 conspicuous pale plates above each arm, the rest of the upper surface covered with scale-like plates. **Arms** up to 4 times disc diameter; arm spines short and conical; ASN 3–4 (4 nearer disc). **Colour** bluish-grey-white. **Habitat** under pebbles, rocks and seaweeds from the shore to 250m. **Similar species** many.

Ophioderma longicauda (Retzius) Disc diameter up to 30mm; granulated, leathery appearance, apparently notched above the origin of each arm. **Arms** up to 4 times disc diameter in length; arm spines short and lying against the arm; ASN 15. **Colour** brownish-green, patterned. **Habitat** sandy and rocky bottoms from the shore down to 70m. **Similar species** many; see references above.

Ophiura texturata Lamarck Disc diameter up to 30mm; rounded above with coarse scales; 2 conspicuous plates above the origin of each arm, do not meet. **Arms** tapering, not quite reaching 4 times the disc diameter; arm spines tapering, shorter than the arm width at that point and lying against the arm itself; ASN 3; plates on underside of arm separated by pores. **Colour** reddish-brown above, pale below. **Habitat** burrowing in sand from the shore down to 200m. **Similar species** about 16 recorded from Europe including *O. albida* Forbes.

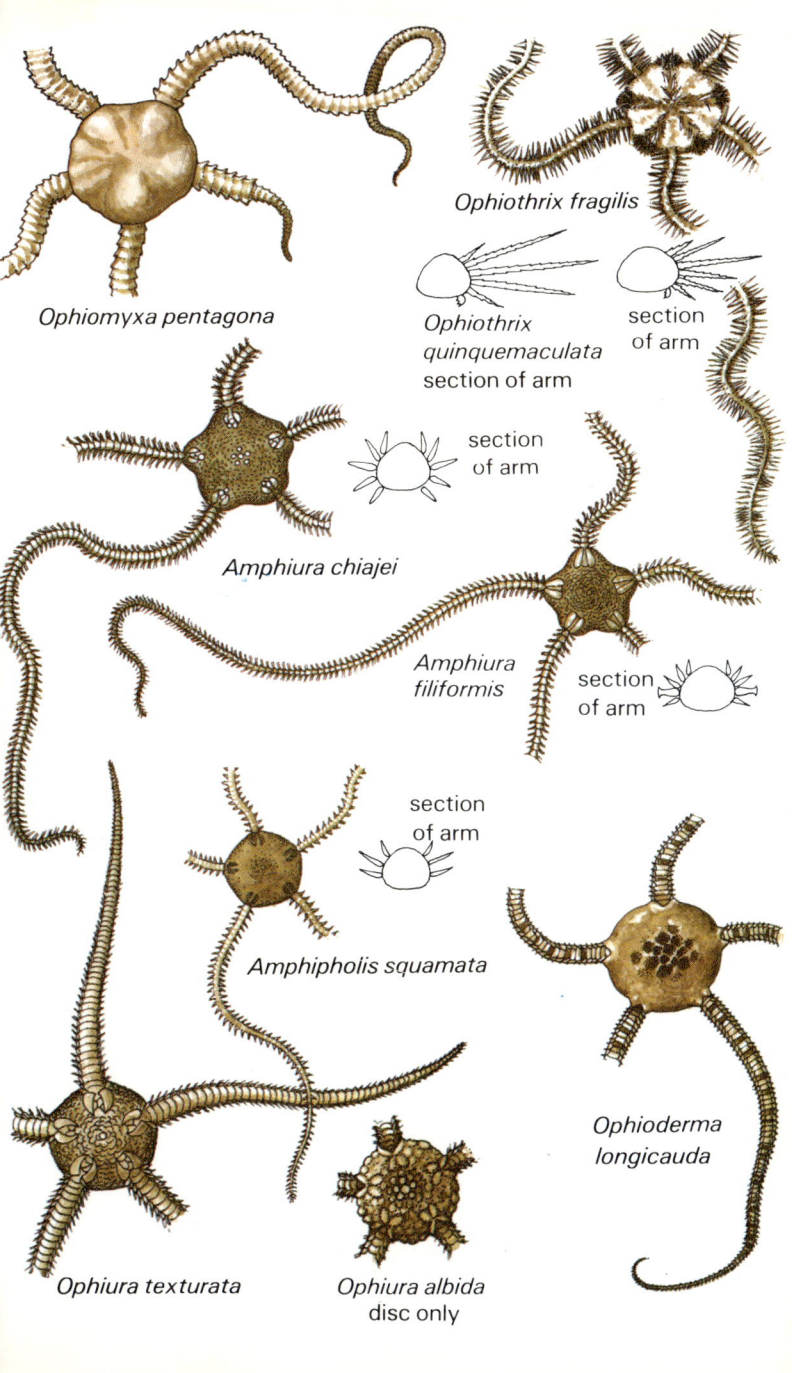

Ophiomyxa pentagona

Ophiothrix fragilis

Ophiothrix
quinquemaculata
section of arm

section
of arm

section
of arm

Amphiura chiajei

Amphiura
filiformis

section
of arm

section
of arm

Amphipholis squamata

Ophioderma
longicauda

Ophiura texturata

Ophiura albida
disc only

Class Echinoidea
Sea-urchins, sand-dollars and heart-urchins

Subclass Perischoechinoidea
Order Cidaroida and Subclass Euechinoidea
Superorders Diadematacea and Echinacea

Spherical echinoderms with a chalky shell-like skeleton (the test) bearing mobile spines mounted externally on small knobs and being perforated by 5 double rows of pores which in life permit fluid to pass into the long tube-feet from within the body. The mouth is on the underside and has chewing teeth, and the anus is at the apex of the test. Mediterranean recorded species = 9.

Cidaris cidaris (Linnaeus) *(=**Dorocidaris papillata**)* Test up to 70mm in diameter. **Spines** large spines reach twice the test diameter and are often encrusted with sponges, hydroids etc; solid when broken; with longitudinal ridges made up of rows of very fine thorns. Many small spines are arranged around the bases of the large spines and on either side of the rows of tube-feet. **Colour** yellow-green-red-brown. **Habitat** various; from 30m down. **Similar species** *Stylocidaris affinis* and *Centrostephanus longispinus*.

Stylocidaris affinis (Philippi) Test up to 40mm in diameter. **Spines** large tapering spines reach a little more than test diameter; solid when broken; with many small thorns. Many small spines are arranged around the bases of the large spines and on either side of the rows of tube-feet. **Colour** orange-brown. **Habitat** on coralline algae and rocks from the shore down to 30m. **Similar species** *Cidaris cidaris* and *Centrostephanus longispinus*.

Centrostephanus longispinus (Philippi) Test up to 60mm in diameter. **Spines** large spines tapering to a sharp point, reaching twice test diameter, long and slender with minute thorns; hollow when broken. Smaller spines are more equally distributed over the test and not arranged around the bases of the larger ones nor on either side of the tube-foot rows. **Colour** of test red-brown; spines banded brown and white. **Habitat** normally below 40m and often in very deep water. **Similar species** *Cidaris cidaris* and *Stylocidaris affinis*. **Note** this species may respond actively to light from an underwater torch by rotating its spines and moving away; the sharp spines may inflict painful wounds.

Arbacia lixula (Linnaeus) **Black Sea-urchin** Test up to 50mm in diameter; this species is difficult to identify with certainty from a distance. **Spines** up to 30mm long, smooth, solid when broken and with sharp tips. **Colour** in life black; the cleaned test is pinkish with characteristic red lines marking the position of the tube-foot pores. **Habitat** on rocks, often among coralline algae from the shore down to 40m. **Similar species** *Paracentrotus lividus*: these two species often live together in large 'flocks' just below the surface. The only reliable way to tell them apart is to examine them closely. *A. lixula* has a large oral opening in the test which in life is overlain by an extensive soft membrane surrounding the mouth. *P. lividus* has a relatively smaller oral opening in the test and a smaller membrane surrounding the mouth. The test of *P. lividus* when cleaned is grey-green. **Note** the sharp spines of *A. lixula* are frequently the cause of painful wounds to bathers. They are difficult to extract from feet and hands but often come out several days after the injuy when subjected to pressure.

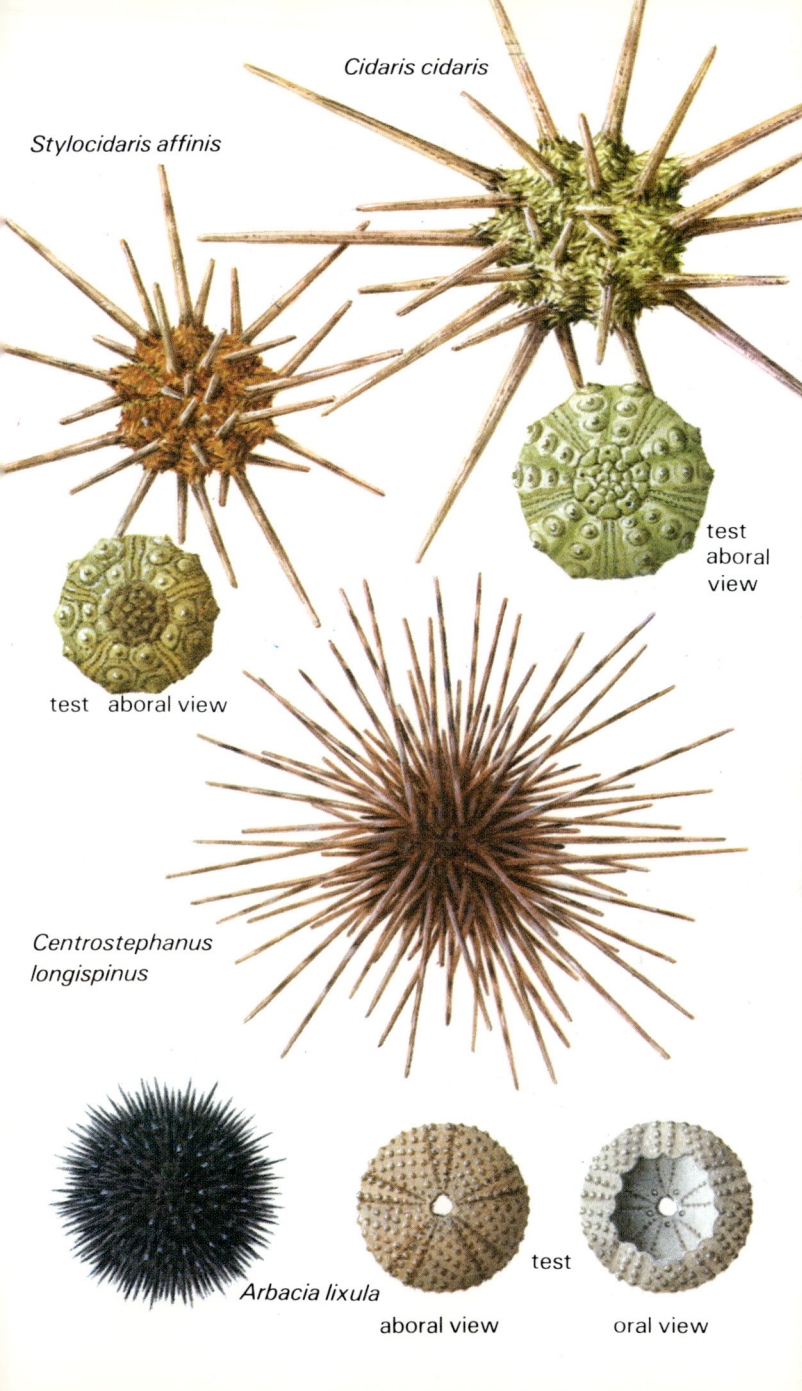

Cidaris cidaris

Stylocidaris affinis

test
aboral
view

test   aboral view

*Centrostephanus
longispinus*

Arbacia lixula

aboral view

test

oral view

Sphaerechinus granularis (Lamarck) **Violet** or **Purple Sea-urchin** Test up to 120mm in diameter. The cleaned test can be recognized by the 10 narrow slits (each about 2mm long) which occur around the oral opening on the underside. **Spines** reach 20mm in length. **Colour** this species may be easily recognized in life by its colours, the relatively short spines having purple shafts with white tips or being entirely white and conspicuous against the purple of the test. **Habitat** on rocks and coralline algae from the shore down to 100m. **Similar species** *Echinus acutus* and *Echinus melo* are of similar size but in each case their coloration is quite different; they also appear more 'bald', having a lower density of spines on the test.

Paracentrotus lividus (Lamarck) **Test** up to 60mm in diameter. This species is difficult to identify with certainty from a distance. **Spines** up to 30mm long, smooth, solid when broken. **Colour** in life variable from green to dark brown; the cleaned test is grey-green all over. **Habitat** on rocks, often among coralline algae from the shore down to about 30m; may be in pools. **Similar species** *Arbacia lixula* and *Psammechinus microtuberculatus*. To distinguish *Paracentrotus lividus* from *Arbacia lixula* examine membrane around mouth, oral opening and test colour (see above under *Arbacia lixula*). To separate *P. lividus* from *Psammechinus microtuberculatus* with certainty, an examination of the test with a lens is necessary. *Paracentrotus lividus* has test plates which bear five pore pairs (one pore pair per tube-foot) whereas *Psammechinus microtuberculatus* has only three pore pairs per plate. Superficially the species may be separated by the generally smaller size of *Psammechinus microtuberculatus*.

Psammechinus microtuberculatus (Blainville) **Green Sea-urchin** Test 350mm in diameter. This species is difficult to identify with certainty at a distance. **Spines** up to 15mm long, fairly slender. **Colour** greenish, spines with reddish tips; test when cleaned grey-green. **Habitat** on rocks and stones from 4 to 400m. **Similar species** *Paracentrotus lividus*: a close examination of the test may be necessary to make positive identifications (see above). N.B. this is a more delicate form of *Psammechinus* than *P. miliaris* (Gmelin) (not illustrated) which occurs in Northwest Europe.

Echinus acutus Lamarck Test up to 160mm in diameter, bulky, pink, conical in profile, almost as tall as broad, not well covered with spines. **Spines** may reach 20mm in length, many much shorter; do not densely cover test. **Colour** reddish-brown above, paler below. **Habitat** often on soft substrates from 20 to 1000m. Not found in Adriatic Sea. **Similar species** *Echinus melo* (below).

Echinus melo Lamarck (Not illustrated) **Test** up to 170mm in diameter, bulky, pink, globular in profile. **Spines** may reach 20mm in length, many much shorter. **Colour** reddish-brown-pink above, paler below. **Habitat** on rocky substrates, 20 to 1000m. **Similar species** *Echinus acutus*. **Note** the status of *E. melo* is doubted by some authorities who believe that it may represent a form of *E. acutus*. *Echinus esculentus* Linnaeus (not illustrated) which is a familiar species in the Atlantic sublittoral, is not found in the Mediterranean.

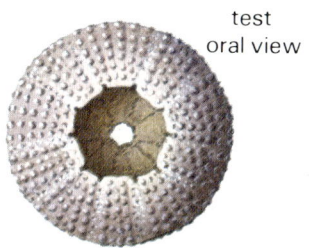

test
oral view

Sphaerechinus granularis

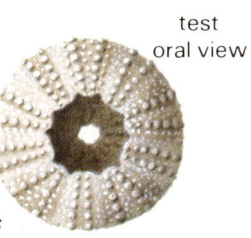

test
oral view

Paracentrotus lividus

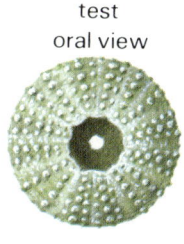

test
aboral view

test
oral view

Psammechinus microtuberculatus

Echinus acutus

test
lateral view

Superorder Gnathostomata
Order Clypeasteroida Sand-dollars

Bilaterally symmetrical disc-shaped echinoids with tube-feet which are mainly restricted to the upper side and are often arranged in a petal-like pattern. The anus is on the underside, sometimes close to the mouth which bears chewing teeth. Mediterranean recorded species = 1.

Echinocyamus pusillus (O.F. Müller) **Pea-urchin** Test up to 15mm in length. **Spines** very short and thickly set, giving a felt-like texture. **Colour** green-grey; test white when cleaned. **Habitat** burrowing in sand and gravel from 1 to 800m; inconspicuous. **Similar species** none.

Superorder Atelostomata
Order Spatangoida Heart-urchins

Bilaterally symmetrical heart-shaped echinoids with most tube-feet borne on the upper side but a few arranged below. The mouth and the anus are situated on the underside towards the anterior and posterior ends respectively. The mouth lacks chewing teeth. The test is thickly covered with fine spines which are moulded to fit the test contours and which give a fur-like appearance. These urchins are highly adapted for burrowing and will not be seen unless dug or dredged for. All species on this page could be confused, especially at a distance. Identification is dependent on specialist knowledge and the use of a reference such as Koehler, R. 1921. Mediterranean recorded species = 8.

Spatangus purpureus O.F. Müller **Purple Heart-urchin** Test up to 120mm long, bearing 5 rows of tube-feet of which the anterior row is the longest and lies in a pronounced but relatively shallow furrow leading towards the mouth. **Spines** mostly short but some on the upper side are longer. **Habitat** red-violet in life, test grey-white when cleaned. **Habitat** burrowing in coarse sand and shell gravel from 5–800m. **Similar species** all on this page, but the purple colour and heart shape are important in this species.

Echinocardium cordatum (Pennant) **Sea-potato** Test up to 90mm in length, bearing 5 rows of tube-feet of which the anterior row is the longest and unlike the rest lies in a deep furrow reaching nearly to the mouth. **Spines** mostly short, but some are long and curved, densely set and pointing backward. **Colour** yellow-brown in life; test yellow-white when cleaned. **Habitat** burrowing in sand from the water line to 200m. **Similar species** three other species of *Echinocardium: E. mediterraneum, E. flavescens* and *E. pennatifidum* have been recorded from the Mediterranean. For a description see Koehler, R. 1921.

Brissopsis lyrifera (Forbes) **Lyre-urchin** Test up to 70mm long, bears 5 rows of tube-feet, the posterior two being shorter than the anterior rows which lie in a frontal notch (not as deep as that of *Echinocardium cordatum)*; the anus is fractionally above the edge of the test so as to appear slightly on the upper side. **Spines** are short, dense and fur-like. **Colour** brown-red in life, yellow-grey when cleaned. **Habitat** burrowing in sand from 5 to 300m. Not recorded from the Adriatic Sea. **Similar species** all on this page, especially *Brissus unicolor.*

Brissus unicolor Klein Test up to 130mm in length, bearing 5 rows of tube-feet all of similar length, but with the front row reaching the mouth and not lying in a furrow; mouth well forward. **Spines** coat the test like fur. **Colour** yellow-brown in life, test greyish when cleaned. **Habitat** burrowing in muddy sand. **Similar species** all on this page but is relatively longer than the others.

Schizaster canaliferus (Lamarck) Test up to 70mm in length, broad at front and tapering towards posterior; bearing 5 rows of tube-feet of which the front row is decidedly the longest, reaching nearly to the mouth and lying in a furrow with a conspicuous indentation at the front of the test. **Spines** mostly short and thick-set, longer at the front. **Colour** in life grey with a pink tinge; cleaned test greyish. **Habitat** burrowing in sand and mud from 9 to 900m. **Similar species** all on this page, but the very long frontal tube-foot row is important as a distinguishing mark.

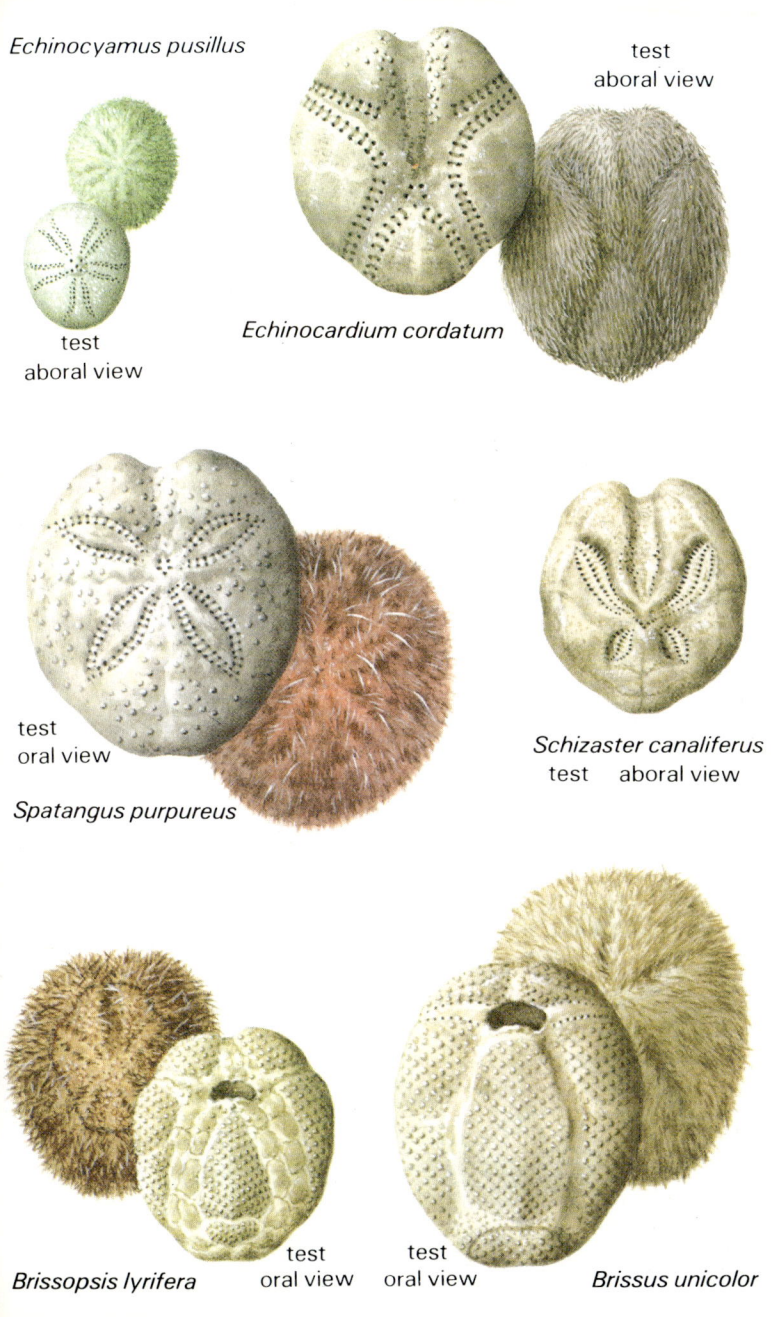

Echinocyamus pusillus

test
aboral view

test
aboral view

Echinocardium cordatum

test
oral view

Spatangus purpureus

Schizaster canaliferus
test   aboral view

Brissopsis lyrifera

test
oral view

test
oral view

Brissus unicolor

Class Holothuroidea Sea-cucumbers

Bilaterally symmetrical echinoderms lacking spines, arms and rays, and usually cucumber-shaped or worm like. Tube-feet present in many species, arranged in 5 rows along the sides of the animal; 3 rows usually in contact with the substrate and equipped with suckers for locomotion; 2 rows facing away from the substrate and often reduced. Mouth is at the anterior, often surrounded by modified branched tentacle-like tube-feet; anus posterior. Skeleton of loosely associated calcareous spicules embedded in the skin. Sea-cucumbers are not always easy to identify with certainty unless laboratory preparations of these spicules are made. For further details reference should be made to Koehler, R. 1921 or Mortensen, T. 1927. Mediterranean recorded species = 27.

Stichopus regalis Cuvier Length up to 300mm. Body with a tough solid-feeling skin; upper surface convex, covered with longitudinally arranged rows of white-bordered papillae; underside flat and sole-like, bearing subterminal mouth (not visible from above) and 3 rows of locomotory tube-feet; when viewed from above conical papillae appear to encircle the whole body. Colour basically yellow-brown above with white markings; underside brownish at edges with reddish middle area. Habitat on sandy substrates and among corals down to 400m. Similar species may be confused with *Holothuria* species; see reference above.

Holothuria forskali Delle Chiaje Length up to 200mm. Body with a thick skin but soft and easily deformable; upper surface convex and covered with about 7 rows of conical papillae which become white as they taper off but are dark at the tips; mouth terminal (visible from above); underside flat, 3 rows of tube-feet, those of the middle row being larger than those of the side rows. Colour black above, sometimes with an iridescence; underside paler. Habitat on mud and sand, some-times in *Zostera* beds (page 44) down to 100m. Note when handled may eject sticky white threads from the Cuvierian organs. Similar species several (see reference) especially *H. tubulosa* Gmelin (not illustrated) and *H.polii* Delle Chiaje (not illustrated). Both are of similar shape and form, reaching 150 to 200mm. The former is coloured deep brown or violet-maroon above and has irregular rows of tubercles and papillae above which are never white; skin is tough. *H. polii* has a velvety-black skin, being lighter brown below; the appendages are closely grouped.

Cucumaria elongata Duben & Koren Length up to 150mm. Body not easily divisible into upper and lower surfaces; often of curved shape and pointed towards the posterior; tube-feet apparently more evenly distributed and arranged in 5 double rows. Colour brown or greyish-brown, including the oral tentacles. Habitat on muddy substrates from 5–150m. Similar species about 9 have been recorded from the Mediterranean; see Koehler, R. 1921. *C. planci* Brandt is common in many parts; it is similar to *C. elongata*, reaching 100mm in length; body bears 10 branching feathery oral tentacles.

Thyone fusus (O. F. Müller) Length up to 200mm. Body plump, tapering to both ends; exterior covered by many irregularly arranged tube-feet; 10 oral tenta-cles may be visible. Colour variable but generally white-pink. Habitat on soft substrates from 10 to 150m. Similar species 3 other species of *Thyone* occur; see Koehler, R. 1921. Also *Phyllophorus urna* Grube (not illustrated) which reaches 200mm long and has an urn-shaped body with 15–20 oral tentacles; tube-feet distributed evenly but not densely over the body.

Leptosynapta inhaerens (O. F. Müller) **Worm-cucumber** (Not illustrated) Length up to 180mm. Body has no tube-feet apart from 12 around the mouth modified for feeding; each of these carries 5–7 pairs of minute finger-like branches; minute anchor-shaped skeletal spicules used in locomotion protrude through the soft skin, making it adhesive. Colour usually pale pink. Habitat on or burrowing in mud and sand from 10–50m. Similar species *Labidoplax digitata*.

Labidoplax digitata (Montagu) Length up to 180mm. Body lacks locomotory tube-feet; oral tube-feet each have 2 pairs of minute finger-like branches. Habitat on mud and sand from the shore down to 70m.

Stichopus regalis

Holothuria forskali

Cucumaria elongata

Thyone fusus

Cucumaria planci

Labidoplax digitata

Phylum Chaetognatha

The chaetognaths are a small group of exclusively marine animals. No more than fifty species have been identified and these are placed in one class. Their bodies are difficult to see unless they are held up to the light in a jar of seawater, and it is necessary to examine them under the microscope against a dark ground in order to make out a lot of detail. Although rarely met with on the shore they will be frequently encountered in plankton samples. About twelve or so species have been recorded from European waters, but because they are small and mainly oceanic only two are mentioned here.

Certain species of chaetognaths are typically associated with particular currents of ocean water, and because of this they are important to oceanographers who need to trace the origin of planktonic communities. Such species are known as 'indicator' species, one of which is *Sagitta setosa*, a common indicator for coastal water.

The body plan of the chaetognaths is quite characteristic. Essentially a small, bilaterally symmetrical, torpedo-shaped body bears paired side fins and a tail fin. The anterior mouth is equipped with strong grasping spines. Circulatory and excretory systems are lacking. The main variations in the body plan are in the number of side fins and the general shape of the body.

Chaetognaths are active predators. Typical items in their diet are small crustaceans called copepods. Chaetognaths appear able to detect the presence of prey by the vibrations that are sent out when swimming. The prey is then captured with the help of the grasping spines around the mouth, and engulfed. Mediterranean recorded species = about 10.

Sagitta setosa J. Müller **Arrow Worm** (see page 247) **Length** up to 15mm. **Body** narrow and transparent without colour; 2 pairs of lateral fins, the anterior being about halfway along the body, the posterior terminating a little in front of the tail. **Habitat** planktonic coastal waters, sometimes stranded in rock pools. **Similar species** about 6; see Riedl, R. 1963.

Spadella cephaloptera (Busch) (see page 247) **Length** up to 8mm. **Body** less elongated than *Sagitta setosa* but broader; 1 pair of lateral fins which are more or less continuous with the tail fin. **Colour** may be tinted yellowish-brown. **Habitat** unlike the majority of chaetognaths, this is a benthic species which attaches itself to rocks and seaweeds by means of suckers. It may be found on the seabed or in rock pools. **Similar species** about 2; see Riedl, R. 1963.

Phylum Hemichordata

The hemichordates are another small group of exclusively marine animals. Until recently they were classified as a subphylum within the Chordata. About 80 species are known and these are divided into 3 classes: Enteropneusta, Pterobranchia and Planctosphaeroidea. Of these the first class only falls within the scope of this book. Hemichordates possess bilaterally symmetrical worm-like bodies divided into three distinct zones, with or without gill slits. A circulatory system is lacking. Tentacles may be on the middle zone of the body. Both solitary and colonial forms occur.

Class Enteropneusta Acorn worms

These solitary marine worms can be quite easily distinguished from the other worms described in this book by the manner in which their bodies are divided into three regions or zones: an anterior proboscis, a short collar lying immediately behind the proboscis and a long abdomen. The relative shapes of the various parts of the body are important in classification. The mouth opens at the junction of the proboscis and collar and gives rise to the gut which passes back through the abdomen. The abdomen itself may be rounded or flattened. At the front of the abdomen are gills which communicate with the gut and facilitate respiration. Tentacles are lacking. In some species the gut forms a number of pockets towards the rear of the body and these can be discerned as small lumps arranged on either side of the posterior part of the abdomen. They are known as hepatic pouches. There is a simple circulatory system and a simple nervous system.

Enteropneusts live in U-shaped burrows in sand or mud and may form characteristic casts reminiscent of polychaete worm casts. Like so many bottom-dwelling invertebrates they are filter feeders, extracting particulate food from seawater. The sexes are separate and the sperms and eggs escape into seawater by rupture of the body wall. A pelagic larva is formed after fertilization. Mediterranean recorded species = about 4. Further information is given by Spengel, I. N. 1893.

Balanoglossus clavigerus Delle Chiaje (See page 247) **Length** up to 300mm. **Body** proboscis short and yellow in colour; small collar region; pale brown abdomen flattened at anterior with the edges folded or frilled; middle gill area not so spread out; rounded long tail. **Habitat** burrowing in sand, mud and clay in shallow and deeper water. **Similar species** all Enteropneusts on this page could be confused unless care is taken; see Riedl, R. 1963.

Glossobalanus minutus (Kowal) (See page 247) **Length** up to 100mm. **Body** proboscis short; small collar; abdomen thicker and more swollen than in *Balanoglossus clavigerus* with the edges not folded or frilled; middle gill area not so spread out, rounded tail. **Colour** delicate, milky-white-yellowish-transparent. **Habitat** burrowing in mud and sand down to 50m, often associated with *Posidonia* beds (see page 44). **Similar species** see note under *Balanoglossus clavigerus*.

Saccoglossus mereschkowskii (Wagner) (Not illustrated) **Length** up to 40mm. **Body** long flesh-coloured proboscis; short reddish collar; abdomen rounded and faint yellow-brown with olive markings, not so readily divisible into 3 regions as in the other species described on this page. **Habitat** soft bottoms, especially if detritus rich, to 40m. **Similar species** see note under *Balanoglossus clavigerus*.

Phylum Chordata

Chordates are animals possessing a single, hollow, dorsal nerve cord. A true body cavity (coelom) is present, as are gills and a notochord. The tail is post anal.

This phylum includes the fishes, which are the most familiar of all sea animals. A more detailed introduction to them is provided on pages 254-255. Far less familiar are the so-called invertebrate chordates (subphylum Urochordata), i.e. those animals whose evolutionary standing places them near the vertebrates yet which as adults lack any trace of the diagnostic notochord or backbone, and which therefore superficially resemble the invertebrates. Two urochordate classes are treated here.

The class Thaliacea includes a number of organisms commonly known as salps. The adults lead a pelagic life, swimming amid their food supply (smaller planktonic organisms) which they filter by means of their gills. The body is surrounded by a number of muscle bands which can contract to force water out of the posterior exhalent opening, thus moving the animal forward. Water with food suspended in it is drawn through the inhalent opening at the anterior end. The body structure is quite complicated, and a good deal of it is associated with the reproductive process. Thaliaceans have a complex life-cycle, and a very small tadpole-like larva with a notochord is formed. Some species are capable of forming associations or colonies. In some cases a species may be both colonial and solitary, and thus exist in two forms. Fig. 40 illustrates the basic characteristics of a solitary salp.

The class Ascidiacea comprises the sea-squirts. Unlike the salps, these are bottom-dwellers as adults, and live attached to rocks or other organisms. They do retain a small free-swimming tadpole-like larva, however. They also filter suspended food particles from the water which passes through the gills supplying the respiratory needs. The body is enclosed in a thick tunic made from cellulose-like material which is often gelatinous. As with the salps there are inhalent and exhalent openings, but these are both generally situated towards the end opposite to the attachment point of the animal. The precise relationship of the two openings (the inhalent

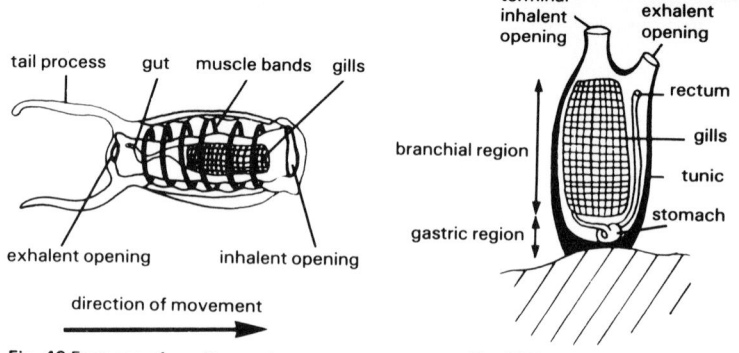

Fig. 40 Features of a solitary salp

direction of movement

Fig. 41 Features of a sea-squirt

one is normally terminal) is important in identification. Another important feature is the proportion of body length which is occupied by the gastric region. Below the gills lie the stomach and reproductive system as well as much of the circulatory system, but for simplicity the latter two systems have been omitted from the diagram. Some sea-squirts, e.g. *Botryllus*, are colonial, and encircle common exhalent canals and are supported in a massive tunic.

Subphylum Urochordata
Class Thaliacea

Pelagic adults resemble miniature floating jelly-like barrels. A hollow dorsal nerve cord and tail are present only in the larva. The nervous system is reduced. Salps and doliolarians are likely to be caught in plankton nets and will probably need careful examination to establish identity precisely. For more details see Riedl, R. 1963 or Godeaux, J. 1973. The life-cycle is complex. Mediterranean recorded species = 10.

Salpa maxima Forskål Solitary. **Length** up to 100mm. **Body** barrel-shaped with 9 muscle bands along it. **Colour** transparent. **Habitat** planktonic. **Similar species** several.

Salpa democratica Forskål **Length** solitary individuals up to 15mm, colonial individuals up to 6mm; colonial groups often have long streamers up to 300mm or more. **Colour** transparent. **Habitat** planktonic. **Similar species** several.

Dolium mülleri Khrohn Solitary. **Length** up to 4mm. **Body** barrel-shaped with small indentations around openings; 10–14 gill slits; about 6 conspicuous muscle bands; U-shaped gut. **Colour** transparent. **Habitat** planktonic.

Dolium denticulatum Grobben (Not illustrated) **Length** up to 9mm. **Body** similar to *D. mülleri* above; about 45 gill slits; up to 6 muscle bands; gut may be curved. **Colour** transparent. **Habitat** planktonic.

Pyrosoma atlanticum Peron (Not illustrated) **Length** up to 100mm. **Body** cylindrical with transparent tentacle-like appendages; consists of many individual animals arranged around a central canal to produce a cylindrical colony the walls of which are made up of stiff jelly-like material in which the individuals are embedded. There is one tentacle for each appendage. The colony opens at one end only where the collective exhalent currents of all the individuals are vented. **Colour** transparent white. **Habitat** planktonic. Similar species other oceanic forms may occur. **Note** this species is luminescent due to the light-generating power of bacteria associated with *Pyrosoma*.

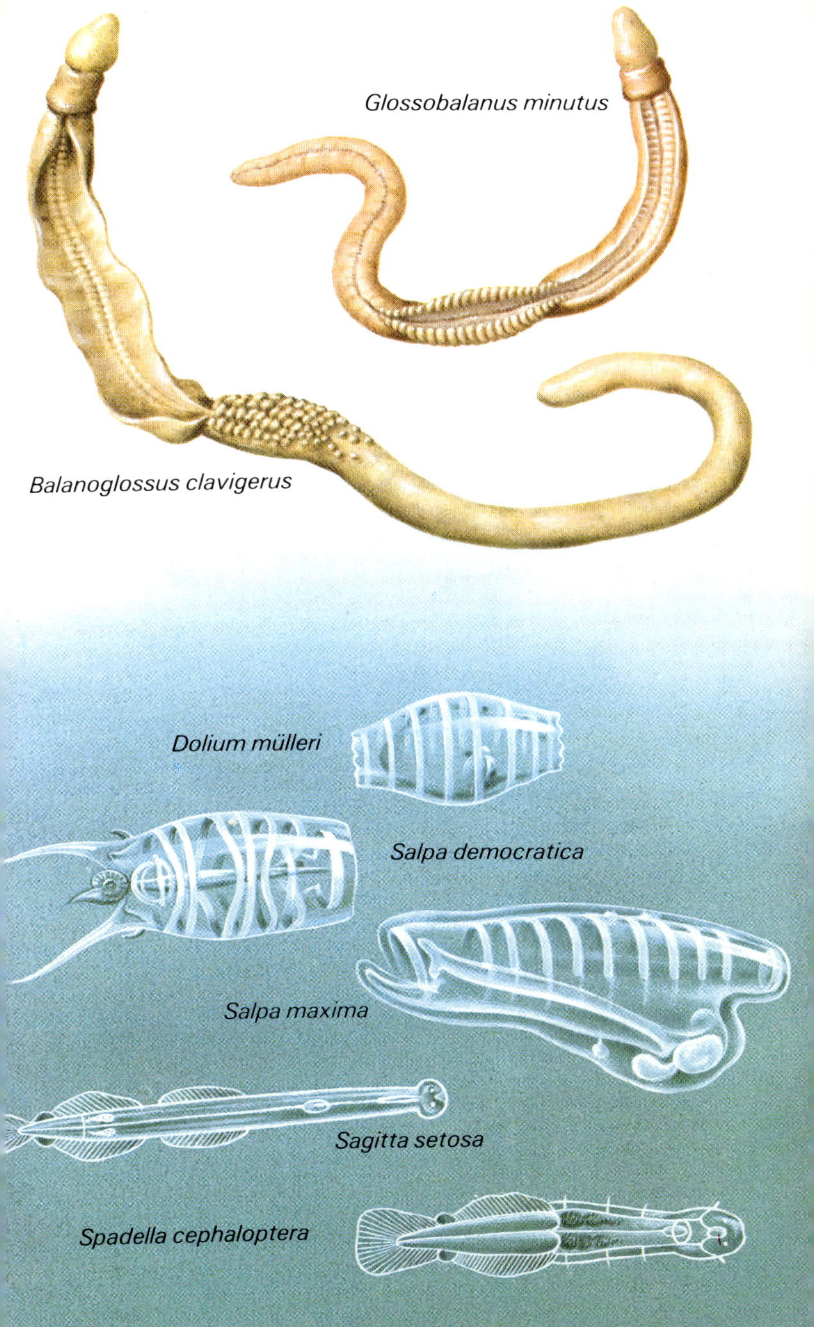

Glossobalanus minutus

Balanoglossus clavigerus

Dolium mülleri

Salpa democratica

Salpa maxima

Sagitta setosa

Spadella cephaloptera

Class Ascidiacea

Adult ascidians are sessile or pelagic chordates bearing no resemblance whatsoever to the vertebrates. Their soft sac-like or tube-like bodies are sheathed with a protective tunic of cellulose; the body wall may contain spicules of calcium carbonate. They may be solitary or colonial, but all are hermaphrodite. Only in the tadpole-like larva is there a hollow dorsal nerve chord and a primitive spine (notochord). Berrill, N.J. 1950, Drasche, R. 1883; Harant, H. and Verniers, P. 1933 and 1938; Lafargue, F. 1970, Millar, R. H. 1970 and Riedl, R. 1963 give more information on sea-squirts likely to be found in the Mediterraean. Identification is difficult. Total Mediterranean species unrecorded, possibly about 60.

Clavelina lepadiformis (O. F. Müller) Colonial, sessile, up to 30mm high; jelly-like individuals joined at the base only by thin stolons; exhalent opening subterminal; branchial region with about 17 gill bars (visible with a lens), shorter than gastric region. Colour transparent with pink, yellow and white marks; a clear fine yellow ring around each opening. Habitat on stones, weeds and shells to 50m.

Distoma adriaticum Drasche Colonial, sessile, up to 90mm high; individuals growing in thick round colonies and not easily distinguished; exhalent opening subterminal, inhalent opening formed into 6 minute lappets ornamented with fine pigment spots (visible with a lens); branchial region with 24 gill bars and shorter than gastric region. Colour whitish-brown. Habitat on stones and shells down to 40m. Similar species several.

Aplidium proliferum (Milne-Edwards) *(=Amaroucium proliferum)* Colonial, sessile, up to 50mm tall; individuals growing in club-shaped fleshy smooth colonies and not easily distinguished, arranged around common exhalent openings. Colour transparent, with red individuals showing through. Habitat on stones, rocks, seaweeds and shells down to 50m. Similar species *A. nordmanni* which forms flat-topped growths, not club-shaped.

Aplidium conicum Olivi *(=Amaroucium conicum)* Colonial, sessile, up to 120mm tall; individuals growing in conical fleshy smooth colonies and not easily distinguished; inhalent openings with 6 small flaps or lappets (visible with a lens). Colour colony orange with patterns of darker pigment. Habitat on sand and shell gravel etc., down to 50m. Similar species none.

Sidnyum turbinatum (Savigny) Colonial, sessile, up to 15mm high; individuals growing in flattish colonies which themselves are sometimes interlinked by stolons; inhalent openings with 8 small flaps or lappets (visible with a lens) grouped around a common exhalent opening. Colour colony orange. Habitat growing on algae, sea-grasses, shells and stones down to 200m. Similar species *Distaplia rosea* Dellavalle (not illustrated) with red-white colonies; a more rounded upper surface; no lappets around inhalent openings.

Didemnum candidum Savigny *(=D. maculosa)* Colonial, sessile, up to 2mm high and 40mm wide; individuals growing in very flat encrusting leathery colonies; 5–8 individuals share a common exhalent opening; crystals of calcium carbonate are set into the colony. Colour orange, brown, grey or violet. Habitat on stones, rocks, seaweeds, shells etc., down to quite deep water. Similar species *Molgula manhattensis* (de Kay) and sea-squirts.

Clavelina lepadiformis
individual

Distoma adriaticum
colony

Aplidium proliferum

individual

colony

Didemnum candidum

Aplidium conicum
colony

magnified
crystal

detail of
inhalent
opening

Sidnyum turbinatum

Ciona intestinalis (Linnaeus) Solitary, sessile, up to 120mm high; exhalent opening close by terminal inhalent opening; both openings with small lappets; outer skin transparent and viscera visible inside; branchial region longer than gastric region. **Colour** yellowish-green, openings tinted yellow. **Habitat** often growing in great numbers on rocks, piers and pilings as well as on weeds, from the shore down to very deep water. **Similar species** *C. edwardsi* which is solitary, sessile and very like *C. intestinalis* but brilliant green in colour.

Diazona violacea Savigny Colonial, sessile, up to 200mm high; individuals growing in rounded or flattish colonies which may be 400mm across; height of individual never as great as colony; each retains terminal inhalent and exhalent opening, and the branchial region and opening protrude from the colony by about 20mm. **Colour** translucent yellow-green. **Habitat** on rocks and stones from 30 to 200m, often in currents. **Similar species** few.

Ascidiella aspersa (O. F. Müller) (Not illustrated) Solitary, sessile, up to 60mm high, sometimes larger; exhalent opening about one-third of the way down the body; both openings have close-grouped lappets; outer skin rough and warty. **Colour** brown, grey or black. **Habitat** sometimes on soft substrates, when it may develop a stalk, or on stones, weeds or pilings, may be encrusted with other sea-squirts. **Similar species** few.

Ascidia mentula O. F. Müller Solitary, sessile, up to 100mm high; exhalent opening more than half-way down the body; both openings slightly puckered; outer skin thick and cartilage-like, covered with low swellings; branchial and gastric regions about equal in size. **Colour** translucent greenish. **Habitat** on rocks and shells down to about 200m. **Similar species** few.

Ascidia virginea (O. F. Müller) Solitary, sessile, up to 80mm high; openings slightly puckered, exhalent opening about one-third of the way down the body; outer skin relatively thin and flimsy. **Colour** translucent milky, with red marks. **Habitat** on stones and shells, usually from 30m down. **Similar species** few.

Phallusia mammillata (Cuvier) Solitary, sessile, up to 120mm high; exhalent opening about halfway down the body; both openings bear lappets; outer skin relatively thick and cartilage-like, bearing smooth swellings, especially on the lower part of the body. **Colour** varies with depth, from white to brown. **Habitat** usually attached to stones buried in mud or clay, from the shore down to about 180m. **Similar species** few.

Diazona violacea

individual

colony

Ciona intestinalis

Ciona edwardsi

Ascidia mentula

Ascidia virginea

Phallusia mammillata

Styela plicata Lesueur Solitary, sessile, up to 70mm tall; exhalent opening up to one-quarter of the body length, away from inhalent opening; thick leathery crinkled skin. **Colour** white to brown. **Habitat** attached to stones and rocks, often in close proximity to others. **Similar species** few.

Distomus variolosus Gaertner Colonial, sessile, rounded individuals up to 10mm high, joined by stolons; exhalent close to inhalent openings; skin rough. **Colour** red-brown. **Habitat** encrusting rocks and holdfasts of algae on the shore and in shallow water. **Similar species** few.

Botryllus schlosseri (Pallas) **Star Ascidian** Colonial, sessile, encrusting gelatinous colonies of various shapes; 3–12 individuals are arranged in star-like groups around a common exhalent opening; several to many groups per colony; individuals about 2mm long. **Colour** variable: brown, yellow, green or red. Colour of 'stars' generally contrasts with background. **Habitat** encrusting stones, rocks, weeds and hydroids and other ascidians, on the shore and in shallow water. **Similar species** *Botrylloides leachi.*

Botrylloides leachi (Savigny) Colonial, sessile, encrusting gelatinous colonies of various shapes; individuals arranged in irregular rows on either side of an extended exhalent cavity, not in 'stars'. **Colour** orange, yellow, grey. **Habitat** encrusting stones and other ascidians. **Similar species** *Botryllus schlosseri.*

Halocynthia papillosa Linnaeus Solitary, sessile, plump rounded individuals up to 60mm tall, occasionally more; openings appear epiciform when viewed end-on and are often surrounded by a fringe of spicules; exhalent opening nearly halfway down body. **Colour** brownish-red to bright red, sometimes paler below. **Habitat** in shallow water, often in sandy places. **Similar species** none.

Microcosmus sulcatus Coquebert *(=**M. vulgaris**)* Solitary, sessile; individuals up to 80mm long or more; exhalent and inhalent openings widely separated; thick leathery folded skin, often encrusted by other organisms. **Colour** brownish-red, siphons with red marks. **Habitat** on sand and shell gravel. **Similar species** few.

Molgula manhattensis (de Kay) Solitary, sessile, up to 30mm high; body rounded and flask-like; exhalent opening near inhalent; soft skin, sometimes covered with sand grains and fibrils. **Colour** bluish-green. **Habitat** on various substrates from the shore down to 100m. **Similar species** few.

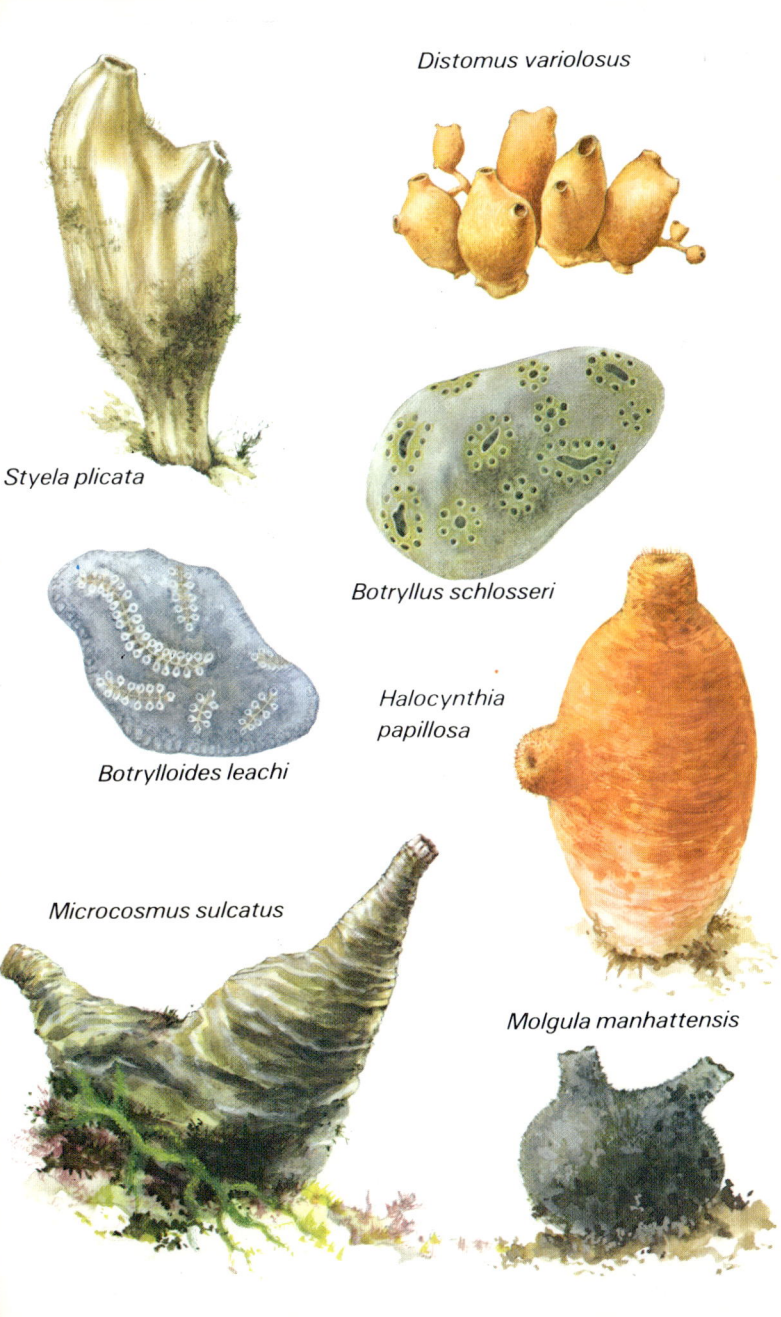

Distomus variolosus

Styela plicata

Botryllus schlosseri

Botrylloides leachi

Halocynthia papillosa

Microcosmus sulcatus

Molgula manhattensis

Subphylum **Vertebrata** Vertebrates

In addition to possessing chordate features, vertebrates have a backbone composed of many articulated body units, or vertebrae, arranged segmentally along with the main body musculature. The distinct head has associated sense organs, and the brain is protected by a brain case of skeletal material. Vertebrates often have paired front (pectoral) and rear (pelvic) appendages.

In terms of known species (about 46,700 species), the vertebrates are but a fraction of the whole animal kingdom. However, they include many of the most familiar organisms, and they range from the lowly lampreys to Man himself. In this book there is room to treat only one section of the vertebrates, the fishes, although seabirds, and mammals such as the porpoise and seal, are also familiar marine organisms. Several classes of vertebrates may be described as fishes. These are the Agnatha (fishes lacking true jaws), which include the lampreys and hagfishes; the Placodermi (an extinct class of primitive jawed fishes); the Chondrichthyes (cartilaginous fishes lacking hard bones), which include the sharks and rays; and the Osteichthyes which includes all the higher bony fishes of which the salmon and the goldfish are examples. The following account will include all these groups apart from the Placodermi.

The class Agnatha are primitive fishes represented today by one living order – the Cyclostomata. This name refers to the circular, sucker-like area armed with rasping teeth which surrounds the mouth. It includes two groups, the lampreys and the hagfishes. Lampreys which exist in fresh water and in the sea live as parasites by attaching themselves to other fishes, rasping a hole in them and sucking out the blood. Hagfishes are entirely marine and feed on dead and dying fishes and a variety of bottom-living organisms. The whole of the hagfish life cycle is passed in the sea, but the sea lamprey enters fresh water to spawn, and may be taken in estuaries. The river lamprey also spawns in fresh water, but spends at least one year in the sea during its growth to maturity, before it returns to fresh water.

Although in general terms the cyclostomes show a low level of vertebrate development (simple, unpaired fins; poorly developed sense organs, etc.), they show special modifications to their way of life. These include the mouth and surrounding teeth, and the arrangement of the gills which may permit respiration even when part of the head is buried in the host's tissues.

The class Chondrichthyes contains some of the largest and most voracious marine animals, and in European waters it is represented by a range of forms including the smaller dogfishes, various skates and rays, and the very large basking sharks. The sharks, unlike the cyclostomes mentioned above, are all rapid and powerful swimmers, often leading lives as voracious carnivores. They frequently prey on shoaling fishes like herring and mackerel. Some of the smaller species are scavengers, while the basking sharks are filter feeders, straining the surface waters for small, planktonic organisms in the manner of many whales. The skates and rays

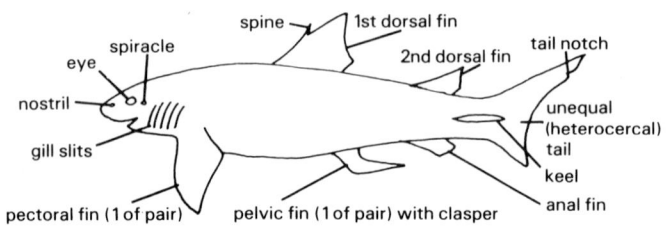

Fig. 42 External features of a generalized male shark

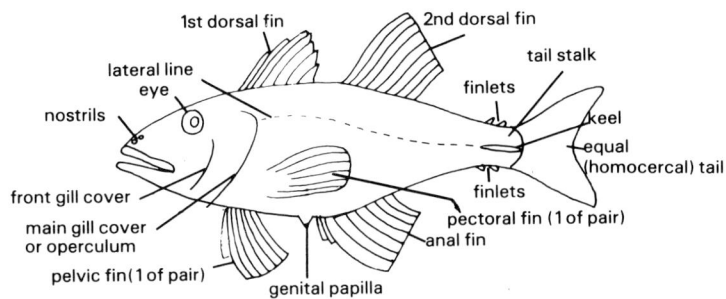

Fig. 43 External features of a generalized teleost

are bottom dwellers, feeding on invertebrates living on or in the sand and mud, as well as preying on flatfishes. Fig. 42 shows the main external features of a male shark. It will be noted that the pelvic fins are modified for use as claspers – structures which allow the transfer of sperms to the female. Females lack claspers. The eggs are normally retained within the female body and, after fertilization, the embryos may develop there until they are sufficiently mature to lead an independent life after birth. In the case of certain species, such as the dogfishes, the embryos are extruded from the female body in special egg cases known as mermaids' purses. These egg cases are attached to seaweeds, and eventually the juveniles hatch from them. Many rays and skates reproduce in a similar fashion. In these species the males can also usually be distinguished by the possession of claspers.

As would be expected from a group of predatory organisms, the senses are well developed, particularly those of vibration and smell which enable the sharks and their allies to track down their prey.

Most of the important food fishes in the world belong to the class Osteichthyes, although in many countries sharks and skates are fished commercially. The Osteichthyes are divided into a number of subgroups of which two concern us here. The first of these is the Chondrostei, which are rather primitive, with unequally developed tail fins. An example is the sturgeon, but there are many fossil forms. The second subgroup is the Teleostei, which includes the dominant fishes of the present day, and of most recent geological eras. They are immediately distinguished externally by the equal tail, and internally by the possession of a swim bladder or buoyancy mechanism which enables them to maintain a particular position in the water without swimming Fig. 43 indicates the principal external features of a teleost. In many cases it is not easy to distinguish between the sexes except by colour at certain times of the year. Most marine teleosts lay thousands of eggs which are fertilized externally by the male. In a number of cases the eggs float, and after fertilization they develop in the plankton as embryos, the juveniles feeding on the yolk supply of the egg. When this is exhausted they can often take small items of planktonic food. The teleosts have evolved to occupy almost all possible niches in the marine habitat, from the open ocean to the rock pool, from the surface waters to the abyssal depths. A great variety will be encountered in the Mediterranean. Readers requiring more information should turn to Hardy, A. C. 1970; Lythgoe, J. and Lythgoe, G. 1971; Marshall, N. B. 1965 or Wheeler, A. 1969, 1978.

Because of variability both in life and after death, references to colour for identification purposes have been kept to a minimum.

Class **Agnatha** Lampreys and hagfishes

Jawless vertebrates with unpaired median fins. Mediterranean recorded species = about 3.

Petromyzon marinus Linnaeus **Sea Lamprey** Length up to 900mm. Head bears small eyes and 7 pairs of gills; oval sucker-like mouth bearing several circlets of small teeth and a large centre tooth with 2 points. Body long, round in section and tapering towards the tail. Skin smooth and scaleless. Fins 2 clearly separated dorsal fins and a tail fin. Habitat from shallow water down to 400m or more, often near river mouths. Similar species *Lampetra fluviatilis* (below).

Lampetra fluviatilis (Linnaeus) **River Lamprey** (Not illustrated) Length up to 500mm, males usually shorter than females. Head, body and fins similar to *Petromyzon marinus* (above); sucker-like mouth with few teeth, 3 pairs on either side; middle of upper side of mouth lacking teeth. Colour dark blue-green dorsally, sides silver, white ventrally. Habitat in shallow water down to about 300m, often near river mouths where it congregates to migrate inland for breeding. Similar species *Petromyzon marinus*.

Myxine glutinosa Linnaeus **Hagfish** Length up to 400mm. Head lacks eyes; 1 pair of gill openings set well back on the body, the left one larger than and behind the right; mouth bears 3 pairs of barbels and has triangular lips on each side; nostrils terminal. Body long, worm-like; skin scaleless and slimy. Fins consist of a narrow strip around tail. Habitat from 20–800m, often in muddy places where it may burrow; may be caught feeding on dead or trapped fish. Similar species none.

Class **Chondrichthyes** Sharks, skates and rays

Vertebrates with jaws and cartilaginous skeletons sometimes reinforced with calcium salts; lacking bony fin rays. The mouth is generally on the underside; the skin normally feels rough when stroked from the tail towards the head and the tail is developed into 2 unequal lobes. Mediterranean recorded species = about 72.

Lamna nasus (Bonnaterre) **Porbeagle** or **Mackerel Shark** Length up to 3.5m. Head with conspicuous point to snout; jaws have triangular teeth with notches or cusps on either side of main point; 1 pair of small spiracles; 5 pairs of gill slits. Body relatively deep. Fins 1st dorsal fin starts where pectoral fin bones terminate; 2nd dorsal fin small, lying above small anal fin; tail stalk with side keels, upper lobe of tail notched and slightly larger than lower lobe. Habitat generally in surface waters, seldom below 150m. Similar species *Isurus oxyrinchus* (below).

Isurus oxyrinchus Rafinesque **Mako** Length up to 4m. Head streamlined; jaws have triangular teeth without notches on either side of the main point; 5 conspicuous gill slits on either side in front of the pectoral fin. Body quite slender. Fins 1st dorsal fin starts immediately behind the pectorals; 2nd dorsal starts just before the anal fin; tail stalk with side keels, upper tail lobe longer and notched. Habitat open water, usually near the surface. Similar species *Lamna nasus* (above).

Cetorhinus maximus (Gunnerus) **Basking Shark** Length up to 15m; huge size makes identification easy. Head elongated point on snout; teeth very small; 5 large gill slits running from near mid-dorsal to near mid-ventral position and equipped with rakers for sieving plankton from the water passing through. Habitat surface waters down to about 150m. Similar species none.

Alopias vulpinus (Bonnaterre) **Thresher** Length up to 4m. Head 1 pair of spiracles; 5 pairs of gill slits; jaws bear small triangular teeth lacking cusps. Fins remarkable extension of upper tail fin gives easy identification. Habitat surface waters. Similar species none.

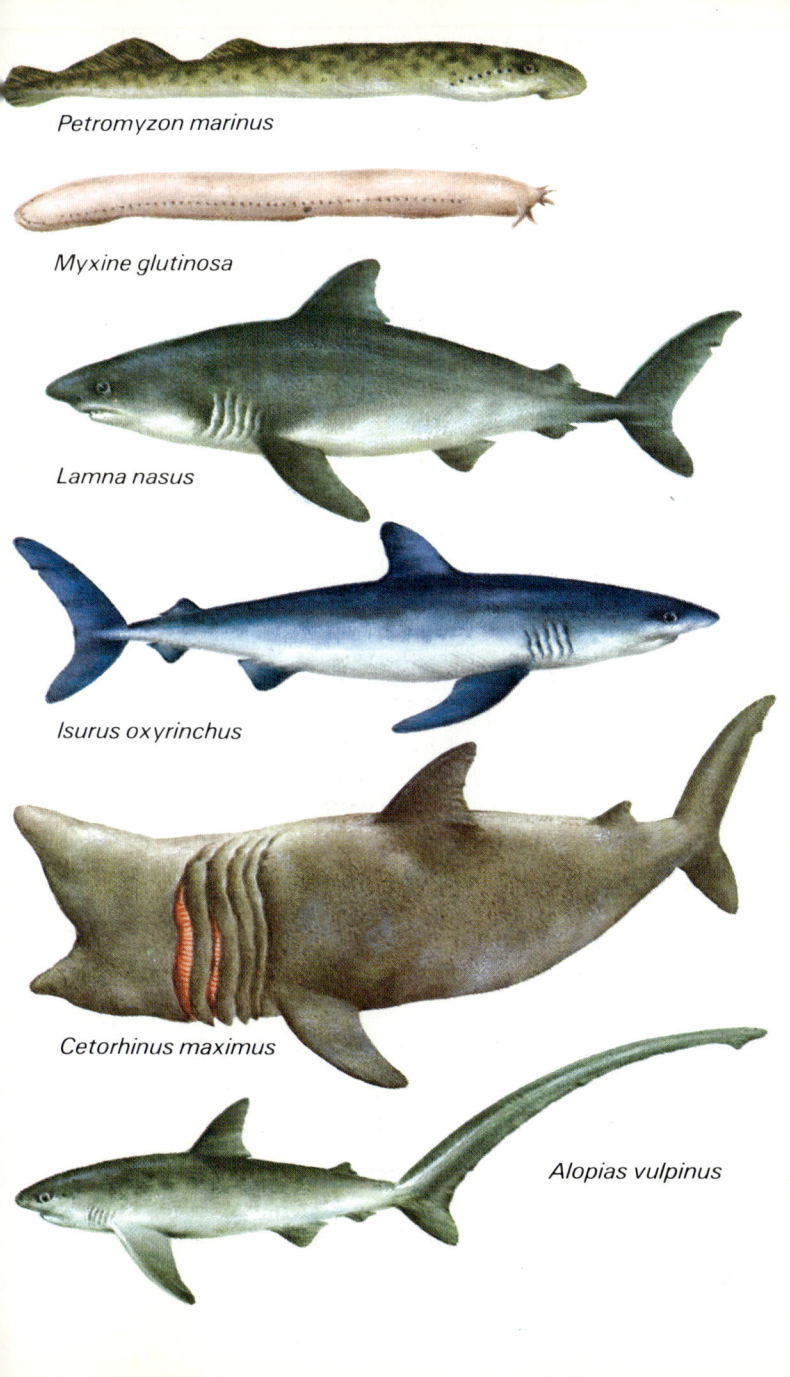

Petromyzon marinus

Myxine glutinosa

Lamna nasus

Isurus oxyrinchus

Cetorhinus maximus

Alopias vulpinus

Scyliorhinus canicula (Linnaeus) **Lesser-spotted Dogfish, Rough Hound** or **Rock Salmon** Length up to 700mm. **Head** blunt, with underside of snout bearing nostrils apparently connected to the mouth on either side by an external groove (see inset diagram); 1 pair of spiracles; 5 pairs of small gill slits, the last 2 pairs overlapping the pectoral fins. **Body** long and tapering to tail. **Fins** 1st dorsal begins behind the pelvics; 2nd dorsal begins level with trailing edge of anal; no keels on tail stalk, lower lobe of tail poorly developed. **Habitat** generally a bottom-dweller in sandy and muddy regions down to 100m. **Similar species** *S. stellaris* and *Galeus melastomus* (below).

Scyliorhinus stellaris (Linnaeus) **Large-spotted Dogfish** or **Nurse Hound** Length up to 1m. **Head** blunt with underside of snout bearing nostril grooves which do not connect nostrils to the mouth but return towards the midline, almost forming a W-shape (see inset diagram). **Fins** 1st dorsal begins just above base of pelvics; 2nd dorsal begins over centre of anal; no keels on tail stalks. **Habitat** often on rocky substrates, especially in sheltered places, down to about 50m. **Similar species** *S. canicula* and *Galeus melastomus.*

Mustelus mustelus (Linnaeus) **Smooth Hound** Length up to 1.5m. **Head** sharply pointed towards snout; jaws carry blunt plate-like teeth of diamond shape in a mosaic arrangement, adapted for crushing prey; 5 gill slits, the last overlapping the leading edge of the pectoral fin. **Body** fairly full, tapering steadily towards tail. **Fins** 1st dorsal begins just after trailing edge of pectorals; 2nd dorsal starts just before anal; tail with notch. **Habitat** on sandy and muddy bottoms from 5–100m. **Similar species** several, especially *Galeorhinus galeus* (below) and *Squalus* species (see page 260).

Mustelus asterias Cloquet **Stellate Smooth Hound** (Not illustrated) Similar to *M. mustelus* (above) but body slightly more massive; colours also similar but with white marks dorsally and on the sides. **Habitat** on soft bottoms down to 150m.

Prionace glauca (Linnaeus) **Blue Shark** Length up to 4m. **Head** streamlined and pointed; jaws carry triangular teeth with serrated edges; no spiracle; 5 small gill slits, the last pair overlapping the leading edge of the pectoral fins. **Fins** 1st dorsal set well back from pectorals; 2nd dorsal lies over anal; upper lobe of tail notched. **Habitat** open surface water, moving inshore in summer. **Similar species** several.

Galeorhinus galeus (Linnaeus) **Tope** Length up to 2m. **Head** characteristically pointed; jaws bear sharply pointed teeth with accessory points on the posterior sides; spiracle; 5 pairs of gill slits, the last overlapping the leading edge of the pectoral fins. **Body** quite deep in the middle regions, tapering towards either end. **Fins** large pectorals; 1st dorsal lies between pectorals and pelvics; 2nd dorsal lies slightly in front of anal; conspicuous notch in upper lobe of tail. **Habitat** on sandy bottoms or gravel in shallow water and down to 250m. **Similar species** several.

Galeus melastomus Rafinesque **Black-mouthed Dogfish** Length up to 800mm. **Head** snout rather flattened and slightly pointed, with nostril grooves (see inset diagram); mouth opening black; 1 pair of spiracles; 5 pairs of gill slits, last pair immediately in front of pectoral fin bones. **Fins** 2nd dorsal begins towards rear of anal; no keels on tail stalks; upper lobe of tail fin appears serrated due to large scales; bold colour pattern. **Habitat** usually in deeper waters from 150 to 400m but sometimes caught in trawls. **Similar species** *S. canicula* and *S. stellaris.*

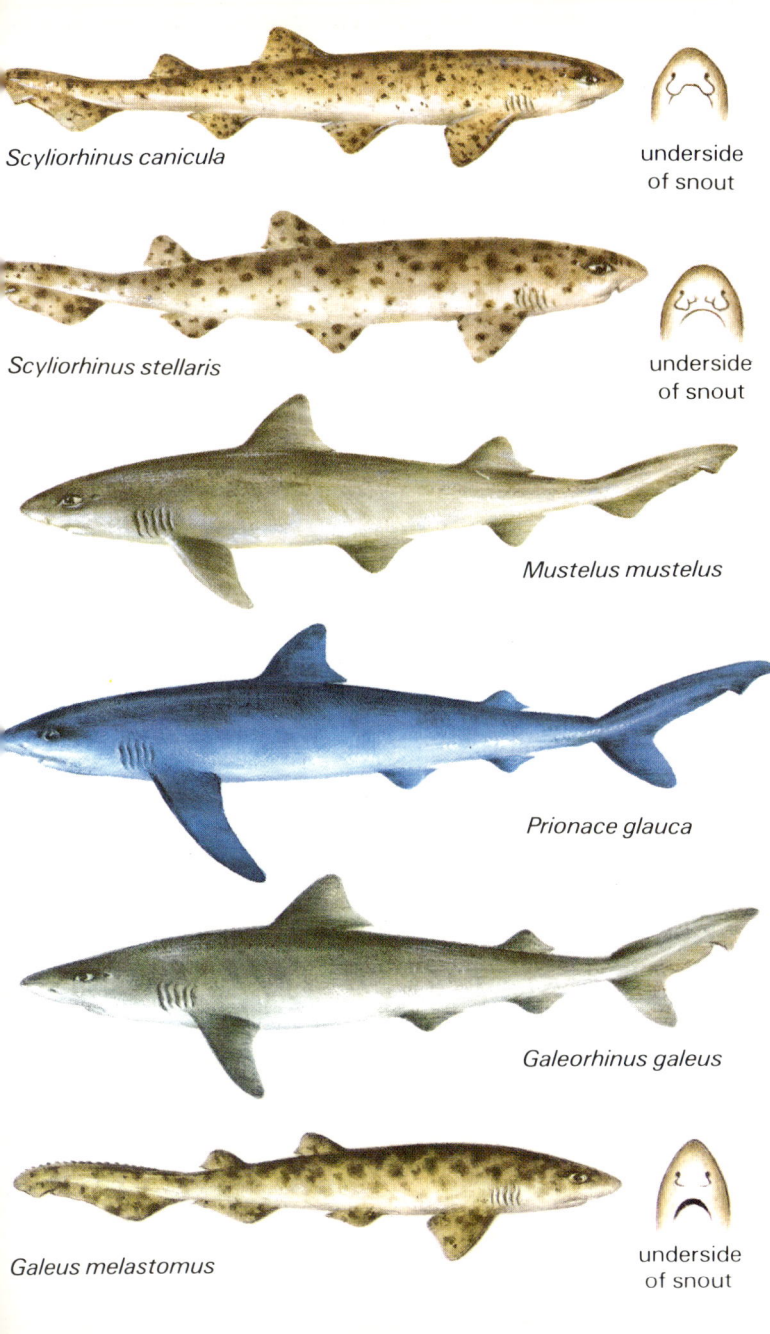

Scyliorhinus canicula

underside
of snout

Scyliorhinus stellaris

underside
of snout

Mustelus mustelus

Prionace glauca

Galeorhinus galeus

Galeus melastomus

underside
of snout

Hexanchus griseus (Bonnaterre) **Six-gilled Shark** Length up to 5m. Head not sharply pointed; jagged teeth; with small spiracle; 6 gill slits, the last in front of the leading edge of the pectoral fin. Fins pectorals triangular; single dorsal set well back between level of pelvics and anal; tail with conspicuously elongated upper lobe bearing a notch. Habitat deep water, 100m or more; may enter coastal waters in summer. Similar species *Heptranchias perlo* (below).

Heptranchias perlo (Bonnaterre) *(=Heptanchus cinereus* Rafinesque*)* **Seven-gilled Shark** Generally resembles *Hexanthus griseus* (above) but reaches only 3m and has 7 gill slits.

Carcharodon carcharias (Linnaeus) **Man-eater** or **White Shark** Length up to 12m. Head massive with pointed snout; jaws bear sharp triangular teeth with serrated edges; 5 deep running gill slits, the last immediately before the origin of the pectoral fins; thick-set body. Fins pectorals large and curved back, bearing a conspicuous posterior lobe near point of origin; 1st dorsal well forward; 2nd dorsal very small, lying between pelvics and anal; tail almost symmetrical; tail keels present. Habitat in deeper water, occasionally near coasts. Similar species two, including *Lamna nasus*.

Sphyrna zygaena (Linnaeus) **Hammerhead Shark** Length up to 3m. Head characteristic processes on either side carrying eyes and resembling a hammer head when viewed from above; slit-like nostrils on leading edge of 'hammer-head'; spiracle lacking; 5 pairs of gills, last pair in front of pectoral fin. Body quite thick-set but tapering towards tail. Fins 1st dorsal well forward, lying over pectorals; 2nd dorsal over anal; dorsal lobe of tail long with reduced notch. Habitat usually in deep water, sometimes entering coastal areas. Similar species one.

Squalus acanthias (Linnaeus) **Spiny Dogfish** or **Spurdog** Length up to 1.2m. Head rounded; spiracle behind eyes; 5 gill slits, the last in front of the pectoral fin. Body streamlined and tapering. Fins 1st dorsal between pectorals and pelvics, 2nd dorsal behind pelvics; both dorsals with a prominent spine at the leading edge but in 2nd it is not as tall as the fin; no anal fin; no keels on tail stalk and notch lacking in tail fin. Habitat over various substrates in coastal waters. Similar species nine in Mediterranean, but especially *S. fernandius* (below).

Squalus fernandius Molina **Blainville's Spurdog** Length up to 700mm; generally similar to *S. acanthias* but head has blunter snout; body not spotted. Fins spine of 2nd dorsal is as tall as the fin. Habitat as above. Similar species nine, but especially *S. acanthias*.

Squatina squatina Linnaeus **Monkfish** or **Angelshark** (Illustrated on page 263) Length up to 2m. Head flat and blunt, with 2 nostrils, each of which carries 2 small barbels; spiracles behind eyes, 5 gills open on underside. Body very flat and ray-like with row of blunt spines along midline. Fins large square pectoral fins, smaller pelvics; 1st and 2nd dorsals behind pelvics; anal fin lacking; upper lobe of tail larger than lower. Habitat on soft substrates, often partly buried; from shallow water down to 100m. Similar species two: *S. aculeata* (Linnaeus) (not illustrated) has conspicuous dorsal spines along the back; and *S. oculata* Bonaparte (not illustrated) has conspicuous black spots on the pectorals and rear body.

Torpedo marmorata Risso **Electric Ray** Length up to 600mm. Head very flat and rounded in outline; merges imperceptibly into body; mouth and 5 pairs of gills opening below; conspicuous eyes and pair of spiracles behind them, these fringed on their inner margins by small protrusions of skin. Body rounded and flat, tapering to tail. Fins 2 small dorsals well back and of equal size; pectorals fringe body; smaller pelvics; anal fins lacking. Habitat on soft substrates, sometimes partly buried. N.B. this fish can give quite powerful electric shocks when discharging its electric organs. Similar species *T. torpedo* (Linnaeus) which is generally brown with 5 blue spots set in dark rings, and has large unfringed spiracles; and *T. nobilana* Bonaparte which is coloured dark grey, dark brown or blackish, has un-fringed spiracles and white edges to the eyes; the first dorsal fin is larger than the second.

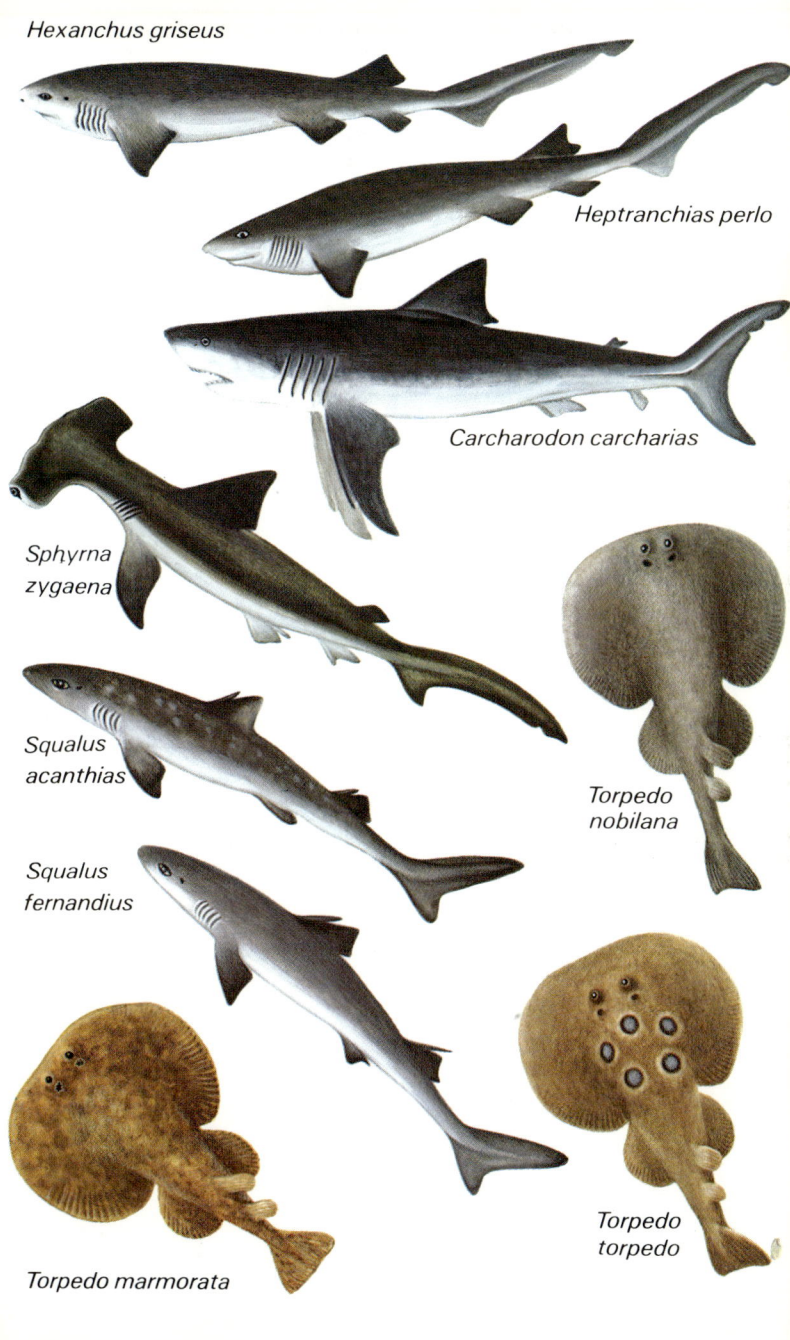

Hexanchus griseus

Heptranchias perlo

Carcharodon carcharias

Sphyrna
zygaena

Squalus
acanthias

Squalus
fernandius

Torpedo
nobilana

Torpedo marmorata

Torpedo
torpedo

Raja clavata Linnaeus **Thornback Ray** Length up to 800mm. Head flat and pointed, merging with body; bears mouth and 5 gill slits on the underside; dorsally eyes and 1 pair of spiracles conspicuous. **Body** flat, drawn out into two 'wings'; tapering tail; conspicuous spines on the upper surface and larger thorns running back to first dorsal fin. **Fins** pectorals form edges of the body; pelvics smaller and set close by; 2 small dorsals with no spines between, set very far back; tail fin reduced. **Habitat** on sandy and muddy substrates down to 100m. **Similar species** sixteen. N.B. male has long ventral claspers reaching halfway to tail.

Raja batis (Linnaeus) **Common Skate** Length up to 2m. Head snout longer than *R. clavata*; 5 gill slits and mouth ventral; eyes and spiracles dorsal. **Body** slightly concave in outline at front; young specimens smooth-skinned but spines develop with age; females have them on front of body, males are spiny on back. **Fins** as *R. clavata* but has spines between 1st and 2nd dorsals. **Habitat** on sandy, muddy substrates to 600m. **Similar species** 16. N.B. Male has claspers.

Raja asterias Delaroche **Starry Ray** (Not illustrated) Length up to 1m. Head snout rounded in outline; 5 gill slits and mouth ventral; eyes and spiracles dorsal. **Body** flat and slightly rounded in outline; edges covered with tiny scales; dorsal surface of tail with 1–3 rows of spines. **Fins** as preceding species. **Colour** yellow-grey above with small white and brown spots. **Similar species** sixteen. N.B. Male has claspers reaching nearly halfway along tail. An endemic species commonly found in the Adriatic region.

Raja miraletus Linnaeus **Brown Ray** (Not illustrated) **Length** up to 600mm. Head with short snout but otherwise typical ray characters. Body and **fins** typical in shape. **Colour** yellowish-grey above, with small black dots and 2 conspicuous 'eye-spots' either side of midline of body, each with a pale blue centre encircled by a black ring and then edge orange or yellow, serving to distinguish this species particularly from *R. asterias* (above). **Habitat** on soft substrates. **Similar species** sixteen, of which two: *R. radula* Delaroche (not illustrated) and *R. naevus* Müller & Henle (not illustrated), also have 'eye-spots'; the former has them with a brown centre and then concentric rings of yellow bounded by brown; the latter with black centres surrounded by yellow. N.B. for further information on Mediterranean species of *Raja* consult Luther, W. and Fiedler, K. 1976; Lythgoe, J. and Lythgoe, G. 1971; or Wheeler, A. 1978.

Dasyatis pastinaca (Linnaeus) **Common Stingray** Length up to 2.3m overall; **Head** snout with a blunt point, flat, with 5 pairs of gills and mouth below, eyes and 1 pair of spiracles above. **Body** merges with head, flat and broad; tail tapering and one and a half times length of rest of body. **Fins** pectorals form margin of body; pelvics well forward; no dorsal or tail fin: dorsal replaced by conspicuous poisonous spine which can cause painful wounds. **Habitat** living on the bottom or swimming in mid-water or at surface; shallow water to 100m. **Similar species** *D. violacea* Bonaparte (not illustrated) which reaches up to 1.5m, and has a violet-coloured body with a flat unpointed snout.

Myliobatis aquila (Linnaeus) **Eagle Ray** Length up to 2m overall. Head reasonably distinct from body; jaws and gills below, eyes and spiracles on side. **Body** broad and drawn out into two wings; tail twice body length and whip-like. **Fins** pectorals fringe body; small pelvics; single dorsal fin immediately in front of conspicuous tail spine. **Habitat** on sandy substrates down to about 100m. **Similar species** *Pteromylaeus bovinus* (Geoffroy Saint-Hilare) which has a more pointed head and a dark striped body.

Chimaera monostrosa Linnaeus **Rabbit-fish** or **Rat-fish** Length up to 1.4m, but often smaller. **Head** jaws with few teeth grouped to give 2 plates in the upper jaw and 1 in the lower; no spiracles; large eyes; 1 gill opening on each side covered by fleshy flap; strange lobe-shaped spiny organ on forehead. **Body** long and tapering. **Fins** 1st dorsal tall with conspicuous anterior spine; 2nd dorsal long and low; large pectorals; smaller pelvics; anal fin separated from long tail fin; males have claspers with 3 small branches. **Habitat** lives on or near bottom in deep water down to 300m. **Similar species** none.

Raja clavata

Raja batis

Dasyatis pastinaca

Myliobatis aquila

Chimaera monostrosa

Squatina squatina

Class Osteichthyes Bony fishes

Fishes with true jaws and bony skeletons.

Subclass Actinopterygii
Infraclass Chondrostei

Primitive fishes with bony rays in the fins, the gill openings covered by a gill cover or operculum, and with unequal tail lobes. Mediterranean recorded species = 4.

Acipenser sturio Linnaeus **Sturgeon** Length up to 3.5m. **Head** pointed, with snout carrying 4 barbels and nostrils, eyes, gill covers and ventral rounded sucker-like mouth. **Body** long and tapering with 5 rows of bony plates. **Fins** dorsal nearly level with anal, pectorals and pelvics; tail with upper lobe better developed. **Habitat** on soft substrates, sometimes in brackish water and penetrating far up rivers to spawn. **Similar species** *Huso huso* (Linnaeus) which reaches up to 5m; **head** with ventral crescent-shaped mouth; **body** with smaller bony plates.

Infraclass Teleostei

More advanced fishes having bony rays in the fins; gill openings covered by an operculum; equally developed tail lobes, and an internal swim bladder for controlling buoyancy. About 460 recorded species in the Mediterranean.

Muraena helena Linnaeus **Moray Eel** Length up to 1.3m. **Head** with pointed snout, powerful jaws, sharp teeth, small gill openings high up. **Body** long, powerful and eel-like; skin slimy and thick; scales lacking. **Fins** no pectorals, no pelvics; dorsal fin starts just behind head and is continuous with tail fin. **Habitat** in crevices, in wrecks and other objects such as amphorae. **Similar species** one other.

Conger conger (Linnaeus) **Conger Eel** Length up to 2m. **Head** upper jaw longer than lower; head itself about two and a half times jaw length. **Body** powerful and cylindrical in section, gradually tapering to tip of tail; no scales. **Fins** dorsal starts just behind pectorals, which themselves are slightly pointed; no pelvics; anal fin merges with tail as does dorsal. **Habitat** in caves and crevices, amongst rocks and wrecks. **Similar species** seventeen others recorded in Mediterranean.

Anguilla anguilla (Linnaeus) **Common Eel** Length up to 1.4m. **Head** lower jaw slightly longer than upper. **Body** more slender than *C. conger* with minute scales. **Fins** dorsal starts well back behind rounded pectorals; no pelvics. There are two forms of common eel: the younger or yellowish type and the silvery form; intermediate forms may be found. Most books on fish describe the migration of this species from the Sargasso Sea to rivers where the eels mature. **Similar species** seventeen; c.f. *Conger conger*.

Sprattus sprattus (Linnaeus) **Sprat** Length up to 150mm. **Head** lower jaw slightly longer than the upper jaw. **Body** with scales, a distinct line of spiny sharp scales running from the throat to the anus forming a sharp keel, which can be felt if a finger is run gently from the vent towards the head. **Fins** dorsal starts behind pectorals; pelvics; anal tail symmetrical. **Habitat** open water, shoals coming inshore in winter. **Similar species** five including *Sardina pilchardus*, *Argentina sphyraena*, and *Alosa fallax*.

Sardina pilchardus (Walbaum) **Sardine** (when small), **Pilchard** (when large) Length up to 250mm. **Head** gill covers have pronounced radiating ridges; jaws of equal length. **Body** with large scales, no sharp keel, and rounded outline. **Fins** last two rays of anal longer than anterior ones. **Habitat** coastal waters in summer, deeper waters in winter. **Similar species** see under *Sprattus sprattus*.

Argentina sphyraena Linnaeus **Argentine** Length up to 270mm. **Head** pointed, with small mouth and large eyes. **Body** long and slender with delicate scales. **Fins** trailing edge of dorsal overlaps pelvics; small adipose fin in front of tail. **Habitat** over muddy bottoms down to 200m. **Similar species** see under *Sprattus sprattus*.

Acipenser sturio

Anguilla anguilla

Muraena helena

Conger conger

Sprattus sprattus

Sardina pilchardus

Argentina sphyraena

Alosa fallax (Lacépède) **Mediterranean Twaite Shad** Length up to 600mm. **Head** large with upper jaw carrying characteristic notch in the midline; radiating ridges weakly developed on gill covers. **Body** quite deep but still herring-like in appearance, with up to 7 dark round spots on the sides behind the gill cover; sharp scales form a toothed keel on the underside. **Habitat** in shallow coastal waters and in rivers, where it migrates to spawn. **Similar species** see under *Sprattus sprattus* (page 264). See Wheeler, A. 1978.

Engraulis encrasicolus (Linnaeus) **Anchovy** Length up to 200mm but often less. **Head** with a characteristically rounded snout and receding lower jaw; mouth stretches back behind the eye. **Body** rounded and slender with delicate scales often abraded with handling. **Habitat** in large schools in the surface waters, moving inshore in summer; an important commercial fish in the Mediterranean area.

Apletodon microcephalus (Brook) **Small-headed Clingfish** Length up to 40mm. **Head** short, flattened and triangular, one-quarter of total length; small mouth and large eyes; small rounded incisor teeth with 1–3 canines. **Body** flattened; males have dark spot on dorsal and anal fins and purple patch on throat. **Fins** contribute to the elaborate sucker arrangement: dorsal set well back (but not joining tail), with 5–6 rays; anal set well back (not joining tail) with 5–7 rays. **Habitat** on the shore among rocks and down to 25m. **Similar species** five.

Diplecogaster bimaculatus (Bonnaterre) (*=Lepadogaster bimaculatus*) **Two-spotted Clingfish** Length up to 60mm; similar to *Apletodon microcephalus* above. **Head** more than one-quarter of total length; teeth uniform and small. **Body** males have 2 purple spots surrounded by yellow rings on the nape. **Fins** contribute to the elaborate sucker arrangement; dorsal set well back but not connected to tail, with 5–7 rays; anal not connected to tail, with 4–6 rays. **Habitat** on the shore among rocks and down to 55m. **Similar species** five.

Lepadogaster lepadogaster lepadogaster (Bonnaterre) **Shore Clingfish** or **Cornish Sucker** Length up to 65mm. **Head** nearly one-third of total length, with a long snout and serrated flap at edge of front nostril; blue 'eye-spots' on top of head surrounded by yellow rings. **Body** tapers to tail. **Fins** contribute to elaborate sucker arrangement; dorsal with 16–19 rays, joining tail; anal with 9–10 rays, joining tail. **Habitat** on the shore amongst rocks and weeds and in shallow water. N.B. this is recognized as a subspecies distinct from the Atlantic variety *Lepadogaster lepadogaster purpurea* (Bonnaterre) (not illustrated). **Similar species** five, especially *L. candollei* Risso, Connemara Clingfish, (not illustrated) which reaches up to 75mm and has dorsal and anal fins reaching almost to tail.

Two other species of clingfish, *Gouania wildenowii* (Risso) and *Opeatogenys gracilis* (Canestrini) exist in the Mediterranean. *Gouania wildenowii* (not illustrated), the Blunt Snouted Clingfish, reaches 60mm and has dorsal, tail and anal fins merged to form a narrow fringe of fin round the posterior of the greyish-yellow to brownish-green body. *Opeatogenys gracilis* (not illustrated), the Pygmy Clingfish, reaches 30mm and has a dorsal fin with 3 rays and an anal fin with 4 rays; body slender with reddish colour above and a spine set in the gill flap.

Merlangius merlangus (Linnaeus) **Whiting** Length occasionally up to 700mm. **Head** longish pointed snout with lower jaw shorter than upper and bearing a minute barbel in young specimens only. **Body** fairly slender. **Fins** 3 dorsal, all connected at their bases; 2 anal, the first beginning in the midline of the 1st dorsal. **Habitat** in shallow water near coasts from 30 to 100m, especially over sandy and muddy bottoms; the young may be associated with jellyfishes. **Similar species** a member of the Gadidae (codfishes); 17 species in the Mediterranean.

Micromesistius poutassou Risso **Blue Whiting** Length up to 450mm. **Head** with lower jaw slightly longer than upper; no barbel. **Body** slender. **Fins** 3 dorsal, all well separated; 1st anal fin very long, beginning in front of 1st dorsal. **Habitat** deeper water away from coasts, between 100 and 300m. **Similar species** see note under *Merlangius merlangus* (above).

Alosa fallax

Micromesistius poutassou

Engraulis encrasicolus

Merlangius merlangus

Apletodon microcephalus

Lepadogaster lepadogaster lepadogaster

Diplecogaster bimaculatus

Phycis blennoides (Brunnich) **Forkbeard** Length up to 750mm but often less. **Head** eyes and nostrils relatively small; barbel on the chin. **Body** deepish with easily detached scales, lateral line curved upwards at anterior. **Fins** 1st dorsal short, triangular and with conspicuous long 3rd ray; 2nd dorsal long; pectorals mid-height of body with long 3-rayed pelvic fins originating in front and stretching to start of anal fin. **Habitat** generally only small immature specimens inshore; adults in deep water down to 350m in schools. **Similar species** a codfish with 16 allied species, but especially *Phycis phycis* Linnaeus (not illustrated) which reaches up to 400mm and which lacks the conspicuous long 3rd ray of 1st dorsal fin; back is coloured brown.

Trisopterus minutus capelanus (Lacépède) **Poor Cod** Length up to 200mm. **Head** with prominent eye; upper jaw longer than lower, which carries a well developed barbel. **Body** fairly elongated. **Fins** 3 dorsals all set close together; 1st anal begins level with or in front of junction of 1st and 2nd dorsal; dark spot on pectoral fin base. **Habitat** in coastal waters from 25 to 300m. **Similar species** sixteen. N.B. this is the Mediterranean subspecies of *T. minutus* Linnaeus occurring in the North Atlantic.

Trisopterus luscus (Linnaeus) **Bib, Whiting Pout, Pout** or **Pouting** Length up to 410mm. **Head** as long as body is deep, conspicuous barbel on lower jaw. **Body** has vertical bands of light and darker colours. **Fins** close-set 3 dorsals, 1st higher; 2 anals, 1st starting in middle of 1st dorsal; pectorals with dark spot at base; pelvics long and stretching back beyond anus when folded up. **Habitat** down to 300m, particularly among rocks, piles and wrecks. **Similar species** sixteen.

Pollachius pollachius (Linnaeus) **Pollack** Length usually up to 500mm, sometimes much larger. **Head** lower jaw longer than upper, lacking barbel; conspicuous large eye. **Body** typically cod-like with lateral lines rising over pectoral fin bases. **Fins** 3 dorsals clearly separated from each other; 2 anals, 1st starting in midline of 1st dorsal. **Habitat** inshore waters among rocks, sometimes, in schools, down to 200m. **Similar species** sixteen.

Gaidropsarus mediterraneus (Linnaeus) *(=Motella mediterraneus* or *Onos mediterraneus)* **Shore Rockling** Length up to 350mm. **Head** bears one barbel on each 1st nostril and one on lower jaw; mouth extends back slightly beyond eye. **Body** long, rounded. **Fins** 1st dorsal low with slender rays apart from conspicuous 1st ray; 2nd dorsal long, running nearly to tail; anal fin not quite as long as 1st dorsal; pectorals with 15 to 17 rays. **Habitat** on the shore, in tide pools and in shallow water among rocks and weeds. **Similar species** three including *G. vulgaris*.

Gaidropsarus vulgaris (Cloquet) **Three-bearded Rockling** Length up to 530mm. **Head** with one barbel on lower jaw and one on each anterior nostril; mouth extends well back beyond eye. **Body** long and rounded. **Fins** similar to *G. mediterraneus* but pectorals with 20–22 rays. **Habitat** generally found offshore in shallow water down to 50m among rocks or over gravel. **Similar species** three.

Molva macrophthalma (Rafinesque-Schmaltz) *(=M. elongata)* **Spanish Ling** or **Mediterranean Ling** (Not illustrated) Length up to 900mm. **Head** with snout about the same length as the eye; lower jaw with moderate barbel and slightly longer than upper jaw. **Body** long and slender, tapering to tail. **Fins** short, 1st dorsal with 10–12 rays, slightly separated from long 2nd dorsal; rounded tail fin; long anal; pelvics in front of pectorals, their longest rays reaching back beyond pectorals. **Colour** back and upper flanks greenish-brown, the lower regions yellow to silver. **Habitat** in deep water from 200–1000m over muddy bottoms, a deep water species but commercially important and often seen in fish markets.

Merluccius merluccius (Linnaeus) **Hake** Length up to 1.8m. **Head** long with large jaws and big teeth; lower jaw longer than upper. **Body** long and tapering, lateral lines straight, not arched over pectoral fins. **Fins** 2 dorsals, 1st roughly triangular, 2nd long and overlying anal; both these last with a slight depression towards the tail. **Habitat** deep water from 150 to 550m. **Similar species** none.

Phycis blennoides

Trisopterus minutus capelanus

Trisopterus luscus

Pollachius pollachius

Gaidropsarus mediterraneus

Gaidropsarus vulgaris

Merluccius merluccius

Cheilopogon heterurus (Rafinesque-Schmaltz) *(=Cypselurus heterurus)* **Atlantic Flying Fish** Length up to 400mm. **Head** herring-like, jaws short but lower jaw fractionally longer than upper. **Body** sideways flattened and herring-like. **Fins** dorsal fin set well back, with 13–14 rays; anal fin starts level with or behind 4th ray of dorsal and has 8–10 rays; lower lobe of tail fin longer than upper; pectorals large, pointed; pelvic fins set quite far back, large and lobed. **Habitat** surface waters where it feeds on plankton and uses its ability to fly by gliding in order to escape predators. **Similar species** five, but especially *Exocoetus volitans* Linnaeus (not illustrated) and *Exonautes rondeleti* (Cuvier & Valenciennes) (not illustrated); both these species reach up to 300mm with short heads and large eyes; in *Exocoetus volitans* the body is grey to blue above and silver to grey below; pectoral fins grey with white periphery; pelvic fins much smaller than in *Cheilopogon heterurus* and set only a little behind pectorals; bluish-white in colour. *Exonautes rondeleti* is bluish-brown above and silvery below with coloured pectoral fins; the pelvics are more like those of *Cheilopogon heterurus*.

Belone belone (Linnaeus) **Garfish, Garpike** or **Greenbone** Length up to 800mm. **Head** long fine tapering snout; lower jaw noticeably longer than upper; teeth well developed. **Body** long and fine. **Fins** dorsal fin set well back over anal fin; pectorals and pelvics small. **Habitat** shallow water inshore during the summer months. The Atlantic, Mediterranean and Black Sea races have been separated as subspecies by some authors. **Similar species** *Scomberesox saurus* (below).

Scomberesox saurus (Walbaum) **Skipper, Saury Pike** Length up to 450mm. Very similar to *Belone belone* but with less conspicuous teeth and several small finlets between ends of dorsal and anal fins and start of tail fin.

Sphyraena sphyraena (Linnaeus) **Barracuda** Length up to 1m. **Head** elongated, large with powerful jaws and strong teeth; lower jaw slightly longer than upper. **Body** long, rounded and tapering. **Fins** pelvics under 1st dorsal which has 5 strong spines; anal under 2nd dorsal. **Habitat** in coastal waters, generally over sandy bottoms, sometimes in shoals. **Similar species** one.

Atherina presbyter Valenciennes **Sand Smelt** Length up to 210mm, but often around 150mm. **Head** short with strongly upturned mouth. **Body** slender; small scales. **Fins** 1st dorsal with 7–8 spines; 2nd dorsal with 2 spiny rays and 10–13 soft branching rays; pectorals high up behind gill covers; small pelvics just in front of 1st dorsal and anal fin with 2 leading spiny rays and 13–16 soft branching rays. **Habitat** in shallow water inshore and in estuaries, often in shoals. **Similar species** *Atherina boyeri* Risso, the Big-scaled Sand Smelt, (not illustrated); differs from *A. presbyter* in having larger head, deeper, slightly shorter body with bigger scales; the anal fin has 2 spiny rays and 11–13 soft branching rays.

Zeus faber Linnaeus **John Dory** Length up to 500mm but often less. **Head** huge with large protrusible mouth. **Body** severely compressed side to side (like head) with conspicuous dark spot on either side. **Fins** characteristic long spiny rays to 1st dorsal and 1st anal. **Habitat** inshore shallow water generally down to 50m, but sometimes deeper. **Similar species** none.

Capros aper (Linnaeus) **Boar-fish** Length up to 160mm. **Head** with relatively large eye and pointed snout; lower jaw longer than upper; mouth protrusible. **Body** laterally compressed. **Fins** 1st dorsal with long spines, 2nd dorsal overlies non spiny region of anal fin; leading edge of this fin has short spines; 1st ray of pelvics is spiny. **Habitat** usually in deep water among corals and rocks, but sometimes taken in less than 100m over sand. **Similar species** none.

Cheilopogon heterurus

Belone belone

Scomberesox saurus

Sphyraena sphyraena

Atherina presbyter

Zeus faber

Capros aper

Macrorhamphosus scolopax (Linnaeus) **Snipe Fish** Length up to 150mm. Head with typical long snout bearing small terminal mouth; conspicuous eyes. Body ovoid, quite deep. Fins 1st dorsal fin has a long anterior spine serrated on the trailing edge; when folded back this spine reaches over the tail fin. Habitat below 25m, usually much deeper; generally over muddy and sandy bottoms. Similar species none.

Syngnathus acus Linnaeus **Greater Pipefish** Length up to 450mm. Head snout more than half head length, long, rounded; tapering from forehead, but flaring slightly at the minute terminal jaws. Body long and tubular; 17–21 conspicuous body rings lying between head and reduced anal fin which lies below anterior of dorsal. Fins well developed pectoral, dorsal and tail fins; other fins reduced; there are 35–46 dorsal fin rays. Habitat often in shallow coastal water among weeds and grasses; males brood eggs in a special pouch in summer. Similar species ten.

Syngnathus typhle Linnaeus **Deep-snouted Pipefish** Length up to 300mm. Head with snout flattened sideways. Body long and tubular with 16–18 body rings between head and reduced anal fin. Fins as for *S. acus* above; dorsal fin with 28–41 rays. Habitat in shallow water, often over sandy bottoms among weeds and sea-grasses. Similar species ten.

Nerophis ophidion (Linnaeus) **Straight-nosed Pipefish** Length up to 300mm. Head relatively small with minute mouth. Body lacking any fins save dorsal which has 33–34 fin rays; 28–32 body segments between head and anus. Habitat in shallow water, particularly amongst long eel-grasses and algae. Similar species ten; for details of other species see Wheeler, A. 1978 or Lythgoe, J. and Lythgoe, G. 1971.

Hippocampus hippocampus Linnaeus **Short-snouted Seahorse** Length up to 160mm. Head with short snout and minute mouth at the tip. Body of typical shape; no 'mane' on neck. Fins pectoral fins with 13–15 rays; dorsal fin 16–18 rays; minute anal fin almost at deepest part of belly; no tail or pelvic fins. Habitat often in shallow water amongst weeds, but rarer than *H. ramulosus*. Similar species *H. ramulosus* (below).

Hippocampus ramulosus Leach **Seahorse** Length up to 150mm. Head with relatively longer snout than *H. hippocampus* and small terminal mouth. Body with 'mane' on neck. Fins 15–18 rays in the pectoral fin and 18–21 in the dorsal fin which overlies the minute anal fin; anal fin does not lie at deepest part of belly, but a little towards the tail. Habitat in shallow water amongst eel-grasses and weeds, uncommon. Similar species *H. hippocampus* (above).

Scorpaena scrofa Linnaeus **Red Scorpionfish** Length up to 510mm. Head scaleless, large in relation to overall size, bearing a number of spines together with a large mouth and moderate eye with no conspicuous flap of tissue above it; other small flaps elsewhere. Body plump with scales. Fins dorsal has anterior part with 11–12 strong spiny rays and a posterior part with 9–10 soft rays; pectorals over pelvics. Habitat over stones or sand from 20 to 100m but not in very shallow water or on the shore. Similar species about nine, including *S. porcus* below. N.B. Poison spines on gill covers and dorsal fin.

Scorpaena porcus Linnaeus **Brown Scorpionfish** (Not illustrated) Length up to 300mm. Similar to *S. scrofa* but head has several flaps of tissue and one conspicuous one over the eye; snout quite short; anterior part of dorsal fin has 12 spiny rays and the posterior part 10–11 softer rays. Colour reddish-brown to brown with darker patches. N.B. Poison spines on gill covers and dorsal fin.

Macrorhamphosus scolopax

Syngnathus acus

Nerophis ophidion

Hippocampus hippocampus

Hippocampus ramulosus

Scorpaena scrofa

Syngnathus typhle

Trigla lyra Linnaeus **Piper** Length up to 400mm, occasionally more. Head large with conspicuous eye and pointed snout; tapering notched plate-like structure on either side of the mouth; various spines on the head and about 5 forward-pointing teeth. **Body** tapers to tail and has conspicuous backward-pointing spine over pectoral fin. **Fins** 1st dorsal with 9–10 spiny rays; 2nd dorsal with 16 softer rays; anterior 3 rays of pectoral fin form separate sensory processes; posterior region of this fin typical in form, long and overlying pelvics. **Habitat** usually in deep water but rarely from 50m down. **Similar species** six.

Eutrigla gurnardus (Linnaeus) **Grey Gurnard** Length up to 450mm. Head large with prominent eye and pointed snout. **Body** tapers to tail. **Fins** 1st dorsal with 7–9 spiny rays; 2nd dorsal with 18–19 softer rays; anterior 3 rays of pectoral form separate sensory processes; posterior region of this fin normal in form but short, not reaching to anus. **Habitat** offshore from 20–50m, usually over sand or mud. **Similar species** six.

Aspitrigla cuculus (Linnaeus) **Red Gurnard** Length up to 400mm. Head as *Eutrigla gurnardus*, but tip of snout carries on each side 3 very small spines; snout slightly dished in profile. **Body** quite stout, lateral line being covered by conspicuously large scales. **Fins** 1st dorsal with 9–10 spiny rays; 2nd dorsal with 17–18 softer rays; pectoral fin divided into 3 sensory rays and posterior 'normal' lobe which stretches just to the anus. **Habitat** in shallow water from 20m down, often caught by commercial trawlers. **Similar species** six.

Dactylopterus volitans Linnaeus **Flying Gurnard** (Not illustrated) Fairly typical gurnard shape, reaching up to 500mm, but with very large 'wing-like' posterior lobes to pectoral fins which are coloured bluish-black with blue stripes and spotting; the fish is greyish-brown above and paler below with pinkish tinges. **Habitat** on sandy and muddy bottoms from 15m.

Serranus cabrilla (Linnaeus) **Comber** Length up to 300mm. Head large and pointed with lower jaw slightly longer than upper; conspicuous eye; main or 2nd gill cover itself covered by 1st gill cover or pre-operculum with small teeth on the lower edge; gill cover bearing 2 flattish spines but no heavy ridge. **Body** somewhat elongated yet ovoid; the dorsal region and flanks are coloured with 7–9 vertical bars and there are greenish-blue stripes along the lower regions of the head and body. **Fins** anterior section of dorsal fin bearing 9–10 spiny rays; the posterior region made up of 13–14 softer rays. **Habitat** from 20m down over rocks and among algae and sea-grasses. **Similar species** *Epinephelus guaza* (page 276), *Serranus scriba* and *S. hepatus* (below).

Serranus scriba (Linnaeus) **Painted Comber** Length up to 250mm; similar to *S. cabrilla* and distinguished from it principally by the divisions of the vertical brown stripes and the single violet or blue patch on the abdomen, and the reddish markings on the head. **Habitat** among sea-grasses and weeds over rocks and sand down to about 30m. **Similar species** *S. cabrilla, S. hepatus* and *Epinephelus guaza* (page 276).

Serranus hepatus (Linnaeus) **Brown Comber** Length up to 130mm; similar to but shorter than the two preceding species; the 4 vertical brown bars on the body and the brownish marking at the junction of the spiny anterior and softer posterior sections of the dorsal fin serve as useful distinguishing features. **Habitat** among rocks, algae and sea-grasses down to about 100m. **Similar species** *S. cabrilla, S. scriba* and *Epinephelus guaza* (page 276).

Anthias anthias (Linnaeus) Length up to 250mm. Head relatively large with conspicuous eye. **Body** superficially like a goldfish. **Fins** anterior section of dorsal fin spiny with 3rd ray twice the length of the rest, posterior part soft; pelvic fins stretch back towards rear of anal fin when folded; tail fin deeply divided into dorsal and ventral lobes. **Habitat** in shallow water, generally where light is low: among rocks and caves. **Similar species** *Apogon imberbis* (page 284).

Trigla lyra

Eutrigla gurnardus

Aspitrigla cuculus

Serranus cabrilla

Serranus scriba

Serranus hepatus

Anthias anthias

Epinephelus guaza (Linnaeus) **Dusky Perch** Length up to 1.4m. Head large with conspicuous eye and jaws; 1st gill cover or operculum has 3 flattened spines but no conspicuous ridge running across it. Body big and quite deep. Fins 1st part of dorsal has 11 spiny rays and the posterior part 13–16 soft rays; tail fin convex. Habitat from 5–200m, usually in crevices or grottos. Similar species *Serranus cabrilla*, *S. scriba* and *S. hepatus* (page 274), *E. alexandrinus* and *P. americanus* (below).

Epinephelus alexandrinus (Cuvier & Valenciennes) *(=**Plectropoma fasciatus**)* (Not illustrated) Length up to 400mm; similar to *E. guaza* but with a concave tail fin; colour brownish-violet with about 5 stripes on the back.

Polyprion americanus Bloch & Schneider **Wreckfish** Length up to 2m. Head massive with powerful jaws; lower jaw protruding; 1st gill cover or pre-operculum has a spiny lower edge and the 2nd gill cover or operculum has a heavy ridge running across it just below the level of origin of the lateral line. Body thick-set and powerful-looking. Fins anterior part of dorsal fin has 11 spiny rays and the posterior part has 12 soft rays. Habitat the young fish are open water dwellers and may be found accompanying flotsam; the older fish live in deeper water around rocks and wrecks down to 200m. Similar species *Epinephelus guaza*.

Dicentrarchus labrax (Linnaeus) *(=**Morone labrax**)* **Bass** Length up to 1m but usually around 600mm. Head with conspicuous eye and jaws; 1st gill cover or pre-operculum has a row of slightly forward-pointing spines on the lower edge; 2nd gill cover or operculum without a ridge. Body elongated, streamlined and powerful with big scales. Fins 1st dorsal with 8–9 spiny rays and a slight gap before 2nd, which has 1 spine and 12–13 soft rays. Habitat coastal water, bays and estuaries; sometimes in schools. Similar species there is a superficial similarity to *Trachurus trachurus* (below), in which the 2nd dorsal and anal fins are longer.

Trachurus trachurus (Linnaeus) *(=**Caranx trachurus**)* **Horse-mackerel, Scad** Length up to 500mm. Head big relative to body size. Body slender and elongated with a row of 69–80 wide scales overlying the lateral line. Fins 1st dorsal with 8 spines and slightly separated from the 2nd dorsal, which has a leading spiny ray and 23–34 soft rays; tail fin deeply forked; long anal fin with 2 leading spiny rays followed by 25–34 softer rays. Habitat often in schools, in surface waters and down to 100m; juveniles often with the bells of jellyfishes in which they may take refuge. Similar species superficially *Dicentrachus labrax*. also **Trachurus mediterraneus (Steindachner)** (not illustrated): length up to 400mm; very similar to *T. trachurus* but with 78–92 less wide scales in the lateral line.

Campogramma glaycos Lacépède *(=**C. vadigo** or **Lichia vadigo**)* **Vadigo** Length up to 650mm. Head jaws large, extending to just behind the eyes; teeth large. Body streamlined; lateral line slightly curved. Fins 1st dorsal has 6 independent short spines; 2nd dorsal has many soft rays; sharply forked tail; anal fin nearly as long as dorsal. Habitat surface waters, but never very close to the shore. Similar species several, including *Lichia amia* (Linnaeus) (not illustrated), reaching 2m long, and with a strongly curved lateral line; also *Seriola dumerili*.

Naucrates ductor (Linnaeus) **Pilot-fish** Length up to 300mm. Head large, streamlined; jaw ends just before eye. Body rounded and streamlined; a distinct keel present on either side of tail stalk. Fins 1st dorsal with 3–6 free spines; 2nd dorsal has a leading spiny ray followed by 26–28 soft rays; sharply forked tail fin; anal fin preceded by 2 short spines. Habitat in open water, often accompanying a large animal such as a shark or turtle. Similar species several, including *Campogramma glaycos* (above) and *Seriola dumerili* (below), but *Naucrates* can be distinguished by dark vertical bands on the body.

Seriola dumerili Risso **Amberjack** Length up to 1m. Head large; powerful jaws ending in midline of the eye. Body streamlined; no keel on tail stalk. Fins 1st dorsal has 5–7 short spines; 2nd dorsal with 36–38 softer rays; tail fin deeply forked; anal fin preceded by 2 spines. Habitat a surface-dwelling open water species which prefers strong currents; the juveniles sometimes associate with jellyfishes. Similar species several.

Polyprion americanus

Dicentrarchus labrax

Epinephelus guaza

Trachurus trachurus

Campogramma glaycos

Naucrates ductor

Seriola dumerili

Trachynotus ovatus (Linnaeus) *(=T. glaucus* or ***Lichia glauca)* Glaucus, Blane** or **Derbio** Length up to 300mm. Head relatively small; jaws terminating immediately in front of eye. Body slender and laterally compressed with 4–6 rounded dark spots on the flanks; no keels on tail stalk. Fins 1st dorsal reduced to 5–6 short spines; 2nd dorsal with 1 spine and 24–25 soft rays; tail fin deeply forked; anal fin preceded by 2 spines; small rounded pelvics. Habitat surface-living in open water, sometimes in schools. Similar species *Campogramma glaycos, Seriola dumerili, Lichia amia* and *Carynx hippos.*

Brama brama (Bonnaterre) **Ray's Bream** Length up to 650mm. Head rounded with large mouth and jaws extending back towards rear of eye; high forehead. Body deep but tapering sharply towards tail. Fins dorsal long with 3–5 spiny rays followed by 30–32 soft rays, the anterior rays being much taller than the posterior ones; anal fin quite long with deeper rays at front. Habitat open sea, often in midwater down to 100m; quite important commercially.

Note on the sea breams The sea breams belong to a family of fish (Sparidae) which is represented throughout the world's seas. They have evolved to lead a variety of life styles and this is reflected in their teeth which have adapted to different types of food. While their teeth show many adaptations their bodies are generally rather similar, being deep and sideways compressed. Many bream species are important commercially. 24 species from the Mediterranean.

Spondyliosoma cantharus (Linnaeus) *(=Cantharus lineatus)* **Black Sea Bream, Old Wife** Length up to 510mm but often less. Head quite small, with small jaws not reaching back to eye and bearing sharp, somewhat curved but small teeth; lips not thick. Body deep. Fins dorsal has 11 spiny and 12–13 soft rays. Habitat often in shallow coastal waters around rocks and wrecks; hermaphrodite: female phase succeeded by male phase. Similar species *Sarpa salpa* (page 280).

Pagellus erythrinus (Linnaeus) **Pandora** (Not illustrated) Length up to 300mm, sometimes longer. Head relatively small; snout pointed; jaws not reaching back to the small eyes; teeth in front of jaw sharp and pointed, small rounded molars. Body deep but tapering strongly to tail. Fins dorsal has 12 spiny rays followed by 10 soft ones. Colour similar to *Spondyliosoma cantharus* but the back is orange-red with pinkish upper flanks and a bluish lateral line; there is a bluish patch above the eye; dorsal fin base bluish; upper edge of operculum reddish; inside of mouth black. Habitat down to 100m over sand and mud; hermaphrodite: female phase succeeded by male phase. Similar species none.

Pagellus bogaraveo (Brünnich) *(=P. centrodontus)* **Red Sea Bream** or **Spanish Bream** Length up to 510mm but often less. Head relatively small; large mouth reaching back to the conspicuous eye; front teeth are sharp and curved but not very long; 2 or 3 molars. Body deep with dark spot above upper limit of operculum. Fins dorsal has 12 spiny rays followed by 12 to 13 soft rays. Habitat over rocks and wrecks from shallow water down to 200m. Similar species small individuals sometimes confused with *P. acarne* (below) which is more slender.

Pagellus acarne Risso **Spanish Sea Bream** Length up to 350mm but often less. Head relatively small but mouth itself quite big and extending back to the eye, which is of moderate proportions; teeth in front of mouth curved and sharp but small. Body slender and not very deep, with a darkish spot at the pectoral fin base. Fins dorsal with 12 spiny rays followed by 11–12 softer rays; anal with 3 spiny and 10 soft rays. Habitat close to the bottom from 20–100m, usually over sand or mud; hermaphrodite: male phase succeeded by female phase. Similar species *P. bogaraveo* (above) and *Lithognathus mormyrus* (page 282).

Sparus auratus Linnaeus **Gilthead** Length up to 700mm. Head large with arching forehead; large mouth reaches back to big eye; large pointed curved teeth in front of jaws, smaller ones behind, and rounded molars behind them. Body deep but tapering to tail; darkish patch at origin of lateral line. Fins dorsal has 11 spiny rays followed by 13-14 softer ones; pectoral fins quite long and pointed. Habitat inshore, often in areas of low salinity such as estuaries and lagoons.

*Trachynotus
ovatus*

Brama brama

*Spondyliosoma
cantharus*

*Pagellus
bogaraveo*

Pagellus acarne

Sparus auratus

Boops boops (Linnaeus) *(=Box boops)* **Bogue** Length up to 360mm. Head small mouth; jaws bearing 1 row of compressed cutting teeth, those of upper jaw bearing several rounded cusps; jaws not reaching back as far as the conspicuous eye. Body long, not deep. Fins dorsal with 14–16 spiny rays followed by 14–15 softer rays. Habitat in very shallow water, often among algae and sea-grasses, and down to 150m; an important commercial fish. Similar species none.

Sarpa salpa (Linnaeus) *(=Boops salpa* or *Box salpa)* **Saupe** Length up to 450mm but often less. Head with small mouth which is just subterminal; lips quite thick; jaws not extending back to eye; teeth in lower jaw pointed with serrations, in upper jaw with notches. Body quite deep. Fins dorsal has 11–12 strong spiny rays with 14–16 softer rays behind. Habitat often in compact shoals among sea-grasses and algae. Similar species *Spondyliosoma cantharus* (page 278) but *Sarpa* is distinguished by its thicker lips and the jaws not reaching as far back as the eye.

Oblada melanura (Linnaeus) **Saddled Bream** Length up to 300mm. Head quite large; lower jaw slightly longer than upper; upward-pointing jaws reaching back to the start of the large eye; lips thin. Body moderately deep relative to length; no vertical lines, but a conspicuous black patch or 'saddle' on tail stalk, bordered on either side by white bands and not lying under trailing edge of dorsal fin. Fins dorsal has 11 spiny rays and 14 soft rays; anal has 3 spiny rays and 13–14 soft rays. Habitat in shallow water over sandy bottoms and among sea-grasses. Similar species none.

Diplodus annularis (Linnaeus) **Annular Gilthead** Length up to 200mm. Head larger than in *Oblada melanura*, with steeper convex forehead; mouth not reaching back to eye; lips quite thick. Body deep, with black ring on tail stalk lacking white edging and not lying under trailing edge of dorsal fin. Colour somewhat variable: the faint vertical or diagonal dark bands are generally not conspicuous against the greyish, silvery or violet background; fins all lack black pigmentation. Habitat among rocks and in shallow water. Similar species *Diplodus sargus, D. vulgaris* and *Puntazzo puntazzo*.

Diplodus sargus (Linnaeus) Length up to 500mm but often less. Head large, with large mouth; high convex forehead; jaws reaching back to start of eye. Body deep and ovoid, tapering to tail; about 8 poorly defined vertical bands of brown with black spot on tail stalk, which begins immediately at end of dorsal fin. Fins pectoral fin bases and pelvic fins blackish. Habitat in shallow coastal waters, often in brackish lagoons and estuaries. Similar species *D. annularis, D. vulgaris* and *Puntazzo puntazzo*.

Diplodus vulgaris (Geoffroy) Length up to 400mm but often less. Head large, with very slightly concave forehead; mouth and jaws large, reaching back to start of eye. Body deep with dark saddle behind head immediately in front of leading spine of dorsal fin; another dark saddle just anterior to tail stalk beginning below trailing rays of dorsal fin; horizontal gold stripes running along body. Fins pelvics black with white periphery. Habitat in shallow coastal waters, often among rocks. Similar species *D. annularis, D. sargus* and *Puntazzo puntazzo*.

Puntazzo puntazzo (Cetti) *(=Charax puntazzo)* Length up to 450mm but often less. Head large (in order specimens) with pointed snout and slightly concave forehead; jaws do not reach as far back as eyes. Body with about 10 vertical dark lines, thicker ones usually alternating with thinner ones and a saddle of dark pigment on the tail stalk, not lying under the trailing part of the dorsal fin. Fins posterior part of dorsal, tail and anal have black peripheral band. Habitat on rocky coasts, generally in shallow water. Similar species *Diplodus annularis, D. vulgaris* and *D. sargus*.

Boops boops

Sarpa salpa

Oblada melanura

Diplodus annularis

Diplodus sargus

Diplodus vulgaris

Puntazzo puntazzo

Lithognathus mormyrus (Linnaeus) *(=Pagellus mormyrus)* Length up to 450mm but often less. Head large, with a slight concavity above the eye. Body not very deep and tapering towards the tail, with about 12 vertical dark bands alternating with paler ones. Habitat generally over sandy bottoms in shallow water. Similar species *Pagellus acarne* (page 278) which lacks the vertical bands and has a dark spot at the pectoral fin base.

Dentex dentex (Linnaeus) *(=D. vulgaris)* Dentex Length up to 500mm, sometimes greater. Head large with steep forehead (humped in old males), with powerful jaws reaching back to the eye; 4–6 conspicuous curved teeth in the front of the jaws, smaller teeth behind these. Body deep, powerful; back and flanks spotted blue-green, sometimes with 4 dark vertical stripes. Fins pelvic and anal fins coloured orange. Habitat in rocky areas down to 200m. Similar species 2 related: *Dentex filosus* Valentin and *D. macropthalmus* Bloch; see Luther, W. and Fiedler, K. 1976.

Argyrosomus regius Asso **Meagre** Length up to 2m but often less. Head large, with jaws reaching well back to the conspicuous eyes. Body long and slender with heavy scales. Fins unlike the breams, the anterior spiny portion is almost separate from posterior soft part. Habitat in sandy areas, often in brackish water; emits a deep rumbling sound under water. Similar species none.

Chelon labrosus (Risso) *(=Mugil chelo* or **Crenimugil chelo)** Thick-lipped Grey Mullet Length up to 700mm, occasionally more. Head broad, with rounded snout and relatively small mouth; lips thick, upper lip still visible when mouth is closed; 3 rows of small warts on lower edge of upper lip as well as close-set small teeth; eye only slightly covered by fatty eyelid. Body long and rounded in section. Fins 1st dorsal has 4 spiny rays; 2nd dorsal 9–10 soft rays. Habitat in shallow coastal waters. Similar species six.

Liza aurata (Risso) **Golden Grey Mullet** Length up to 440mm. Head upper lip quite thin, lacking small warts and not visible when mouth is shut; small teeth on upper lip conspicuous; small thick fatty eyelids. Body golden spot on gill covers. Fins 1st dorsal has 4 spiny rays, 2nd dorsal has 3 spiny and 7–9 soft rays. Habitat coastal waters, lagoons and estuaries. Similar species six.

Liza ramada (Risso) **Thin-lipped Grey Mullet** Length up to 600mm; very similar to *L. aurata* but with minute, almost invisible warts on upper lip. Fins 1st dorsal with 4 spiny rays, 2nd dorsal with 8–9 soft rays. Similar species six; for other species of mullet see Riedl, R. 1963.

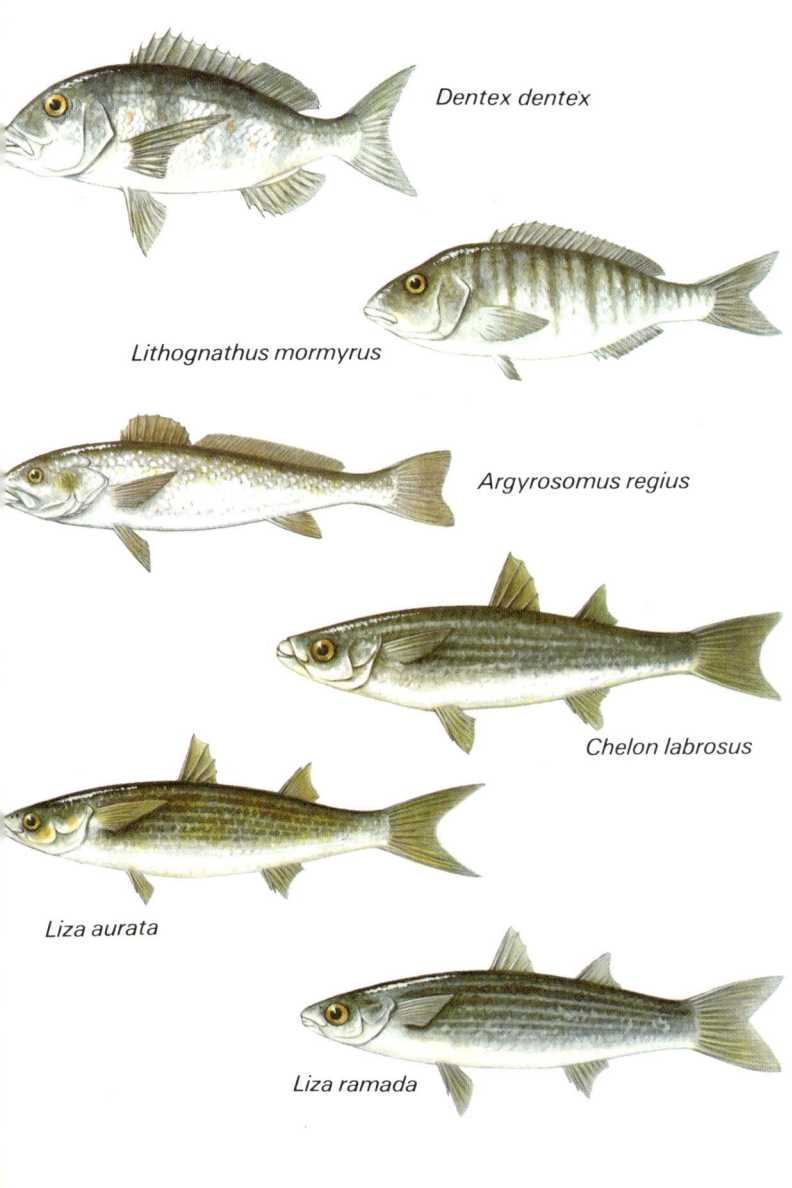

Dentex dentex

Lithognathus mormyrus

Argyrosomus regius

Chelon labrosus

Liza aurata

Liza ramada

Apogon imberbis (Linnaeus) **Cardinal Fish** Length up to 150mm. Head relatively large, with jaws reaching back to rear of conspicuous eye, which has two cross bands; 2nd gill cover or operculum has teeth on free edge. **Fins** short and small; 1st dorsal spiny; 2nd soft apart from leading ray; tail fin only slightly forked. **Habitat** in caves and rock crevices or under rocky overhangs where the illumination is low. **Similar species** *Anthias anthias* (see page 274) which has a deeply forked tail fin.

Maena maena (Linnaeus) *(=M. vulgaris)* Length up to 200mm. Head not large; snout pointed; small mouth with jaws not reaching back to eye. **Body** ovoid with a rectangular black mark in the mid-flank edged in white; background coloration normally blue-grey above and grey-white below but varies according to sex and season; breeding males have conspicuous blue stripes and spots. **Fins** dorsal fin uniform in height. **Habitat** over sandy bottoms and among sea-grass beds. **Similar species** three, but especially *M. sineris* and *M. chryselis*.

Maena smaris (Linnaeus) Length up to 200mm, males usually bigger than females. **Head** usually pointed and relatively smaller than in other species of *Maena*; jaw reaches back to eye. **Body** slender and not very deep, with an almost square black spot on flank. **Colour** normally greyish-brown above, but breeding males have about 4 bluish stripes along the body and a curved blue-green mark between the eyes. **Fins** dorsal fin of uniform height, dorsal and anal fins spotted bluish-green. **Habitat** over muddy and sandy bottoms, often associated with sea-grasses. For details of reproductive behaviour see Luther, W. and Fiedler, K. 1976.

Maena chryselis Valenciennes **Picarel** Length up to 200mm. Head pointed with terminal jaws reaching back to large eye. **Body** grey-brown or yellow-brown above; grey, white or silver below with indistinct diagonal stripes; black spot on the flank a little further forward than in *M. smaris*; breeding males have 4 undulating blue lines and a shiny blue-green stripe across the forehead reaching back to the eyes. **Fins** with blue spots; dip in middle of dorsal fin. **Habitat** over sand or coarse gravel. **Similar species** sometimes confused with *Boops boops* (page 280).

Sciaena umbra Linnaeus *(=Corvina nigra, Johnius umbra)* **Corb** or **Brown Meagre** Length up to 700mm but often less. Head large rounded snout; moderate jaws of equal length reaching back to the large eye. **Body** deep, ovoid. **Fins** 1st dorsal spiny and just connected to 2nd soft dorsal; 2nd ray of anal fin has a conspicuous spine; tail almost straight-ended. **Habitat** among rocks and weeds in shallow water, often in shoals. N.B. juveniles are brownish with brown pelvic fins and brown anterior rays to the anal fin, which is otherwise transparent. **Similar species** two.

Umbrina cirrosa (Linnaeus) **Bast Umber** or **Corb** Length up to 700mm but often less. Head large, with lower jaw shorter than upper and bearing truncated barbel; jaws reach back to large eye. **Body** slightly less deep than *Sciaena umbra* and coloured by diagonal wavy lines of yellow alternating with blue. **Fins** as for *Sciaena umbra*, but no conspicuous anal fin spine. **Habitat** in shallow water over sand and mud, often near estuaries. **Similar species** three.

Coryphaena hippurus Linnaeus **Dolphin (Fish)** No relation to the marine mammal of the same name. Length up to 1m. Head with large jaws reaching back to small eye; forehead steep. **Body** long and massive, tapering slowly to the tail. **Fins** dorsal long and ribbon-like, stretching from top of head almost to tail stalk; deeply forked tail. **Habitat** open water, often far out to sea. **Similar species** none.

Apogon imberbis

Maena maena

Maena smaris

Maena chryselis

Sciaena umbra

Umbrina cirrosa

Coryphaena hippurus

Mullus surmuletus Linnaeus **Red Mullet** Length up to 400mm. Head quite large with snout sloping up to prominent eye; 2 large scales below eye; jaws set low on head and bearing 2 long barbels. **Body** fairly slender and tapering to tail. **Fins** 1st dorsal yellow with a darkish patch; tail fin forked. **Habitat** among rocks and over sand down to 100m. **Similar species** *M. barbatus*. N.B. not a close relative of the Grey Mullets (see page 282).

Mullus barbatus Linnaeus **Red Mullet** Length up to 300mm; very similar to *M. surmuletus* but with very steep profile, 2 large scales and 1 much smaller below the eye. **Fins** 1st dorsal with a pink membrane and yellow markings; tail fin forked. **Similar species** *M. surmuletus* (above).

Cepola rubescens Linnaeus **Red Hand Fish** Length up to 700mm but often less. **Head** small with large upward-pointing mouth; many sharp curved teeth, and conspicuous eye. **Body** long and eel-like. **Fins** dorsal ribbon-like, starts above the gill cover and runs to the tail; anal starts a little further back and runs likewise. **Habitat** among sea-grasses and on sandy bottoms, sometimes burying itself in mud; down to 200m. **Similar species** none.

Chromis chromis (Linnaeus) **Damselfish** Length up to 150mm, often less. Head small, with small thin-lipped mouth. **Body** quite deep. **Fins** anterior part of dorsal spiny, posterior part soft; tail deeply cleft; pectorals quite large. **Colour** juveniles bluish-violet, adults brown. **Habitat** shallow water among rocks, piers.

Sparisoma cretense (Linnaeus) *(=Euscarus cretensis* or *Scarus cretensis)* **Parrotfish** Length up to 400mm. **Head** quite big; snout blunt; small mouth with powerful jaws and grinding teeth. **Body** with large scales. **Habitat** among rocks and weeds, often in shallow water. **Colour** two forms are known: the one illustrated is probably the female and the grey form with a black patch behind the gill cover probably the male. **Similar species** none.

Note on the wrasses Wrasses form a diverse and interesting family, exhibiting features such as nest building and aggressive behaviour. Some are remarkable for their association with other species of fish, working as cleaners removing parasites. The Mediterranean has about 20 wrasse species and their identification is not always easy. A species may take on different coloration according to sex and season. A few species show sex reversal with primary males and secondary males that are initially female and then change sex. Consequently alternative descriptions of colour have to be given in most cases, although there may be a persistent general identification feature. In addition to the usual features the presence and colour of the genital papilla may be important. The head is usually pointed with the mouth at the tip of the snout. The jaws do not normally reach as far back as the eye. The first gill cover or pre-operculum is often toothed and the main (second) gill cover or operculum never bears spines. There are usually a few spiny rays at the front of the dorsal fin.

Labrus mixtus Linnaeus *(=L. bimaculatus)* **Cuckoo Wrasse** Length up to 400mm. **Head** large, pre-operculum without teeth. **Body** long and not very deep. **Colour** breeding males orange-red with bright blue heads and flanks, blue extending to front of dorsal fin; white head is courtship coloration; immature males orange-red; females orange-red with 3–4 black patches on back and tail stalk; hermaphrodite: male succeeding female phase. **Habitat** rocks from 10m down.

Labrus bergylta Ascanius *(=L. maculatus)* **Ballan Wrasse** Length up to 450mm. **Head** large; lips thick; forehead slightly humped; gill covers without teeth or spines. **Body** deep; scales large. **Colour** background may be green, red or brown with red-brown netted pattern due to scale coloration on lower flanks. **Habitat** among rocks and weeds down to about 20m. Hermaphrodite: male succeeds female phase.

Labrus merula Linnaeus **Brown Wrasse** Length up to 450mm. **Head** small; snout pointed; lips thick; gill covers without teeth or spines. **Fins** tail usually has a small 'snip' out of it in the midline. **Colour** usually olive or brownish. **Habitat** shallow water among rocks, often solitary. Hermaphrodite: male succeeds female

Mullus barbatus

Mullus surmuletus

Cepola rubescens

Sparisoma cretense

Chromis chromis

juvenile

breeding ♂

mature ♂

Labrus mixtus

Labrus turdus

Labrus merula

Labrus bergylta

phase. Similar species *Labrus turdus* which has horizontal silvery lines.

Labrus turdus Linnaeus **Green Wrasse** (Illustrated on page 287) **Length** up to 450mm. **Head** snout pointed; forehead not steep. **Colour** variable, most commonly reddish to light green, sometimes with a silvery or white horizontal stripe; alternatively olive above with reddish colouring on the gill cover and belly; there is also a basically red variety without green but having silver and red on the belly. **Habitat** among sea-grasses such as *Posidonia* and in rocky places. Hermaphrodite: male succeeds female phase. **Similar species** *L. merula*, which lacks white horizontal stripe.

Crenilabrus mediterraneus (Linnaeus) **Axillary Wrasse** **Length** up to 170mm. **Head** small, with pointed snout; lips whitish and conspicuous; lower jaw shorter than upper; pre-operculum toothed. **Body** not very deep; lateral line curved like top of body. **Fins** dorsal fin drawn out into a slight lobe at trailing edge. **Colour** basically red or pinkish in breeding male, but may be brown; female coffee-brown with white vertical stripes, but always with a conspicuous dark mark on base of pectoral fin (blue in male set in a gold ring; brown in female); dark mark in both sexes above lateral line on tail stalk; genital papilla white in male, black in female. **Habitat** often in transitional zone between rocks and sand in shallow water.

Crenilabrus melops (Linnaeus) **Corkwing Wrasse** **Length** up to 200mm. **Head** greenish or blue crescentic spot behind eye. **Body** deeper than in most wrasses; lateral line curved but straightening out towards rear of dorsal fin; a blackish spot on the tail stalk over or below the lateral line; 3 spiny rays at front of anal fin. **Colour** background is variable: breeding males may be yellowish with 5 transverse bands; females olive-brown with 2 faint longitudinal bands. **Habitat** in shallow water in weeds. **Similar species** *C. quinquemaculatus* (Bloch & Schneider) (not illustrated) but this lacks the crescentic spot behind the eye and has 5 dark spots: 2 near the spiny rays of the dorsal fin and 3 near the soft rays.

Crenilabrus cinereus (Bonnaterre) **Grey Wrasse** **Length** up to 110mm. **Head** moderate with pointed snout and sometimes slightly concave forehead. **Body** with 1 black, blue or brown spot in front of dorsal fin and another on underside of tail stalk. **Colour** pale brown to grey, sometimes green-grey or yellowish; breeding male has a blue-green stripe on the gill cover and a dark spot above the base of the pectoral fin, the periphery of the dorsal and anal fins is bluish; female pale brown to green with horizontal brown stripes and a conspicuous black genital papilla in breeding state. **Habitat** shallow water, especially in sea-grass meadows.

Crenilabrus scina (Forskål) **Long-mouthed Wrasse** **Length** up to 120mm. **Head** thin lips, long snout and long jaws; shallow forehead with distinct concavity over eye. **Body** pale spot in midline of tail stalk; both sexes with a genital papilla. **Colour** green to red-brown or brown; breeding males reddish and breeding females with gold to yellow belly. **Habitat** in shallow water amongst rocks, sea-grasses and algae, more common in summer.

Crenilabrus ocellatus (Forskål) **Ocellated Wrasse** **Length** up to 120mm, males longer than females. **Head** pointed. **Body** deep in male, female slighter; small black spot in midline of tail stalk in both sexes. **Colour** breeding male greenish with a bluish-violet spot on rear of gill cover, horizontal rows of blue and green spots on flanks and belly; background coloration of breeding female green-brown with 2 brown stripes along the flanks, white genital papilla but spot on gill cover not clearly defined. **Habitat** very common amongst weeds and rocks, piers and harbour walls as well as in sea-grass meadows.

Crenilabrus tinca (Linnaeus) **Painted Wrasse** **Length** up to 300mm. **Head** quite large. **Body** deeper in male. **Colour** breeding male lemon-yellowish flanks with red horizontal stripes, fins spotted red and bordered in dark blue; black spot above pectoral fin base; female without blue and red patterns, short white genital papilla, abdomen silvery; ripe females and defeated males have 2 dark horizontal stripes. **Habitat** common in rocky places and in sea-grass meadows.

Crenilabrus mediterraneus

Crenilabrus melops

Crenilabrus cinereus

Crenilabrus scina

Crenilabrus ocellatus

breeding ♂

Crenilabrus tinca

♀

Coris julis (Linnaeus) **Rainbow Wrasse** Length up to 250mm. Head mouth and eyes small. **Body** shallow and relatively elongated. **Fins** dorsal long: in male first 3 spiny rays taller than the rest, black spot at leading edge. **Colour** male background blue above, silvery below; a short horizontal black mark behind pectoral fin and a zig-zag golden-yellow horizontal band; female drab by comparison and with brownish and white horizontal bands. **Habitat** in shallow water among rocks, especially during summer; hides in sand at night. Hermaphrodite: some drab females turning to colourful males; primary males occur with similar coloration to the female and not becoming colourful until 1 year old. **Similar species** none.

Thalassoma pavo (Linnaeus) **Peacock Wrasse, Turkish Wrasse** or **Rainbow Wrasse** Length up to 200mm. **Head** not large. **Body** long, a little deeper than *Coris julis*. **Colour** a black spot on back at the mid-point of the dorsal fin; background green with paler vertical stripes or a red and blue stripe running from the anterior of the dorsal fin to the pelvic fin. **Fins** large males have elongated tips to the tail fin, making it look crescentic. **Habitat** in rocky coastal areas during the summer, down to about 20m; hides in sand at night; hermaphrodite: males succeed females.

Trachinus draco Linnaeus **Greater Weever Fish** Length up to 400mm. **Head** small with eyes almost on top; mouth relatively large, upward-pointing; large venomous spine on each gill cover. **Body** elongated and laterally compressed; scales small; brown and blue stripes alternate on flanks. **Fins** 1st dorsal with 6–7 poisonous spiny rays and long 2nd dorsal with 29–32 soft rays; trailing edge of tail fin slightly curved. **Habitat** sandy bottoms; often partially buried so as to lie in wait for prey such as small crustaceans and fish. **Similar species** three, especially *Trachinus radiatus* Delaroche (not illustrated) which has black ring-like marks on the flanks.

Echiichthys vipera (Cuvier) *(= Trachinus vipera)* **Lesser Weever** Length up to 200mm. **Head** eyes almost on top; mouth relatively large, upward-pointing; large venomous spine on each gill cover. **Body** grey-yellow above, white below. **Fins** similar to *Trachinus draco*; trailing edge of tail fin straight and black. **Habitat** sandy bottoms in shallow water down to 50m. **Similar species** three.
N.B. poisonous 1st dorsal spines can cause severe pain.

Uranoscopus scaber Linnaeus **Stargazer** Length up to 250mm. Head large, jaws large and upward-pointing; eyes on top of head; spine on upper side of gill cover. **Body** long and tapering, rounded in section. **Fins** 1st dorsal short with 4 spiny rays; 2nd dorsal longer with 13–14 soft rays; large pectoral fins. **Habitat** usually buried in sand and mud with eyes and mouth visible. **Similar species** none.

Note on the blennies A common element of the fish fauna. The head is large and the mouth generally also big. They either have very small scales on their bodies or no scales at all; the body is usually slimy and compressed sideways. The dorsal and anal fins are elongated and the pelvic fins occur below the gills in front of the pectorals. The pelvics have 1 spiny and 1–3 softer rays and their forward position distinguishes the blennies from the gobies (pages 296–302) which are superficially similar. Most blennies occur in shallow rocky areas. About 21 species are recognized from the Mediterranean.

Blennius pavo Risso **Peacock Blenny** Length up to 120mm. Head large, with steep forehead and minute tentacles; mouth small; oval dark spot surrounded by blue ring behind the eye. Males have an orange crest over the eye; females lack head crest. **Body** elongated and compressed sideways; background coloration yellow to green, anterior with dark vertical bands edged in bright blue; at posterior these disperse into spots. **Fins** dorsal with 12 spiny rays and 21–23 soft rays. **Habitat** in shallow water on soft substrates, often near rocks.

Blennius ocellaris Linnaeus **Butterfly Blenny** Length up to 170mm. Head large with small mouth, steeply rising forehead bearing a pair of small branched tentacles above the eye. **Body** green-brown; tapering; with about 6 transverse

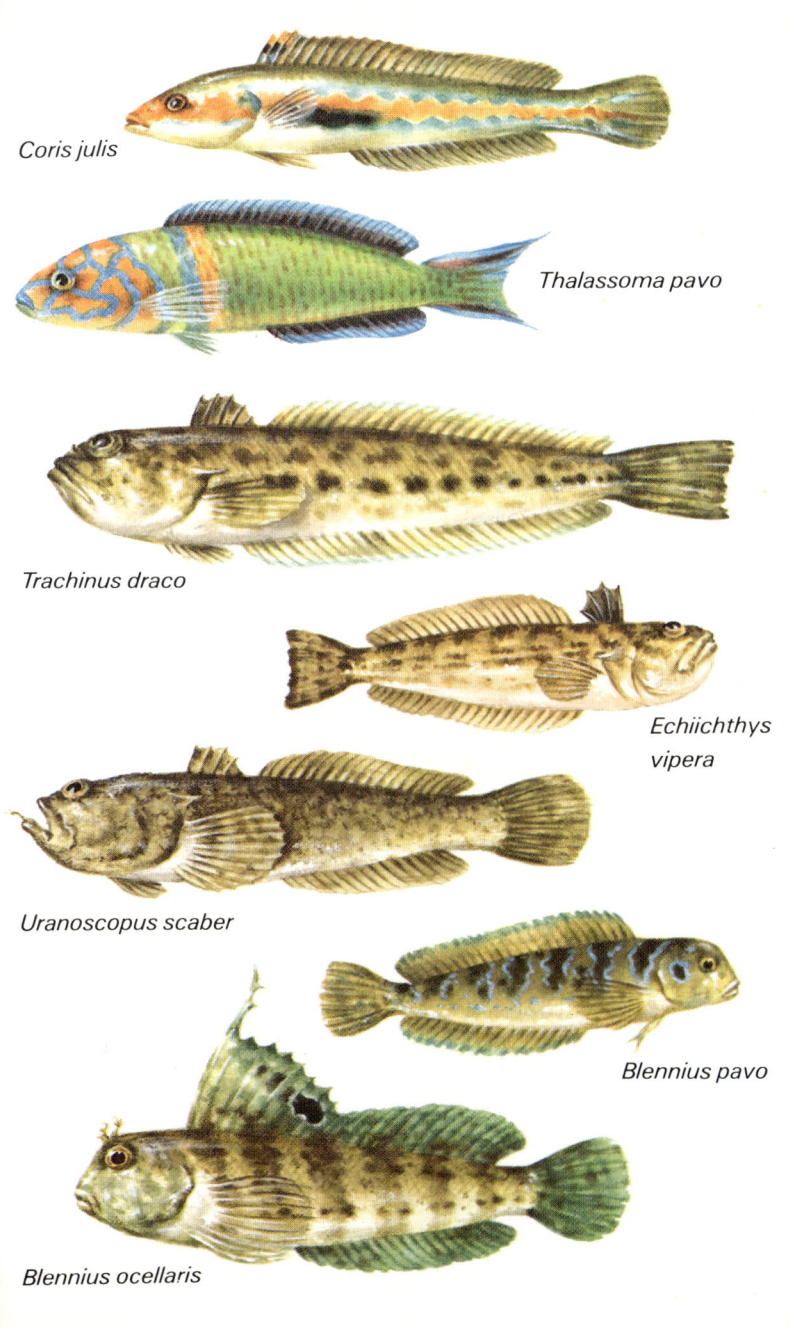

Coris julis

Thalassoma pavo

Trachinus draco

*Echiichthys
vipera*

Uranoscopus scaber

Blennius pavo

Blennius ocellaris

bands of darker pigment. **Fins** dorsal divided into anterior lobe bearing a blue-black spot edged with white around the sixth ray; 10–12 spiny rays and 14–16 softer rays; 2 very small tentacles on either side of 1st dorsal fin ray. **Habitat** on various hard substrates down to 20m.

Parablennius gattorugine (Linnaeus) *(=Blennius gattorugine)* **Tompot Blenny** Length up to 250mm. **Head** large; snout slopes steeply up to eye; conspicuous branching tentacle above each eye, less tall than 1st ray of dorsal fin and branched in all directions. **Body** tapering. **Fins** junction of spiny and smooth sections of dorsal fin marked by a slight dip; 12–14 spiny rays and 17–20 soft rays. **Colour** about 6 vertical dark bands running from dorsal fins over flanks to anal fin. **Habitat** on rocks in shallow water.

Blennius tentacularis Brünnich **Horned Blenny** Length up to 150mm. **Head** large; tentacles conspicuous above eye with branches on their trailing edges only, and as tall as 1st ray of dorsal fin; separate small tentacles associated with nostrils. **Body** tapering. **Fins** dorsal not clearly divided into spiny and soft sections, 12–14 spiny rays, 18–25 soft rays. **Habitat** on sandy substrates where there are a few stones; and in brackish places. **Similar species** *Parablennius gattorugine* but distinguished from it by eye tentacle form.

Blennius sanguinolentus Pallas **Red Speckled Blenny** Length up to 150mm, sometimes more. **Head** forehead rounded; jaws small with powerful conical teeth; very small tentacles above eye. **Body** stout with fat stomach but compressed sideways towards the head. **Fins** dorsal of equal height along its length with 12–13 spiny rays and about 20 soft rays; a black spot overlying the 1st and 2nd spiny rays. **Habitat** in shallow rocky places, especially in crevices.

Blennius sphinx Valenciennes Length up to 80mm. **Head** forehead steep; small eyes located well up and small mouth well down; conspicuous unbranched tentacle above each eye; bluish-grey spot behind eye surrounded by red ring. **Body** compressed sideways. **Fins** dorsal with high lobed anterior region of 12 spiny rays and a lower posterior region of 16–17 soft rays. **Habitat** in shallow water among rocks and weeds.

Blennius rouxi Cocco **Striped Blenny** Length up to 70mm. **Head** forehead steeply rising; large eyes with a tentacle of about 5 unbranched threads over each eye as well as a tiny tentacle on the lower nostril. **Body** elongated, grey to white and bearing a conspicuous horizontal black stripe. **Fins** dorsal of about equal height along its length and a dark spot often between 1st and 2nd fin rays; 11 spiny and 24 soft rays. **Habitat** in shallow water down to about 10m among rocks with sand.

Blennius zvonimiri Kolombatović **Zoanimir's Blenny** **Head** forehead steeply rising; mouth small; lips thick and not reaching to behind eye; conspicuous branching tentacles over eyes and smaller ones over nostrils. **Body** flattened sideways; greyish to reddish brown. **Fins** dorsal divided in two with 12 spiny and 17 soft rays. **Habitat** in deep water, often among red coralline algae.

Blennius nigriceps Vinciguerra **Black-necked Blenny** Length up to 40mm. **Head** with black patterns; forehead steep; eyes located well up and jaws well down, lacking any tentacles. **Body** long, belly slightly swollen, background red. **Fins** spiny section of dorsal with 12 rays and almost separated from soft section of 15 rays by 'notch'. **Habitat** amongst rocks, especially near entrances to dark crevices; often associated with encrusting algae and the similar species *Tripterygion minor* (see page 294) but this is distinguished by the presence of 3 dorsal fins.

Blennius canevae Vinciguerra Length up to 70mm. **Head** forehead steep but profile slightly rounded; eye small and set well up, mouth well down and not reaching far back; lacking tentacles and with golden cheeks. **Body** slightly tapering, background coloration dark brownish with a tracery of lighter marks. **Fins** dorsal with 13 spiny rays and 16 soft rays separated by a 'notch'. **Habitat** on lower shore and in extreme shallow water amongst rocks.

Parablennius gattorugine

Blennius tentacularis

Blennius sanguinolentus

Blennius sphinx

Blennius rouxi

Blennius zvonimiri

Blennius nigriceps

Blennius canevae

Blennius dalmatinus Steindachner and Kolombatović Length up to 40mm. **Head** forehead very steep; eye prominent, lacking tentacles above it; very inconspicuous tentacle present over lower nostril. **Body** slender, compressed sideways; background colour olive above and green to gold below, with about 9 brownish bands on the back. **Fins** anterior spiny section of dorsal with 12 rays which are somewhat shorter than the posterior 15–16 soft rays. **Habitat** in very shallow water on rocky substrates.

Blennius trigloides Valenciennes Length up to 90mm. **Head** steep forehead and conspicuous eye, entirely lacking tentacles. **Body** quite full but tapering to tail; background coloration green to yellow-brown with darker patches. **Habitat** lower shore and in shallow water amongst rocks and stones.

Coryphoblennius galerita (Linnaeus) *(=Blennius galerita* or *B. montagui)* **Montagu's Blenny** Length up to 80mm. **Head** forehead not steep, lacking tentacle above eye but with flap of fringed filaments running across forehead between eyes, from the midline of which a line of up to 7 small spines or filaments extends back towards the 1st dorsal fin. **Body** elongated, coloration olive to grey-brown with blue and white spots on head and back. **Fins** dorsal with 12–17 spiny rays separated from 17–18 soft rays by a conspicuous 'notch'; spiny section not as tall as soft section. **Habitat** on lower shore and in shallow water, on sand and among stones and shells.

Cristiceps argentatus Risso *(=Clinus argentatus)* Length up to 100mm, superficially like a true blenny. **Head** relatively small. **Body** with scales, rounded in section. **Fins** 1st dorsal triangular in shape with 3 spiny rays; 2nd dorsal with 25–28 spiny rays and 3–4 soft ones; tail may appear forked due to the lack of pigment in the middle of the membrane but is actually not so; pelvic fins lobed. **Habitat** shallow water among weeds and in crevices. **Similar species** none.

Tripterygion tripteronotus (Risso) *(=T. nasus)* **Black-faced Blenny** Length up to 80mm. **Head** with more tapering snout than the usual fairly steep forehead of the true blennies; tentacles over eyes. **Body** deepest below 2nd dorsal fin; background coloration depends on sex and maturity: breeding male has black head and blue spots on the gill cover, rest of the body red with dark vertical bands; non-breeding males may be brownish above, yellowish below and have diagonal stripes. **Fins** 1st dorsal has 3 spiny rays, 2nd dorsal has 17–18 spiny rays and 3rd dorsal has 12–13 soft rays; all just separated; in the male there is a black spot on the trailing edge of 1st dorsal fin and the 2nd dorsal has elongated leading rays and there are blue edges to the 2nd and 3rd dorsals; the female lacks a spot on the 1st dorsal and the long rays of the 2nd. **Habitat** on rocky substrates from extremely shallow water down to about 10m. **Similar species** *Tripterygion minor* (Risso) which reaches only 50mm and has a less steep snout; the eye is lower on the head and there is a barbel on the lower lip; body bears 3 conspicuous medium-sized white spots on the back between each of which is a single smaller spot; see also *Blennius nigriceps,* in whose company this species is often found.

Carapus acus (Brünnich) *(=Fierasfer acus)* **Pearlfish** (Not illustrated) Length up to 200mm. **Head** smaller; jaws reaching back to eye. **Body** long and eel-like and compressed sideways. **Fins** dorsal and anal meet at tip of tail. **Habitat** in the respiratory trees and body cavities of sea-cucumbers such as *Holothuria* and *Stichopus* species (see page 242), where they feed on the reproductive organs of the sea-cucumbers.

Blennius dalmatinus

Blennius trigloides

Coryphoblennius galerita

Cristiceps argentatus

Tripterygion minor

Tripterygion tripteronotus

♂

♀

Callionymus reticulatus Valenciennes Length males up to 100mm, females up to 80mm. **Head** has 3 spines on the front gill cover or pre-operculum, all of which point backwards and upwards. **Body** very slender; mature males have background of orange-brown above and creamy underparts, together with dark round spots and pale blue spots on the flanks and dark brown-orange saddle-like marks across the back; female lacks dark spots on the body but has 'saddles'. **Fins** 1st dorsal has 4 spiny rays and 2nd dorsal always has 10 soft rays; males have blue-white spots and lines on dorsal fins. **Habitat** on sandy bottoms and shell gravel from 20 to 40m.

Callionymus lyra Linnaeus **Common Dragonet** Length male up to 300mm, female up to 200mm. **Head** large and flat with a long snout; mouth on lower surface; upper jaw longer than lower; lips thick; eyes borne high up; front gill covers or pre-opercula bear 4 spines which can cause injuries; 1 points forward and the others point back or upwards. **Body** lacking scales; flat and slender; tapering to tail, where it is rounded in cross-section; breeding males have brown-yellow background with blue-violet or green markings; females and immature males generally pale brownish with darker brown marks. **Fins** 1st dorsal has 4 spiny rays, 2nd dorsal 9 soft rays; in male 1st ray of 1st dorsal when folded down reaches back to base of tail fin, successive rays shorter; all rays of 2nd dorsal fin about equal in height; female has 1st ray of 1st dorsal about same height as rays of 2nd dorsal fin. **Habitat** common on sandy and muddy substrates from 20 to 100m. **Similar species** *C. festivus*.

Callionymus maculatus Rafinesque *(= C. dracunculus)* **Spotted Dragonet** Length male up to 140mm, female up to 110mm; generally similar to *C. lyra* but smaller. **Head** flattish and has first gill cover or pre-operculum with 1 forward-pointing and 3 upward-pointing spines; silvery coloration. **Body** tapering to tail, which is rounded in section; background coloration brownish-yellow above and paler yellow below, with 2 horizontal rows of brownish spots on flanks and 4 irregularly-shaped dark saddle-like marks on the back; there may also be silvery as well as bluish spots on the flanks. **Fins** 1st dorsal has 4 spiny rays, 2nd dorsal normally has 9 soft rays (rarely 10); in mature males dorsal fins have 4 rows of dark spots interspersed with bluish-white spots; also the 1st dorsal fin, when folded down, reaches to the trailing edge of the 2nd dorsal. In females and immature males the dark dorsal fin spots are less distinct, and there are 2 rows in the female. **Habitat** on sandy substrates from 70 to 300m. **Similar species** *C. belenus*. (Risso) (Not illustrated) Similar to *C. maculatus*; **length** up to 80mm, but this species has only 3 spines on the first gill cover or pre-operculum; 1st dorsal fin with 3 spiny rays, 2nd dorsal with 8–9 soft rays.

Callionymus festivus Pallas Length males up to 150mm, females up to 100mm. **Head** flat with long rounded snout; mouth small; lips thick; 1st gill cover or pre-operculum carries 3 spines. **Body** without scales; coloration grey to sandy or brown; the male has blue stripes and spots on the flanks and the head; the female is not so brightly coloured. **Fins** 1st dorsal has 4 spiny rays; 2nd dorsal has 7 soft rays; in the male the 2nd dorsal rays are taller than those of the first and they protrude beyond the fin membrane; the fins are yellowish with undulating blue lines edged in brown; lower parts of anal and tail fin blackish; in the female the 1st dorsal fin is black and the protruding rays of the 2nd dorsal fin are lacking. **Habitat** in shallow water on sandy bottoms, often partly buried.

Note on the gobies The gobies are a large group of small, superficially similar fishes with about 45 species in the Mediterranean. They seldom get longer than 200mm and have a large head with conspicuous eyes and pouting cheeks. The body is round in section and has scales. There are two dorsal fins and the large pectoral fins lie about level with the pelvics, which are normally fused to form a weak funnel-like sucker. The anterior margin of this is formed by a membrane which stretches from the spiny rays of the pelvic fin on one side across the underside of the body of those of the other. The precise form of this horizontal membrane is important in identification because it varies in form, sometimes having lateral lobes and sometimes having minute tubercles or villi.

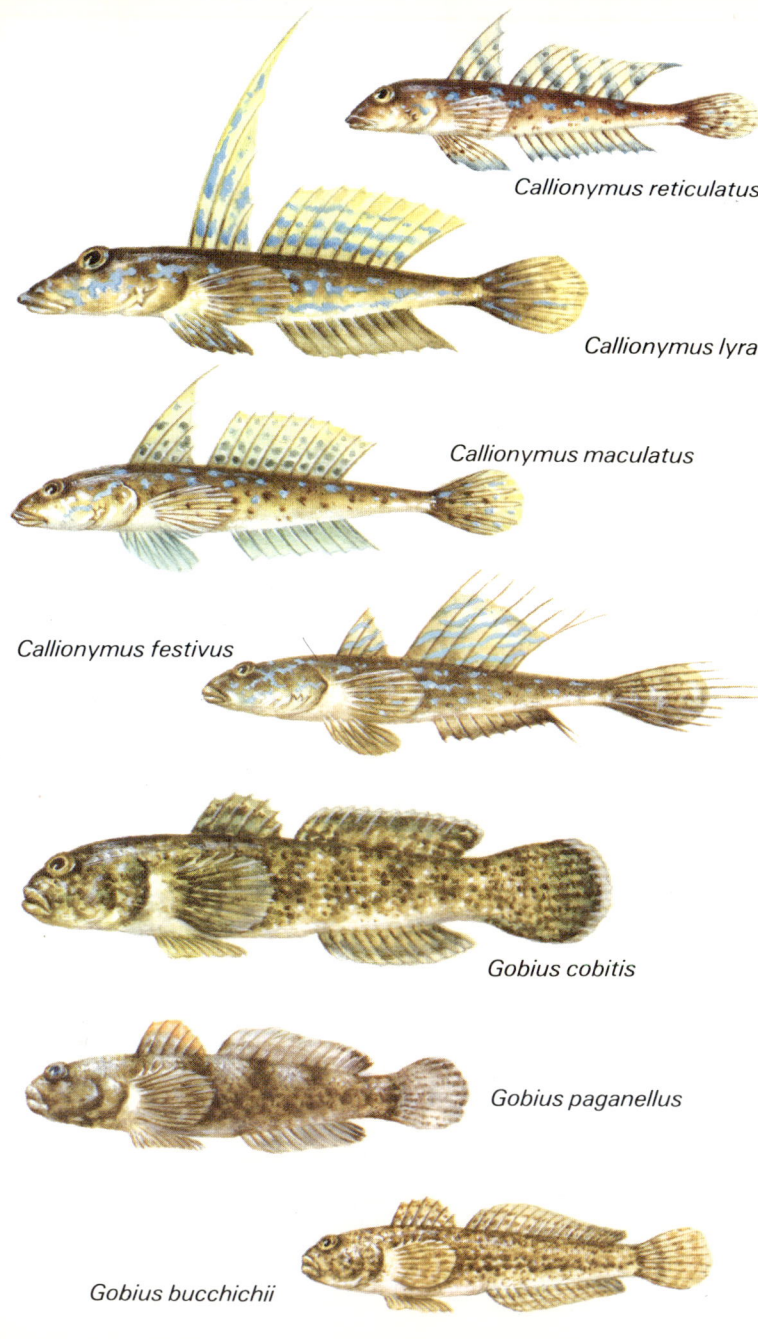

Callionymus reticulatus

Callionymus lyra

Callionymus maculatus

Callionymus festivus

Gobius cobitis

Gobius paganellus

Gobius bucchichii

Gobius cobitis Pallas **Giant Goby** (Illustrated page 296) **Length** up to 270mm; probably the largest goby in the Mediterranean. **Head** large, with large wide lips; jaws not reaching back to eye; nostrils have a divided tentacle. **Body** typical shape; scales on back above gill cover. **Fins** 1st dorsal has 6 spiny rays; 2nd dorsal has 1 spiny and 13 soft rays; pectorals have free upper rays; leading membrane of pelvics has conspicuous lateral lobes; 'sucker' rounded and short. **Habitat** in shallow rocky places and among weeds.

Gobius paganellus Linnaeus **Rock Goby** (Illustrated page 296) **Length** to 120mm. **Head** 1st nostril has tentacle with about 6 branches at tip. **Body** scales on back above gills and cheek; background coloration pale brown with dark markings; breeding males are brown-purple. **Fins** 1st dorsal with 6 spiny rays, 2nd dorsal with 1 spiny and 13–14 soft rays; normally upper edge of 1st dorsal has pale band along it but in breeding males this is orangish; free rays of pectorals well developed and often reaching up to start of 1st dorsals; leading membrane of pelvics conspicuous, normally no lateral lobes. **Habitat** among rocks and stones in shallow water, harbours, muddy places. **Similar species** *G. niger.*

Gobius bucchichii Steindachner **Bucchich's Goby** (Illustrated page 296) **Length** up to 100mm. **Head** relatively small; nostrils have tentacles which are sometimes folded. **Body** has scales on neck above gill covers. **Fins** 1st dorsal has 6 spiny rays, 2nd dorsal has 1 spiny and 13–14 soft rays; highest rays of pectorals free at tips; leading membrane of pelvics lacking lateral lobes. **Habitat** on soft substrates, often hides in *Anemonia viridis* (page 74) tentacles.

Gobius niger Linnaeus *(=****G. jozo)*** **Black Goby** **Length** up to 150mm. **Head** large with conspicuous eyes. **Body** has scales on neck above gill covers; background brownish, darker in breeding males, with darker patches. **Fins** 1st dorsal 5–7 spiny rays; 3rd, 4th and 5th elongated in male and when folded reaching back beyond leading edge of 2nd dorsal, which has 1 spiny and 11–13 soft rays; leading corners of both dorsals with black spots; pectorals with short protruding free upper rays; 'sucker' rounded; tail fin rounded. **Habitat** on soft bottoms in sea-grass beds and down to about 75m. **Similar species** *G. paganellus.*

Gobius geniporus Valenciennes **Slender Goby** **Length** up to 160mm. **Head** has anterior nostril with a small lobe on the edge. **Body** slender; background coloration brown with dark patches on the flanks. **Fins** 1st dorsal with 6 spiny rays, 2nd dorsal with 1 spiny and 12–13 soft rays; pectorals with only extreme tips of upper rays protruding; leading membrane of pelvics large but no lateral lobes; 'sucker' has straight rear edge. **Habitat** shallow water, over mud and sea-grasses.

Gobius cruentatus Gmelin **Red Mouthed Goby** **Length** up to 180mm. **Head** 1st nostril bears an unbranched tentacle. **Body** has scales on neck above gill cover, on gill cover itself and on hind edge of cheeks; coloration brown-red with patches on flanks, lips and cheeks bright red; black sensory spots on head. **Fins** 1st dorsal has 6 spiny rays, 2nd dorsal 1 spiny and 14 soft rays; pectorals have some free rays; leading membrane of pelvics large but lacking lateral lobes; 'sucker' split a little on hind edge. **Habitat** rocks, sandy places, sea-grass beds to 40m.

Chromogobius quadrivittatus Steindachner *(=****Relictogobius kryzhanovskii)*** **Banded Goby** **Length** up to 65mm. **Head** anterior nostrils without tentacle; posterior nostrils extended as a short tube; dark black spot on gill cover, head patterned. **Body** no scales on neck above gill covers; coloration brownish with about 12 upright dark stripes on the flanks, paler saddle-like marks at front and rear of 2nd dorsal fin; wide pale mark across neck from one pectoral fin to the other, the rear edge strikingly edged with black. **Fins** 1st dorsal has 6 spiny rays, 2nd dorsal 1 spiny and 8–11 soft rays. **Habitat** rocky shore and shallow water.

Thorogobius ephippiatus (Lowe) *(= Gobius forsteri)* **Leopard-spotted Goby** **Length** up to 130mm. **Head** nostrils lack tentacles. **Body** bears no scales on neck above the gill covers; coloration pale brownish with reddish patches. **Fins** 1st dorsal has 5–7 spiny rays and a black spot in the hind corner; 2nd dorsal 1 spiny and normally 11 soft rays; pectorals lack free rays; leading membrane of pelvics lacks lateral lobes; tail fin rounded. **Habitat** in rocky places down to 40m.

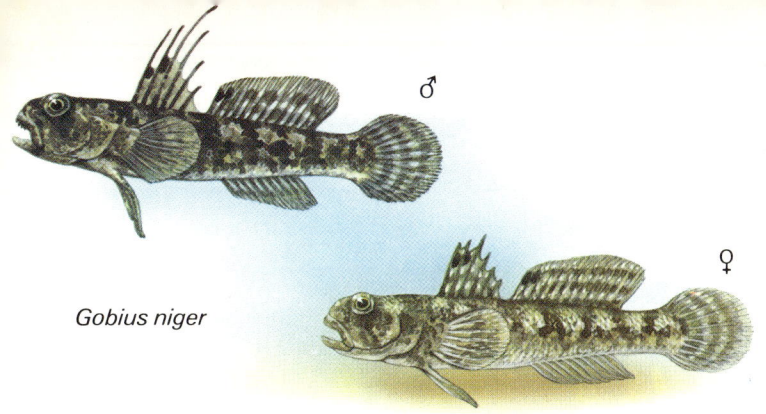

Gobius niger

♂

♀

Gobius geniporus

Gobius cruentatus

Chromogobius quadrivittatus

Thorogobius ephippiatus

Zosterissor ophiocephalus

Zosterissor ophiocephalus (Pallas) *(=Gobius lota)* **Grass Goby** (Illustrated on page 299) **Length** up to 250mm. **Head** with nostrils lacking tentacles. **Body** has scales on neck above gill covers. **Fins** 1st dorsal has 1 spiny and 14–15 soft rays; pectorals lack free rays and often have a dark spot at the upper end of the base; a small dark spot at the tail fin base; pelvics leading membranes but no lateral lobes; coloration yellow-green with irregular patterns on flanks. **Habitat** often in brackish water, in lagoons, estuaries and in muddy places or near sea-grass.

Lesueurigobius friesii (Malm) **Fries's Goby** **Length** up to 120mm. **Body** large, slender; scales on neck over gill covers; basic coloration pale brown with many small yellow and gold spots. **Fins** 1st dorsal fin with 6 elongated spiny rays; 2nd dorsal 1 spiny ray and 13–16 soft rays; pectorals lack free rays; leading membrane of pelvics lacks lateral lobes; sucker complete; short tail stalk and tailfin, lance-shaped; fins have yellow/gold spots. **Habitat** on muddy ground, often below 50m and sometimes associated with *Nephrops norvegicus* (page 204) in whose burrows it may hide.

Gobius flavescens (Fabricius) **Two-spotted Goby** **Length** 60mm or more. **Head** eyes large, set wide apart; anterior nostrils bear tubular opening. **Body** flattened sideways; lacks scales on back and above gill covers; coloration red-greenish-brown with pale saddle-like marks on back and on back of neck as well as dark patterns; pale and dark bluish marks on flanks; large black spot on flanks below 1st dorsal fin. **Fins** 1st dorsal has 7–8 spiny rays; 2nd dorsal has 1 spiny and 9–10 soft rays; pectorals lack free rays, pelvics with leading membranes and 'sucker'. **Habitat** coastal waters to 16m and among algae and sea-grasses.

Aphia minuta (Risso) **Transparent Goby** **Length** up to 51mm. **Head** nostrils lack tentacles or other processes; eyes situated on the side of the head; mature males have large canine teeth, females have very small teeth. **Body** flattened sideways; no scales on neck above gills; body is transparent with very small dots of dark colour along some of the fin bases and on the head. **Fins** generally smaller in the female, but 1st dorsal has 5 spiny rays, 2nd dorsal has 1 spiny ray and 11–13 soft rays; pectorals lack free rays; pelvics with a leading membrane and a 'sucker': this is smaller in female. **Habitat** in shallow areas in midwater and *not* on bottom; often swimming over sand and sea-grasses.

Crystallogobius linearis (Von Düben) **Crystal Goby** **Length** up to 47mm, female smaller. **Head** bears eyes on the sides; male has conspicuous lower jaw with big canine teeth; female has smaller toothless jaws and pigment spots on chin. **Body** lacks scales, is flattened sideways. **Fins** 1st dorsal of male with 2 spiny rays (vestigial in the female) 2nd dorsal 1 spiny and 14–15 soft rays; pectorals lack free rays; pelvics form a deep tubular 'sucker' in the male and are vestigial in the female. **Habitat** deep water down to 400m over shells, mud or sand.

Pomatoschistus marmoratus (Risso) **Marbled Goby** **Length** up to 65mm. **Head** with no flap-like membrane on 1st nostrils; darkish pigment below chin to below gill covers; male has dark spot under chin. **Body** no scales on back above gill covers; background coloration is sandy-brown with darker patterns and saddle-like marks; males often have 4 upright stripes on flanks. **Fins** 1st dorsal with 6 spiny rays and black spots towards rear; 2nd dorsal with 1 spiny and 8–9 soft rays; pectorals lack free rays; leading membrane of pelvics with villi and 'sucker' darker than other fins, which apart from tail and anal have reddish bands. **Habitat** in shallow sandy places. **Similar species** several.

Pomatoschistus pictus (Malm) **Painted Goby** **Length** up to 57mm. **Head** 1st nostrils lack flap-like membrane. **Body** scales lacking on back above gill covers; background coloration sandy-brown to fawn with dark patterns and big saddle-like marks reaching halfway down flanks with 4 double black spots along the midline. **Fins** 1st dorsal with 6 spiny rays and 2nd dorsal with 1 spiny and 8–9 soft rays; both these fins have tracts of dark spots overlain by pink pigments; pectorals lack free rays; pelvics with leading membranes lacking villi. **Habitat** in coastal waters down to 50m on sand and gravel; rarely in rock pools.

Lesueurigobius friesii ♀

♂

Gobius flavescens

♂

♀

Aphia minuta

Crystallogobius linearis ♂

Pomatoschistus marmoratus

Pomatoschistus pictus

Pomatoschistus minutus (Pallas) **Sand Goby** Length up to 95mm. Head front nostrils lack a flap on the rim. **Body** slender and tapering; scales on neck above gill covers; background coloration sandy-brown with darker spots and small saddle-like marks on the back; 4 conspicuous upright stripes of darker pigment on the flanks. **Fins** 1st dorsal 6–7 spiny rays; 2nd dorsal 1 spiny ray and 10–12 soft rays; in males a dark blue mark on 1st dorsal fin between 5th and 6th spiny ray; pectorals have no free rays; pelvics with leading membrane and 'sucker' darker than the other fins. **Habitat** near coasts, often over sand from the shore down to 20m. **Similar species** several.

Pomatoschistus microps (Krøyer) **Common Goby** Length up to 64mm. **Head** slightly orange; 1st nostril lacks a flap-like membrane; dark colour below chin to below gill openings. **Body** lacks scales on back above gill covers; background coloration grey to pale-brown, darker saddle-like marks on back; about 10 vertical bars running up flanks. **Fins** 1st dorsal has about 7 spiny rays with dark mark on membrane between rays 5 and 6; 2nd dorsal about 9 soft rays; both fins red-brown; pectorals without free rays; leading membrane of pelvics lacks villi; sucker dark, tail fin rounded. **Habitat** near coasts and in estuaries, salt flats and pools, even on the shore. **Similar species** several.

Scomber scombrus Linnaeus Length up to 660mm. **Head** with jaw set well down, large eye and transparent vertical eyelids. **Body** elongated, rounded and tapering to tail with characteristic mackerel stripes along back; no corselet of scales around head and pectoral fin base. **Fins** 1st dorsal with 10–13 spiny rays, not lying in a groove and widely separated from 2nd dorsal which has 11–13 soft rays; there are 5 finlets between it and the tail, and 5 finlets between the anal fin and the tail; a smaller keel at the base of the tail fin lobes on either side. **Habitat** a surface water fish often living in huge shoals which are migratory, often entering coastal waters. **Similar species** *S. japonicus.*

Scomber japonicus Houttuym *(=S. colias)* **Spanish Mackerel** or **Chub Mackerel** Length up to 400mm, occasionally longer; very similar to *S. scombrus* but head larger and has 9–10 spiny rays in 1st dorsal fin; a corselet of scales runs around head and pectoral fin base; body coloration blue-green, blue flanks with whitish belly; a golden-yellow strip running from head to tail and with numerous dark spots running the length of the body. **Habitat** surface, open and coastal waters. **Similar species** *S. scombrus.*

Thunnus thynnus (Linnaeus) **Blue-fin Tunny** or **Tuna** Length up to 3m or occasionally more. **Head** large and pointed; mouth almost in midline but jaws relatively small, reaching back to the eye; small conical sharp teeth; well developed keel on either side of the tail stalk. **Fins** 1st dorsal with 13–15 spiny rays, the leading rays being long and the posterior ones short, set close to 2nd dorsal fin which has 1 spiny and 13–15 soft rays; 8 to 10 finlets lie between it and the tail-fin while there are 8–9 finlets between the anal and the tail; pectoral fin short and about the same length as the snout. **Habitat** open surface waters in schools, seldom found deeper than 100m; migratory; historically an important Mediterranean food fish now under heavy pressure from fisheries. **Similar species** several.

Thunnus alalunga (Bonnaterre) (Not illustrated) **Long-finned Tunny** or **Albacore** Length up to 1m. **Head** large. **Fins** 1st dorsal 13–14 spiny rays, 2nd dorsal 13–14 soft rays and 7–8 finlets lying between it and the tail; 7 finlets between anal fin and tail; pectoral fins markedly longer than snout. Fins clear yellow; an important commercial fish. Coloration clear blue above with yellow on the flanks and silvery on the belly. **Similar species** *T. thunnus.*

Auxis thazard (Lacépède) *(=A. rochei)* **Frigate Mackerel** Length up to 610m, occasionally more. **Head** with small eyes placed well forward. **Body** like the true mackerels; fleshy keels on either side of tail stalk as well as larger ones on each side of the tail fin itself. **Fins** 1st dorsal has 10–11 spiny rays and is well separated from the 2nd which has 11–12 soft rays and 8–9 finlets between this and the high thin-lobed tail fin; 7–8 finlets between anal fin and tail. **Habitat** in shoals in surface waters, coming inshore in summer.

Pomatoschistus minutus

Pomatoschistus microps

Scomber scombrus

Scomber japonicus

Thunnus thynnus

Auxis thazard

Euthynnus alletteratus (Rafinesque) **Little Tunny** or **Bonito** (Not illustrated) **Length** up to 800mm. **Head** large with quite long jaws, but these do not reach back to rear of eye. **Body** typical tunny shape; background coloration dark blue to blackish above, much paler below; there is an area of mackerel-like lines and dots on the upper rear part of the body beginning behind the pectoral fin, and sometimes more dark spots in the lower flank region behind the pectorals; keels either side of tail stalk well developed. **Fins** 1st dorsal with 13–16 spiny rays, the leading ones tall and the remainder getting progressively shorter; 2nd dorsal has 11–14 soft rays behind which are 8 short finlets; tail crescentic; pectoral fin short; origin of anal fin is level with the 1st dorsal finlet; anal fin has 12–14 soft rays. **Habitat** quite common in schools near the surface.

Euthynnus pelamis (Linnaeus) **Skipjack Tuna** or **Ocean Bonito** (Not illustrated) **Length** up to 1m. **Head** large; jaw not reaching back to rear of eye. **Body** typical tunny shape, moderately deep, keels either side of tail stalk; background colour dark blue above, flanks green, belly silvery and with up to 6 wide stripes running from the belly upwards towards the tail. **Fins** 1st dorsal with 15 spiny rays, the anterior ones much the longest and becoming progressively shorter towards the rear and separated from the 2nd dorsal of 11–16 soft rays by a very short space; 8 finlets between the 2nd dorsal and the tail fin and 7 finlets between the anal fin and the tail fin; pectoral fin considerably shorter than snout; anal fin originating below trailing edge of 2nd dorsal; anal fin has 11–16 soft rays. **Habitat** an open water species usually in deep waters, occasionally close inshore.

Xiphias gladius Linnaeus **Swordfish** **Length** up to 4–9m, often less. Unmistakeable fish: head bears jaws drawn out into long flattened sword-like blades. **Fins** 1st dorsal short with 3–4 spiny rays; no pelvics. **Habitat** on surface waters down to 600m. No very similar species, but the Sail Fish *Istiophorus platypterus* can be distinguished from it by the upper jaw which is only about twice the length of the lower jaw and the long sail-like 1st dorsal fin which runs for about three-quarters of the body length. The marlins, genus *Tetrapturus* have 'swords' which are not as long as those of *Xiphas* and are rounded in section, not flattened; they also have a long 1st dorsal. Two species occur in the Mediterranean: *T. belone* Refinesque and *T. albidus* Poey.

Arnoglossus lanterna (Walbaum) **Scaldfish** **Length** up to 190mm. Flatfish lying on its right side when adult. **Head** quite small, body ovoid. **Fins** dorsal begins in front of eyes with 87–93 rays, the first few being partly free of membrane but not elongated anal fin has 65–74 fin rays. **Habitat** on sandy bottoms from 10 to 60m. **Similar species** two closely related species in the Mediterranean: *A. imperialis* (Rafinesque-Schmaltz) and *A. thori* Kyle, but these both have elongated, partially free anterior dorsal fin rays; the number of rays in their dorsal and anal fins is respectively 96–106/74–82 and 81–91/62–67.

Scophthalmus rhombus (Linnaeus) **Brill** **Length** up to 750mm. Adult lying on right side. **Head** quite large, mouth curved. **Body** ovoid. **Fins** first few rays of dorsal fin partially free of membrane, having 73–83 rays in all, anal fin with 56–62 rays. **Habitat** on sandy bottoms from the shore down to about 75m; quite important commercially. **Similar species** *S. maximus*.

Scophthalmus maximus (Linnaeus) **Turbot** **Length** up to 1m. **Head** large. **Body** broad and with scattered bony tubercles. **Fins** dorsal with 57–71 rays, none free of membrane even for part of their length; anal fin with 43–52 rays. **Habitat** on sandy and gravelly bottoms.

Pleuronectes platessa Linnaeus **Plaice** **Length** up to 910mm, often less. Adult lies on left side. **Head** with relatively small jaws; a row of 4–7 bony tubercles running from between the eyes to the top of the gill cover. **Fins** anal with 48–59 rays. **Habitat** sandy and muddy bottoms from the shore down to 200m.

Platichthys flesus (Linnaeus) **Flounder** **Length** up to 510mm. Similar to *Pleuronectes platessa* but lacks bony knobs on the head. **Body** bears sharp spines along the inside edge of the dorsal and anal fins; anal fin with 35–46 rays. **Habitat** on soft substrates down to 50m.

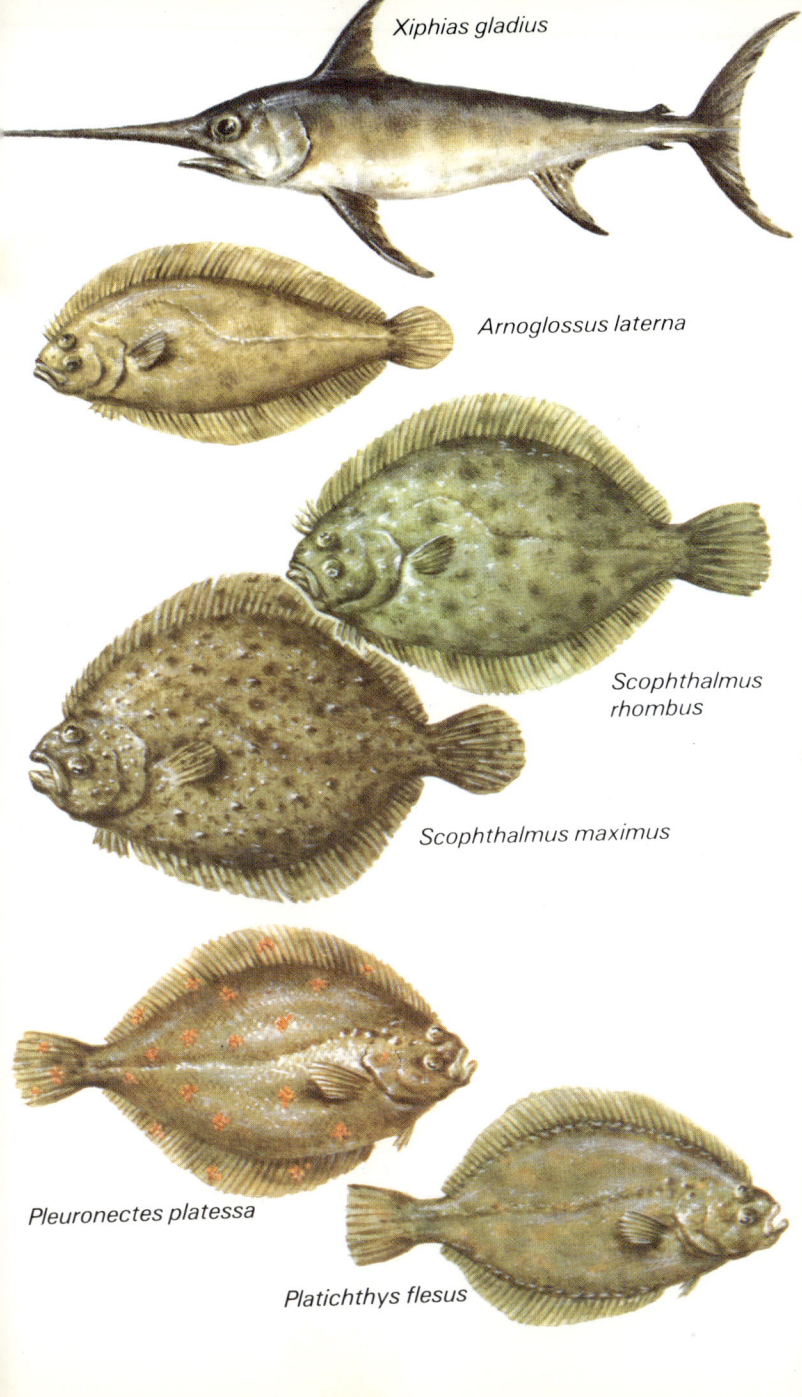

Xiphias gladius

Arnoglossus laterna

Scophthalmus rhombus

Scophthalmus maximus

Pleuronectes platessa

Platichthys flesus

Bibliography

There are relatively few English language guides to groups of organisms occurring in the Mediterranean Sea. The following list of reference works, which is by no means exclusive, includes a variety of monographs, fauna lists and guides to specific areas as well as a few scientific papers published in journals. The author has referred to these books in the course of preparing the present volume. In order to guide the reader further, * denotes references which are recommended for assisting with identification of particular groups of organisms; † denotes those which are fauna lists for particular areas; ‡ denotes those which are other important books of the field guide nature; and § denotes references of historical interest. It will be noted that a number of the books referred to in this bibliography relate principally to British waters. Many species described therein occur in the Mediterranean, and in the absence of more specific English language guides to Mediterranean groups, these are included.

Admiralty Tide Tables *Vol 1 European Waters.* Published annually by the Hydrographic Department, Admiralty, Ministry of Defence, Taunton, UK. Contains tidal predictions for Mediterranean ports.
* Arrecgros-Déjean, J. 1966 *Coquillages Marins: Petits Atlas Payot Lausanne No 33.* Editions Payot, Lausanne.
* Ax, P. 1956 Les Turbellariés des étangs cotiers du littoral Méditerranéen de la France Méridionale *Vie et Milieu, Supplement* 5 1–215.
Baltzer, F. 1917 Monographie der Echiuriden des Golfes von Neapel. *Fauna und Flora Neapel 34.*
Berquist, P. R. 1978 *Sponges.* Hutchinson University Library, London.
* Berrill, N. J. 1950 *The Tunicata, with an account of the British Species.* Ray Society, London.
† Best, M. B. 1969 Étude systématique . . . des madréporares de Banyules *Vie et Milieu 20* 2A 293–375.
Bibliographica Oceanographica 1954. A bibliography for marine science relating to the Mediterranean.
† Biggs, H. E. J. and Wilkinson, C. 1966 Marine Mollusca from Malta. *J. Conch* 26, 52–65.
Bouvier, E. L. 1940 *Faune de France 3 – Décapodes Marcheurs* Paul Lechevalier, Paris.
Brayfield, A. E. 1978 *Life in Sandy Shores: Studies in Biology 89.* Arnold, London.
Brinkmann-Voss, A. 1970 Anthomedusae/Athecatae (Hydrozoa, Cnidaria) of the Mediterranean, Pt. I Capitata. *Fauna und Flora Neapel 39.*
* Brown, G. H. and Picton, B. E. 1979 *Nudibranchs of the British Isles – a colour guide.* Underwater Conservation Society, Manchester.
* Burger, O. 1895 Die Nemertinen des Golfes von Neapel *Flora und Fauna Neapel 22.*
Burton, M. 1963 *Revision of Classification of Calcareous Sponges.* British Museum (Natural History), London.
Campbell, A. C. 1976 *The Hamlyn Guide to the Seashore and Shallow Seas of Britain and Europe.* Hamlyn Group, London.
* Carpine, C. and Grasshoff, M. 1975 Les Gorgonaires de la Méditerranée *Bull. Inst. Océanogr. Monaco 71,* 1430, 1–140.
† Chardy, P. 1970 Écologie des crustacés péracarides des fonds rocheaux de Banyules-sur-mer Amphipodes, Isopodes, Tanaidaces, Cumacés, infra et circalittoraux *Vie et Milieu 21,* 3B 657–728.
* Chevreaux, E. and Fage, L. 1925 *Faune de France 9: Amphipodes.* Paul Lechevalier, Paris.
* Christiansen, M. E. 1969 Crustacea. Decapoda (brachyura) in *Marine Invertebrates of Scandinavia* No 2. Universitets Forlaget, Oslo.
* Chun, C. 1880 *Die Ctenophoren des Golfes von Neapel.* Wilhelm Engelmann, Leipzig.
Cornelius, P. F. S. 1975 a. The Hydroid species of *Obelia* Coelenterata Hydrozoa: Campanulariidea), with notes on the medusa stage. *Bull. Br. Mus. nat. Hist. (Zool.)* 28, 249–293.
Cornelius, P. F. S. 1975 b. A Revision of the Species of *Lafoeidae* and *Haleciidae* (Coelenterata Hydrozoa Hydroiida recorded from Britain and nearby seas. *Bull. Br. Mus. nat. Hist. (Zool.)* 28, 373–426.
Cornelius, P. F. S. 1979 A Revision of *Sertulariidae* (Coelenterata Hydrozoa: Hydroiida) recorded from Britain and nearby seas. *Bull. Br. Mus. nat. Hist. (Zool.)* 34, 243–321.
Cornelius, P. F. S. 1982 Hydroids and medusae of the family Campanulariidae recorded from the eastern North Atlantic with a world synopsis of genera. *Bull. Br. Mus. nat. Hist. (Zool.) 42* 27–144.
Crisp, D. J., Southward, A. J. and Southward, E. C. 1981 On the Distribution of the Intertidal Barnacles *Chthamalus stellacus, Chthalamus montagui* and *Euraphia depressa. J. Mar. Biol. Ass. U.K. 61* 359–380.
* Cuenot, L. 1922 *Faune de France 4: Sipunculiens, Echiurens, Priapuliens.* Paul Lechevalier, Paris.
Dales, R. P. 1967 Annelids (2nd edn.). Hutchinson University Library, London.
* Darwin, C. R. 1851–1854 *A Monograph of the sub-class Cirripedia.* 2 vols. Ray Society, London.
Darwin, C. R. 1859 *The Origin of Species.* John Murray, London.
† Demetropoulos, A. and Neocleous, D. 1969 The fishes and crustaceans of Cyprus. *Min. of Agriculture and Nat. Resources Fish Dept. Fish. Bull. 1,* 1–21.
† Demetropoulos, A. 1969 Marine Molluscs of Cyprus Part A; Placophora, Gastropoda, Scaphopoda, Cephalopoda. *Min. of Agriculture and Nat. Resources Fish Dept. Fish Bull. 2,* 1–15.
† Demetropoulos, A. 1971 Marine Molluscs of Cyprus Part B: Bivalvia: some additions to part A: A check list of marine molluscs. *Min. of Agriculture and Nat. Resources Fish Dept. Fish Bull. 3,* 1–24.
† Demetropoulos, A. and Hadjichristophorou, M. 1976 Echinodermata of Cyprus and some additions to the knowledge of the Macrofauna of Cyprus. *Min. of Agriculture and Nat. Resources Fish Dept. Fish Bull. 4,* 1–84.
Den Hartog, C. 1970 Sea-grasses of the world. *Verbrandelingen Koninkluke Nederlandase Akademie Van Wettenschappen, Afd. Natuurkunde 59 Tweede Reeks,* 1–275.
* Dixon, P. S. and Irvine, L. M. 1977 *Seaweeds of the British Isles Vol. 1 Rhodophyta Pt. 1. Introduction, Nemaliales, Gigartinales.* British Museum (Natural History), London.
* Dobson, F. 1979 *Lichens, An Illustrated Guide.* Richmond Publishing Company.
Döhrn, A. 1881 *Die Pautopoden des Golfes von Neapel.* Fauna und Flora des Golfes von Neapel 3, Monographie.
* Drasche, R. 1883 *Die Synascidien der Bucht von Rovigno,* Vienna (deals with compound ascidians of the northern Adriatic).

* Duncan, U. K. 1959 *A Guide to the Study of Lichens*. T. Buncle & Co., Arbroath.
* Duncan, U. K. 1970 *Introduction to British Lichens*. T. Buncle & Co., Arbroath.
Eltringham, S. K. 1971 *Life in Mud and Sand*. English Universities Press, London.
* Ering, C. C. 1979 British and other Phoronids *Synopses of the British Fauna (New Series) No 13*. The Linnaean Society of London: Estuarine and Brackish Water Association. Academic Press, London and New York.
* Fauvel, P. 1923 *Faune de France 5: Polychètes errantes*. Paul Lechevalier, Paris.
* Fauvel, P. 1927 *Faune de France 16: Polychètes sedentaires*. Paul Lechavalier, Paris.
* Ferry, B. W. and Sheard, J.W. 1969 Zonation of Supralittoral Lichens on Rocky Shores around the Dale Peninsula. Pembrokeshire (with key for their identification). *Field Studies 3*. 1, 41–67.
* *Fiches d'Indentification du Zooplankton* Conseil Permanent International pour l'exploration de la mer. 1939 to current date. A useful series of sheets providing keys and figures for the identification of zooplankton.
§ Forbes, E. 1859 *Natural History of the European Seas*. J. Van Voorst, London.
* Forest, J. 1978 Le genre *Macropodia* Leach dans les eaux atlantiques européenes (Crustacea, Brachyura, Majidae) *Cahiers de Biol. Marine 19*, 323–342.
Fretter, V. and Graham, A. 1962 *British Prosobranch Molluscs*. Ray Society, London.
† Fretter, V. and Graham, A. 1976 The Prosobranch Molluscs of Britain and Denmark: Part 1 Plurotomariacea, etc. *Supplement 1 The Journal of Molluscan Studies*. Malacological Society of London.
† Fretter, V. and Graham, A. 1977 The Prosobranch Molluscs of Britain and Denmark: Part 2 Trochacea. *Supplement 3 The Journal of Molluscan Studies*. Malacological Society of London.
† Fretter, V. and Graham, A. 1978 a. The Prosobranch Molluscs of Britain and Denmark: Part 3 Neritacea, etc. *Supplement 4 The Journal of Molluscan Studies*. Malacological Society of London.
† Fretter, V. and Graham, A. 1978 b. The Prosobranch Molluscs of Britain and Denmark: Part 4 Marine Rissoacea. *Supplement 6 The Journal of Molluscan Studies*. Malacological Society of London.
† Fretter, V. and Graham, A. 1980 The Prosobranch Molluscs of Britain and Denmark: Part 5 Marine Littorinacea. *Supplement 7 The Journal of Molluscan Studies*. Malacological Society of London.
* Ghisotti, F. *Schede Malacologische del Mediterraneo*, Milan: Società Malacologica Italiana. A series of papers giving illustrations and notes on the distribution and synonymy of Mediterranean molluscs.
* Gibbs, P. E. 1977 *Synopses of the British Fauna (New Series) No. 7 British Sipunculans*. The Linnaean Society of London, Academic press, London and New York.
Gibson, R. 1972 *Nemerteans*. Hutchinson University Library, London.
† Godeaux, J. 1973 A contribution to the knowledge of the Thaliacean faunas of the eastern Mediterranean and the Red Sea. *Israel J. Zoology 22*, 39–50.
Graham, A. 1971 *Synopses of the British Fauna (New Series) No 2. British Prosobranch and other operculate gastropod molluscs*. The Linnaean Society of London, Academic Press, London and New York.
Guiterman, D. 1979 *Sponge recognition guide*. Underwater Conservation Society, Manchester.
‡ Haas, W. de and Knorr, F. 1966 *The Young Specialist Looks at Marine Life*., Burke, London.
* Harant, H. and Vernieres, P. 1933 *Faune de France 27: Tuniciers*. Paul Lechevalier, Paris.
* Harant, H. and Vernieres, P. 1938 *Faune de France 33: Tuniciers*. Paul Lechevalier, Paris.
Hardy, A.C. 1970 *The Open Sea: Part 2 Fish and Fisheries* (new edition). New Naturalist Series, Collins, London.
Hardy, A. C. 1971 *The Open Sea: Part 1 The World of Plankton* (revised edition). New Naturalist Series, Collins, London.
* Hayward, P. J. and Ryland, J. S. 1979 *Synopses of the British Fauna (New Series) No. 14 British Ascophoran Bryozoans*. The Linnaean Society of London: Estuarine and Brackish Water Association. Academic Press, London and New York.
Hincks, T. 1868 *A History of British Hydroid Zoophytes* 2 vols. J. Van Voorst, London.
Hincks, T. 1880 *A History of British Marine Polyzoa* 2 vols. J. Van Voorst, London. This book has been largely superseded by the works of Hayward and Ryland, and Prenant and Bobin.
Hiscock, S. 1979 A field key to the British brown seaweeds (*Phaeophyta*) in *Field Studies V.* 1–44.
* Ingle, R. 1980 *British Crabs*. British Museum (Natural History) and Oxford University Press.
* Jones, N. S. 1976 *Synopses of the British Fauna (New Series) No. 7 British Cumaceans*. The Linnaean Society of London, Academic Press, London and New York.
* King, P. E. 1974 *Synopses of the British Fauna (New Series) No. 5 British Sea Spiders: Arthropodal; Pycnogonida*. The Linnaean Society of London. Academic Press, London and New York.
Knight-Jones, E. W. 1954 Notes on invertebrate larvae observed at Naples during May and June. *Publ. Staz. Zool. Napoli,* 25 135–144.
* Koehler, R. 1921 *Faune de France 1: Echinodermes*. Paul Lechevalier, Paris.
Kosswig, C. 1956 Beitrat zur Faunengeschichte des Mittelmeeres *Publ. Staz. Zool. Napoli, 28* 78–88. An account of the origins and history of the Mediterranean fauna.
† Lafargue, F. 1970 Peuplements sessiles de l'archipel de Ienan 1. Inventair Ascidies. *Vie et Milieu 21* 3B 729–742.
* Lewinsohn, C. and Manning, R. B. 1980 Stomatopod Crustacea from the Eastern Mediterranean *Smithsonian Contributions to Zoology No. 305* Smithsonian Institution Press, Washington.
* Lincoln, R. H. 1979 *British Marine Amphipoda: Gammaridea*. British Museum (Natural History), London.
‡ Luther, W. and Fiedler, K. 1976 *A Field Guide to the Mediterranean Sea Shore*. Collins, London.
* Lythgoe, J. and Lythgoe G. 1971 *Fishes of the Sea*. Blandford, London.
* Manuel, R. L. 1980 *The Anthozoa of the British Isles – a colour guide*. Underwater Conservation Society, Manchester.
* Manuel, R. L. 1981 *Synopses of the British Fauna (New Series) No. 18: British Anthozoa*. The Linnaean Society of London: Estuarine and Brackish Water Association. Academic Press, London and New York.
Marshall, N. B. 1965 *The Life of Fishes*. Weidenfeld and Nicholson, London.
Matthews, G. 1953 A key for use in the identification of British Chitons. *Proceedings of the Malacological Society, London, 29*, 241–248.
† Médioni, A. 1970 Les peuplements sessiles des fonds rocheux de la region de Banyuls-sur-mer Ascidies Bryozoaires (Premiere Partie). *Vie et Milieu 21* 3B 591–656.

† Micallef, H. and Evans, F. 1968 *The Marine Fauna of Malta*. Malta University Press.
* Millar, R. H. 1970 *Synopses of the British Fauna (New Series) No 1. British Ascidians: Tunicata: Asadiacea*. The Linnaean Society of London. Academic Press, London and New York.
Morten, J. E. 1979 *Molluscs* (5th edn.). Hutchinson University Library Series, London.
* Mortensen, T. 1927 *Handbook of the Echinoderms of the British Isles*. Oxford University Press. Reprinted by Johnson Reprints.
‡ Muus, B. J. and Dahlstrøm, P. 1974 *Collins Guide to the Sea Fishes of Britain and Europe*. Collins, London.
* Naef, I. 1921–1928 Die Cephalopoden. *Fauna und Flora Neapel 35.*
* Naylor, E. 1972 *Synopses of the British Fauna (New Series) No. 3. British Marine Isopods*. The Linnaean Society of London, Academic Press, London and New York.
Newell, G. E. and Newell, R. C. 1973 *Marine Plankton* (revised edition). Hutchinson, London.
Nichols, D. 1969 *Echinoderms* (4th edition). Hutchinson University Library, London.
* Nordsieck, F. 1968 *Die Europäischen Meeres – Gehäuseschnecken (Prosobranchia) vom Eismeer bis Kapverdarn und Mittelmeer*. Gustave Fischer, Stuttgart.
* Nordsieck, F. 1969 *Die Europäischen Meersmuscheln (Bivalvia) vom Eismeer vis Kapverden, Mittelmeer und Schwarzes Meer* Gustave Fisher, Stuttgart.
Parke, M. and Dixon, P. S. 1976 Checklist of British Marine Algae – third revision. *Journal of the Marine Biological Association of the United Kingdom 56*, 1–11.
* Oax, F. and Müller, I. 1962 Die Anthozoenfauna der Adria *Fauna et Flora Adriatica 3*. Institute za Oceanografiju i Ribarstvo, Split.
Pérès, J. M. 1967 The Mediterranean Benthos *Oceanogr. Mar. Biol. Ann. Rev. 5* 449.
Polunin, O. and Huxley, A. 1967 *Flowers of the Mediterranean* Chatto and Windus, London.
Prenant, M. and Bobin, G. 1956 *Faune de France 60: Bryozaires pt. 1.* Fédération Française des Sociétés de Sciences Naturelles, Paris.
Prenant, M. and Bobin, G. 1966 *Faune de France 68: Bryozaires pt. 2* Fédération des Sociétés de Sciences Naturelles, Paris.
* Pruvot-Fol, A. 1954 *Faune de France 58. Mollusques Opisthobranches*. Paul Lechavalier, Paris.
† Quignard, J. P., Raibaut, A. and Trilles, J. P. 1962 Contribution a la faune ichthyologique Sétoise *Naturalia Monspelensia, Ser. Zool., vol. 4* 61–85.
‡ Riedl, R. 1963 *Fauna und Flora der Adria* Paul Parey, Hamburg and Berlin. A major field guide, specifically relating to the Adriatic but of great use throughout the Mediterranean.
* Russell, F. S. 1953 *The Medusae of the British Isles Vol. 1* Cambridge University Press.
* Russell, F. S. 1970 *The Medusae of the British Isles Vol. 2* Cambridge University Press.
Russell, F. S. and Yonge, C. M. 1975 *The Seas* (revised edition) Warne & Co., London.
* Ryland, J. S. 1962 The biology and identification of intertidal Polyzoa *Field Studies 1*, 4 33–51.
Ryland, J. S. 1970 *Bryozoans*. Hutchinson University Library, London.
* Ryland, J. S. 1974 A Revised key for the identification of inter-tidal *Bryozoa* (Polyzoa) *Field Studies 4*, 1, 77–86.
* Ryland, J. S. and Hayward, P. J. 1977 *Synopses of the British Fauna (New Series) 10 British Anascan Bryozoans*. The Linnaean Society of London. Academic Press, London and New York.
* Schmidt, H. 1972 Prodromus zu einer Monographie der mediterranen Aktinien *Zoologica 42*, 2, 121, 1–120
* Selbie, C. M. 1914 The decapoda reptantia of the coasts of Ireland pt. 1: Palinura, Astaeura and Anomura (except Paguridea). *Fisheries, Ireland, Sci. Invest., 1914 1.*
* Selbie, C. M. 1921 The decapoda reptantia of the coasts of Ireland pt. 2: Paguridea *Fisheries, Ireland, Sci. Invest., 1921 1.*
* Smaldon, G. 1979 *Synopses of the British Fauna (New Series) No. 15: British Coastal Shrimps and Prawns*. Linnaean Society of London: Estuarine and brackish water association. Academic Press, London and New York.
* Southward, E. 1972 Keys for the identification of echinodermata of the British Isles: *Echinoderm Survey, Marine Biological Association of the U.K.*
* Spengel, I. N. 1893 Die Enteropneusten des Golfes von Neapel *Fauna und Flora Neapel 18.*
Stazione Zoologica de Napoli 1928 *Guide to the aquarium of the Zoological Station of Naples*, 1–132. Naples. (English version).
Stebbing, A. R. D. 1979 An experimental approach to the determinants of biological water quality. *Phil. Trans. Roy. Soc. Lond. B 286* 465–481.
* Stephenson, T. A. 1928 and 1935 *The British Sea Anemones* 2 vols. Ray Society, London.
* Tattersall, W. M. and Tattersall, O. S. 1951 *The British* Mysidacea. Ray Society, London.
* Tebble, N. 1976 *British Bivalve Seashells* (2nd edition). British Museum (Natural History) Publications, London.
Theodor, J. 0165 *Vie sous-marine*. 1: *Petits Atlas Payot, Lausanne, No. 46–47*. Editions Payot, Lausanne.
§ Thompson, Sir D'Arcy W., 1947 *A Glossary of Greek Fishes*. Oxford University Press.
* Thompson, T. E. 1976 *Biology of Opisthobranch Molluscs* vol. 1. Ray Society, London.
Thompson, T. E. and Brown, G. H. 1976 *Synopses of the British Fauna (New series) No. 8 British Opisthobranch Molluscs*. The Linnaean Society of London, Academic Press, London and New York.
Vosmaer, G. C. H. 1935 *The Sponges of the Bay of Naples, Porifera Incalcerea* 3 vols. Martinus Nijhoff, the Hague. An extensive but very technical account of non-calcareous sponges.
Westblad, E. 1955–56 Marine Alloecolls (Turbellaria) from North Atlantic and Mediterranean coasts. *Arkiv foor zoologie* Bd 7 & 9.
* Wheeler, A. 1969 *The fishes of the British Isles and Northwest Europe*. Macmillan, London.
* Wheeler, A. 1978 *Key to the fishes of Northern Europe*, Warne, London.
Yonge, C. M. and Thompson, T. E. 1976 *Living Marine Molluscs*. Collins, London.
* Zariquiez-Alvarez, R. 1969 Crustaceos, Decapodos, Ibericos *Investigacion Pesquera* Tome 32. 1–510.

Glossary

Aboral describes the surface of the body opposite that which bears the mouth.

Ambulacrum (of echinoderms) usually a groove, with a row of tube-feet on either side: generally five per animal.

Antenna usually a long, slender, sensory appendage on the heads of some anthropods and some annelids.

Asexual describes organisms which reproduce without sexual processes.

Asymmetrical without symmetry, being irregular or unequal: used to describe the growth form of some animals, e.g. certain sponges.

Benthic dwelling in or on the seabed.

Bilateral symmetry symmetry of an organism (e.g. a fish) which can be divided into two equal and complementary left and right halves, but which has dissimilar front and hind ends.

Biserial describes parts arranged in a double series.

Brackish describes water usually containing less, but occasionally more, salt than is usually found in the sea.

Byssus hair-like filaments which attach some bivalves to rocks or plants.

Calcareous being made of calcium carbonate or chalk.

Calyx refers either to the outer whorl or leaves making the outermost covering of a flower whilst in the bud or alternatively to the skeletal cup protecting a coral polyp.

Caruncula small, fleshy protuberance on the heads of some polychaetes of the family *Amphinonidae*.

Cell smallest functional unit of a plant or animal, consisting of a nucleus surrounded by cytoplasm and bounded by a membrane, and sometimes a cell wall.

Cephalothorax region combining the head and thoracic segments of advanced crustaceans.

Chaeta bristle of polychaetes.

Chela leg of crustaceans which bears pincers or nippers.

Chermoreceptor sense organ for detecting chemical stimuli as in smell or taste.

Chitin organic consitutent of cuticle, as found in arthropods.

Chitinous made of chitin.

Chordate animal with at least a simple form of backbone (the notochord) at some stage in the life cycle: includes the vertebrates.

Cilia minute, filamentous structures which, by beating, may create a current and provide locomotion, visible only under the high power of a microscope.

Cirrus small, tentacular or finger-like appendage found in certain arthropods and polychaetes.

Class major subdivision of a phylum.

Coelom fluid-filled cavity, formed within the middle cell layer of animals.

Commensal organism of one species which lives in close association with one or more different species.

Crenulate having the edge cut into very small scallops.

Cuticle exterior skeleton of chitin and protein; may be tanned as in insects.

Detritus particles of decaying organisms accumulating, for example, on the seabed; forms the food of many invertebrate animals.

Disc (of an anemone) either the mouth disc which bears the tentacles, or the basal, suckerlike attachment disc; (of an ophiuroid) body excluding the arms.

Dorsal upper side of a bilaterally symmetrical animal (cf Ventral).

Ectoparasite parasite living on the outer surface of another organism.

Epiphyte plant which grows on the outer surface of another organism.

Epizoic describes an animal which grows on the outer surface of another organism.

Evert turn inside out, often applied to the process of extending the proboscis of worms.

Exhalent breathing out, applied to respiratory streams of water in organisms or the anatomical structures by which they are conveyed.

Foot (of molluscs) organ on the underside of the body used in gastropods for creeping and in bivalves for various functions including secretion of byssus, digging and burrowing.

Free tooth tooth not attached to jaws; as teeth on the proboscis of polychaetes such as *Nereis*.

Free living living unattached to any other structure.

Frond (of alga) all of the plant except the holdfast.

Gamete sperm or egg.

Gametophyte (of plants) generation which produces sperms and eggs.

Genus group of related species; many genera may form one order.

Growth line recognizable line or mark on a shell which indicates the start or end of a period of shell growth.

Hermaphrodite organism which has reproductive organs of both sexes and thus produces sperms and eggs.

Heteromorphic (of plants) condition where the gametophyte and sporophyte generators are dissimilar in form (c.f. Isomorphic).

Holdfast attachment organ of seaweeds.

Inhalent breathing in; applied to respiratory streams of water in organisms or the anatomical structures by which they are conveyed.

Invertebrate without a backbone.

Isomorphic (of plants) condition where the gametophyte and sporophyte generations are similar in form (c.f. Heteromorphic).

Lamella thin, plate-like structure or layer.

Larva developmental phase of an organism which usually does not resemble the adult or lead a way of life similar to it; a phase often associated with an entirely different manner of feeding from the adult and which provides a dispersive mechanism in many sedentary marine species; always terminates with the process of metamorphosis.

Lusitanian applied to water masses and plankton originating from the Mediterranean and Atlantic region of Portugal.

Mandible jaw, especially as applied to arthropods.

Mangrove refers either to trees or shrubs of the genus *Rhizophora* (or a related one) growing on the seashore or shallow water, or alternatively to the complex community formed by the mangrove plant.

Mantle special region of the body wall, particularly of molluscs, which secretes the shell and encloses the mantle cavity.

Medusa the jellyfish phase in the life cycle of hydrozoan and scyphozoan cnidarians.

Metamorphosis the act of transformation of a larva into an adult.

Neap tide tide with the smallest range between high and low water.

Nekton (c.f. Plankton) swimming animals which are able to determine their position in the sea.

Nematocyst special cell of cnidarians which discharges threads to sting or ensnare the prey.

Nephridium excretory organ of many invertebrates.

Niche limiting resources and habitat of a species; determined by its interaction with a wide variety of biological, physical and chemical environmental factors.

Notochord skeletal tube running from front to back in some simple chordates; forerunner of the backbone of vertebrates.

Ocellus simple light receptor.

Oral relating to the mouth; in echinoderms that side of the body on which the mouth is situated (c.f. Aboral).

Order major subdivision of a class.

Papilla small outgrowing structure on the surface of an organism.

Paragaster main cavity inside a sponge.

Parapodium segmental, flap-like appendage of a polychaete annelid; usually bears chaetae or bristles.

Parasitism condition whereby one organism, the parasite, lives on or in another; its host, at the expense of the latter.

Pelagic inhabiting the surface waters of the sea.

Pentamerism five-fold symmetry found in echinoderms.

Perisarc thin, tubular, skeletal structure investing the outer surface of many hydroid polyps.

Peristalsis form of motion resulting from the interaction of circularly and longitudinally arranged muscles in organs like intestines and in whole animals like worms.

Pharyngeal relating to the pharynx.

Pharynx anterior region of the alimentary canal, it adjoins the gills in chordates.

Phylum major division of the animal kingdom which includes those animals thought to have a common evolutionary origin.

Phytoplankton planktonic plants (generally microscopic).

Plankton drifting organisms or swimming organisms which are not able to determine their position in the sea.

Planula simple larva (for instance) of cnidarians resembling a ball of cells, usually ciliated and hence able to move.

Pneumatophore organ of flotation containing gas, or a modified individual in a siphonophoran colony which subserves this function.

Polymorphism the occurrence of different forms of the same species, for instance in a life cycle (polymorphism in time), or in a colony (polymorphism in space).

Polyp sedentary, individual cnidarian such as *Hydra*, basically with a sac-like body opening only by the mouth which is generally surrounded by tentacles.

Proboscis special structure at the anterior end of some animals; in nemertines it is generally everted through the mouth but is itself not part of the alimentary canal; in polychaetes it is everted through the mouth and is part of the alimentary canal.

Propodial refers in sea slugs to the leading tentacles at the front of the foot.

Radial symmetry symmetry of an organism (e.g. a cnidarian) in which the body parts are equally arranged around a median vertical axis which passes through the mouth; lacking definite front and rear ends and hence left and right sides.

Radula small, horny, tongue-like strip bearing teeth used by many molluscs for rasping food.

Rhizoid root-like structure.

Rostrum pointed projection at the extreme anterior end of the crustacean head.

Salinity measure of the salt concentration of water.

Segment one of a (generally fixed) number of functional units of the body and normally bearing a pair of appendages; applied particularly to annelids and arthropods.

Sessile commonly meaning living attached to a structure such as a rock or a shell of another organism.

Siphon tube leading into or out of the bodies of invertebrates and used for conducting water currents, found especially in molluscs and sea-squirts.

Species reproductively isolated group of interbreeding organisms; usually defined by morphological characteristcs.

Spicule minute fragment or crystal of skeletal material.

Spiracle vestigial gill slit found in fishes such as sharks.

Splash zone zone on the shore above the highest point to which the tides flow but which is under the influence of spray and salt.

Spore minute reproductive germ or particle produced by the asexual generation of plants i.e. the sporophyte.

Sporophyte (of plants) that generation which produces asexually reproductive spores, alternates with the gametophyte.

Spring tide tide with the greatest range between high and low water.

Stolon root-like structure found in some animals and linking up individuals in a colony; in plants a horizontal branch which produces its own roots and subsequently a new individual.

Sublittoral biologically defined zone on the seashore which lies below the highest point to which laminarians grow, and only uncovered at the lowest tides and extending down from the shore to the shallow seabed; equivalent to the term *lower* shore as used in this book.

Symbiotic describes an organism of one species which lives in close association with one of another species and to the advantage of both.

Telson terminal flap-like appendage of many crustaceans.

Test shell of a sea-urchin or starfish which, strictly speaking is an internal skeleton.

Thallus entire body of a lower plant, such as an alga or lichen.

Theca cup-shaped skeleton of a hydroid or coral polyp; the cup-shaped test of a feather-star (crinoid).

Thecate possessing a theca.

Thixatropic condition in which sand contains much water and becomes slushy when compressed and so is able to flow.

Torsion twisting of the body, particularly applied to an event in the development of certain larvae, e.g. gastropods and holothuroids.

Tube-foot hydraulic appendage of echinoderms and part of the water vascular system.

Tunic the proteinaceous coat surrounding the bodies of sea-squirts and salps.

Umbilicus aperture in the central pillar of a snail shell.

Umbo part of the shell of a bivalve mollusc.

Ventral underside of a bilaterally symmetrical animal (c.f. dorsal).

Vertebrate animal with a backbone made up of vertebrae.

Viscera organs inside the body cavity, especially intestines, heart, liver, etc.

Visceral hump part of the gastropod body where most of the internal organs are housed.

Water vascular system hydraulic system unique to the echinoderms comprising a system of vessels and organs like the tube-feet, and fulfilling various functions, especially locomotion.

Zoecium casing surrounding an individual ectoproct zooid.

Zonation separation of plants and animals into discrete zones or communities on the shore related to the tidal levels.

Zooid individual animal in a colony; usually applied to the Cnidaria and Ectoprocta.

Zooxanthellae microscopic algae, probably related to dinoflagellates, which live in carrier-cells of animals such as the cnidaria and the ascidia.

Zooplankton animals of the plankton.

Index

Page numbers in **bold** type refer to illustrations. Many marine organisms have alternative scientific names as well as a common name and these appear in the index; however, subsidiary scientific and common names are not included on the illustration pages and in these cases the reader should refer to the first page listed, where they will be found under the most widely used scientific name.